Metagovernance fc

The 17 Sustainable Development Goals (SDGs) which were adopted by the United Nations in September 2015 are universally applicable in all 193 UN member states and connect the big challenges of our time, such as hunger and poverty, climate change, health in an urbanized environment, sustainable energy, mobility, economic development and environmental degradation. Sustainability has the characteristics of a 'wicked problem', for which there are no one-size-fits-all solutions.

This book tests the hypothesis that the implementation of sustainable development, and in particular the 2015 SDGs, requires tailor-made metagovernance or 'governance of governance'. This is necessary to develop effective governance and high-quality and inclusive public administration and to foster policy and institutional coherence to support implementing the SDGs. Based on the growing literature on governance and metagovernance, and taking into account the specificities of societal factors such as different values and traditions in different countries, the book presents a framework for the design and management of SDG implementation. It shows how hierarchical, network and market governance styles can be combined and how governance failure can be prevented or dealt with. The book presents an overview of fifty 'shades of governance' which differ for each governance style and a sketch of a concrete method to apply sustainability metagovernance.

Metagovernance for Sustainability is relevant to academic and practitioner fields across many disciplines and problem areas. It will be of particular interest to scholars, students and policy makers studying Sustainable Development, Governance and Metagovernance, Public Management and Capacity Building.

Louis Meuleman is member of the United Nations Committee of Experts on Public Administration (CEPA) (2018–2021), and works at the European Commission in Brussels as coordinator of the Environmental Implementation Review and the greening of the 'European Semester'. He is affiliated with the universities of Leuven (Belgium), Massachusetts Boston (USA) and Wageningen (the Netherlands).

Routledge Studies in Sustainable Development

This series uniquely brings together original and cutting-edge research on sustainable development. The books in this series tackle difficult and important issues in sustainable development including: values and ethics; sustainability in higher education; climate compatible development; resilience; capitalism and de-growth; sustainable urban development; gender and participation; and well-being.

Drawing on a wide range of disciplines, the series promotes interdisciplinary research for an international readership. The series was recommended in the *Guardian*'s suggested reads on development and the environment.

Sustainable Development Policy
A European Perspective
Michael von Hauff and Claudia Kuhnke

Creative Practice and Socioeconomic Crisis in the Caribbean
A Path to Sustainable Growth
Kent J. Wessinger

Engineering Education for Sustainable Development
A Capabilities Approach
Mikateko Mathebula

Sustainable Pathways for our Cities and Regions
Planning within Planetary Boundaries
Barbara Norman

Land Rights, Biodiversity Conservation and Justice
Rethinking Parks and People
Edited by Sharlene Mollett and Thembela Kepe

Metagovernance for Sustainability
A Framework for Implementing the Sustainable Development Goals
Louis Meuleman

For a full list of titles in this series, please visit www.routledge.com

Metagovernance for Sustainability
A Framework for Implementing the Sustainable Development Goals

Louis Meuleman

First published 2019 by Routledge

2 Park Square, Milton Park, Abingdon, Oxfordshire OX14 4RN

52 Vanderbilt Avenue, New York, NY 10017

Routledge is an imprint of the Taylor & Francis Group, an informa business

First issued in paperback 2020

Copyright © 2019 Louis Meuleman

The right of Louis Meuleman to be identified as author of this work has been asserted by him in accordance with sections 77 and 78 of the Copyright, Designs and Patents Act 1988.

All rights reserved. No part of this book may be reprinted or reproduced or utilised in any form or by any electronic, mechanical, or other means, now known or hereafter invented, including photocopying and recording, or in any information storage or retrieval system, without permission in writing from the publishers.

Notice:
Product or corporate names may be trademarks or registered trademarks, and are used only for identification and explanation without intent to infringe.

British Library Cataloguing-in-Publication Data
A catalogue record for this book is available from the British Library

Library of Congress Cataloging-in-Publication Data
A catalog record for this book has been requested

ISBN: 978-0-8153-7016-1 (hbk)
ISBN: 978-0-367-50046-7 (pbk)

Typeset in Goudy
by Apex CoVantage, LLC

Responsibility for the information and views set out in this book lies entirely with the author.

Contents

List of figures vii
List of tables viii
List of boxes ix
Foreword x
Acknowledgements xi
List of definitions xii
List of abbreviations xv

PART 1
What and for what is governance? 1

1 Introduction: the problem with sustainability governance 3

2 Three governance styles and their hybrids 21

3 Governance failures and their causes 45

4 Introducing metagovernance: governance of governance 72

PART 2
Features of metagovernance 107

5 Fifty shades of governance: a toolbox 109

6 How values, traditions and geography shape the feasibility of governance approaches 160

7 Mind-sets and mental silos: rise and fall of simple switches 173

PART 3
Metagovernance for sustainability 191

8 Metagovernance challenges for the 17 Sustainable Development Goals 193

9 Metagovernance: sketching a method 224

10 Metagovernance, public sector reform, coherence promotion and capacity building 247

11 Conclusions: metagovernance as framework for SDG implementation 277

Index 297

Figures

1.1	Sustainable development and metagovernance: holistic goals meet holistic governance	5
1.2	The UN Sustainable Development Goals	14
2.1	Governance and metagovernance: the conceptual crowd revisited	26
3.1	A model of governance failure occurrence in the policy cycle	67
4.1	Metagovernance as coordination of hierarchical, network and market governance	75
4.2	Academic publications/year using the term meta(-)governance 1997–2017	85
5.1	Breaking the ice? How to move towards collaboration	149
8.1	Framework for clustering the SDGs	194
9.1	Strength analysis	229
9.2	Stakeholder roles in the 'rings of influence'	231
9.3	Problem types	237
9.4	Human behaviour and appropriate intervention types	239
10.1	Policy and institutional coherence for the SDGs: horizontal, vertical and inclusive	256
11.1	Transgovernance: the quest for sustainable development	283
11.2	Governance frameworks on energy transition in three European countries	286

Tables

2.1	Current approaches to SDG governance	37
3.1	Policy failure and governance failure	48
3.2	Causes of governance failure linking to governance capacity	56
3.3	Causes of governance failure linking to governance design	60
3.4	Causes of governance failure linking to governance management	65
4.1	Examples of metagovernance research: various locations and topics	84
4.2	(Meta)governance principles: an overview	92
5.1	Hierarchical, network and market governance: three distinct logics	110
5.2	Governance style differences (1): vision, strategy and orientation	112
5.3	Governance style differences (2): institutions, instruments and tools	123
5.4	Challenges of strategic environmental assessment and typical governance style reactions	136
5.5	Three governance strategies and accountability relationships	138
5.6	Governance style differences (3): processes, people and partnerships	139
5.7	Differences between project and process management	141
5.8	Governance style differences (4): problems, solutions and their linkages	150
6.1	Governance styles and Hofstede's cultural dimensions	164
6.2	Four cultural dimensions: examples across countries	165
6.3	Tentative relations: governance styles and Meyer's cultural scales	166
9.1	Sketch of a metagovernance method in seven steps	226
9.2	Model for a SWOT analysis of existing policy and governance frameworks	236
9.3	Model for analyzing the complexity of the problem	237
9.4	Principles and priorities for SDG integration	241
9.5	Design of a governance framework: principles	242
9.6	Possible template for the design of a governance framework: combining governance styles	243
10.1	Intervention types for promoting SDG coherence: examples of successful practice	257

Boxes

9.1	The case of the EU Environmental Implementation Review (EIR)	226
9.2	Stakeholder mapping for the Environmental Implementation Review	231
9.3	Values and traditions and the EU Plastics Strategy	233
9.4	Institutions relevant for the Environmental Implementation Review	235
9.5	Ex-ante evaluation, the Environmental Implementation Review and the EU Plastics Strategy	236
9.6	Problem definition of the Environmental Implementation Review (EIR)	238
9.7	Contextual goal definition of the Environmental Implementation Review (EIR)	240
9.8	EU Environmental Implementation Review (EIR) governance	243

Foreword

"Ultimately, it's all about governance" – This is what I said to the United Nations General Assembly in September 2015. And indeed, even with the best policies, if we don't have effective governance it will be extremely difficult to implement the Agenda 2030 and the Sustainable Development Goals. Effective governance means using all the institutions, tools and mechanisms at our disposal, and adjusting our approach to the context and the challenges in each country. There is no one-size-fits-all model, and it is imperative that we think outside the box to get our approach right. The SDGs are indivisible and their implementation requires coordination, integration and ultimately effective and coherent policies and institutional design.

This book describes the interesting concept of metagovernance - 'governance of governance'. It is right to say that when we have a problem, we should not always think of a law as the first and only solution. There may be other tools that lead to better results. The range of governance options is so much wider. This book offers a broad collection of ideas and examples of effective governance for sustainability and, in particular, for the implementation of the SDGs. The toolbox provides interesting food for thought for public authorities around the world as they roll out their own local plans for implementing the SDGs.

The questions and possible answers for sustainability governance are based on the author's practical experience. Exact recipes are not prescribed, because the context matters more than the theory. But I think the book will be relevant for practitioners in governments at all levels as well as in civil society and the private sector - our indispensable partners in achieving sustainability. This book makes an important and positive contribution to stimulating the discussion on effective public administration and public services for delivering the SDGs by 2030.

Governance is about delivering results for people; its legitimacy depends on its overall effectiveness, and the effectiveness of its policies. Citizens expect both industry and politicians to deliver results when it comes to sustainability. People are ready for change, they yearn for change, and we have got to enable and provide this change.

Frans Timmermans
First Vice-President of the European Commission
Brussels

Acknowledgements

Writing this book in my free hours outside my full-time job has been quite a challenge, but it was worth it, as it helped me structuring my thoughts and giving my experience and observations a place.

I would like to thank those who have supported me with comments, among others my colleagues Jonathan Parker and Christian Buechter at the European Commission. Special thanks go to Geert Boeckaert, Bob Jessop and Roel in 't Veld who have inspired me for many years. I apologize to all authors who have written, often better than me, about aspects of metagovernance and sustainability governance, for not having given them all the credits they deserve. I could only touch upon their work and I am sure I have forgotten to refer to some of you.

My last and not least thanks go to Inge, for her constructive comments, for enduring my writer's solitude and neglecting other things during this year. Without your support and patience this book would not have been written, and I promise, again: no more book projects (for the moment . . .).

Definitions

Governance	The totality of interactions in which government, other public bodies, private sector and civil society participate (in one way or another), aimed at solving public challenges or creating public opportunities (Meuleman, 2008).
Governance Capacity	The resources and skills a government requires to steer a governance mode so as to make sound policy choices and implement them effectively (Howlett and Ramesh, 2014).
Governance Capacity Failure	Governance failure resulting from the mismatch of governance style and governance capacity. In this case, the chosen governance style may be suitable to address a specific problem type, but governmental actors do not possess the necessary capacity (i.e. competences, skills, capabilities) to bring about results.
Governance Design Failure	Governance failure resulting from the mismatch of problem context and governance style. In this case, the governance style (combination) is incapable to address successfully a specific problem type.
Governance Environment	The whole of rules and laws, institutional setting, policy instruments, division of tasks and roles of governments and societal stakeholders, within a particular country or other boundary or transboundary units (Meuleman, 2015, p. 6).
Governance Failure	The ineffectiveness of governance goals, a governance framework or the management thereof, to achieve policy goals (adapted from Bovens et al., 2001).
Governance Framework	The totality of instruments, procedures and processes designed to tackle a societal problem. They should be adapted to legal, cultural and physical conditions of the problem environment and internally consistent; the normative assumptions (values, hypotheses) should be clear (Meuleman, 2014).

Governance, Hierarchical	Governance based on enforcement by means of legitimate authority, either through an employment relation (i.e. vertical integration) or a detailed contractual arrangement that provides decision-making authority in certain areas (after Steenkamp and Geyskens, 2012).
Governance Management Failure	Governance failure resulting from ineffective management of a governance framework.
Governance, Market	A governance style, which is driven by market and business ideas, treats government organizations as if they were business units, prefers using market-based instruments like taxes, and focuses on principles such as efficiency, competition, devolvement and empowerment.
Governance, Network	The 'management' of complex networks, consisting of many different actors from the national, regional and local government, from political groups and from societal groups (pressure, action and interest groups, societal institutions, private and business organizations (Kickert et al., 1997, p. 735).
Metagovernance	Metagovernance or 'governance of governance' is a means by which to produce some degree of coordinated governance, by designing and managing sound combinations of hierarchical, market and network governance, to achieve the best possible outcomes from the viewpoint of those responsible for the performance of public sector organizations: public managers as 'metagovernors' (Meuleman, 2008).
Policy Failure	The failure to reach a policy objective.
Sustainable Development	Sustainable development is development that meets the needs of the present without compromising the ability of future generations to meet their own needs (World Commission on Environment and Development (Brundtland et al., 1987)).

References

Bovens, M., 't Hart, P. & Peters, B. G. 2001. Analysing governance success and failure in six European states. In: *Success and Failure in Public Governance: A Comparative Analysis*. Cheltenham, Glos: Edward Elgar Publishing.

Brundtland, G., Khalid, M., Agnelli, S., et al. 1987. *Our Common Future* ('Brundtland report'). Oxford: Oxford University Press and United Nations World Commission on Environment and Development.

Howlett, M. & Ramesh, M. 2014. The two orders of governance failure: Design mismatches and policy capacity issues in modern governance. *Policy and Society*, 33, 317–327.

Kickert, W. J. M., Klijn, E. H. & Koppenjan, J. F. M. 1997. *Managing Complex Networks: Strategies for the Public Sector*. London: Sage.

Meuleman, L. 2008. *Public Management and the Metagovernance of Hierarchies, Networks, and Markets*. Heidelberg: Springer.

Meuleman, L. 2014. Governance frameworks. *In:* B. Freedman (ed.) *Global Environmental Change*. Dordrecht: Springer.

Meuleman, L. 2015. Owl meets beehive: How impact assessment and governance relate. *Impact Assessment and Project Appraisal*, 33, 4–15.

Steenkamp, J. -B. E. M. & Geyskens, I. 2012. Transaction cost economics and the roles of national culture: A test of hypotheses based on Inglehart and Hofstede. *Journal of the Academy of Marketing Science*, 40, 252–270.

Abbreviations

ABC	Administration, Business, Civil Society (partnership)
BATNA	Best Alternative to a Negotiated Agreement (in MGA)
CBA	Cost-Benefit Analysis
CBDG	Common But Differentiated Governance
CBDR	Common But Differentiated Responsibility
CEA	Cost-Effectiveness Analysis
CEPA	Committee of Experts on Public Administration (UN)
CoR	Committee of the Regions (European Union)
CSO	Civil Society Organization
EEA	European Environmental Agency
EESC	European Economic and Social Committee
EIA	Environmental Impact Assessment
EIR	Environmental Implementation Review (European Commission)
EPR	Enviromental Performance Review (OECD)
EPSC	European Political Strategy Centre (of the European Commission)
EU	European Union
EUPACK	European Public Administration Country Knowledge (project)
GCT	Garbage Can Theory
GDP	Gross Domestic Product
HDI	Human Development Index
HLPF	High-Level Political Forum
IA	Impact Assessment
IASS	Institute for Advanced Sustainability Sudies (Potsdam)
ICT	Information and Communication Technology
IMC	Inter Municipal Cooperation
IMPEL	EU Network for the Implementation and Enforcement of Environmental Law
MDG	Millennium Development Goal
MGA	Mutual Gains Approach
MSP	Multi-Stakeholder Partnership
NASA	National Aeronautics and Space Administration (USA)
NPG	New Public Governance
NPM	New Public Management

OECD	Organization of Economic Cooperation and Development
OMC	Open Method of Coordination (of the European Commission)
OWG	Open Working Group (on the SDGs)
P2P	Peer-to-Peer
PPP	Public-Private Partnership
SAMOA	SIDS Accelerated Modalities of Action
SD	Sustainable Development
SDG	Sustainable Development Goal
SDSN	Sustainable Development Solutions Network
SEA	Strategic Environmental Assessment
SIA	Sustainability Impact Assessment
SIDS	Small Island Developing State
SPP	Sustainable Public Procurement
SWOT	Strengths, Weaknesses, Opportunities and Threats
TAIEX	Technical Assistance and Information Exchange instrument (European Commission)
TED	Technology, Entertainment, Design (media organization)
TIAS	There Is Always a Solution
TINA	There Is No Alternative
UK	United Kingdom
UNDESA	United Nations Department for Economic and Social Development
UNDP	United Nations Development Programme
UNECE	United Nations Economic Commission for Europe
UNEP	United Nations Environment Programme
USA	United States of America
VNR	Voluntary National Review

Part 1
What and for what is governance?

Chapter 1. Introduction: the problem with sustainability governance
Chapter 2. Three governance styles and their hybrids
Chapter 3. Governance failures and their causes
Chapter 4. Introducing metagovernance: governance of governance

1 Introduction
The problem with sustainability governance

> Ultimately, this is all about governance. About inclusiveness: societies will only accept transformation if people feel their voices have been heard. And it's about breaking out of silos. Sustainable development is not just an economic or social challenge, or an environmental problem: it's all three – and our efforts on each need to reinforce rather than undermine one another.
> – First Vice-President of the European Commission Frans Timmermans at the United Nations Summit on the 2030 Agenda for Sustainable Development in New York, 27 September 2015

1.1 Holistic governance meets holistic goals

Ten years ago, I wrote a research book showing that successful public managers applied a flexible, situation-adapted approach to achieve their objectives; five case studies in different countries underpinned this conclusion. For these public managers, such situational governance was a logical thing to do. It was similar to the approach coined decades ago as 'situational leadership' of employees (Hersey and Blanchard, 1988), which still figures in most leadership training. From experience, they had learned that each policy or implementation challenge requires a situational and dynamic approach, just as employees need guidance adapted to personality and circumstances, to get the best out of them. Many scholars call the application of such a context-specific approach to policy challenges 'metagovernance' (i.e. governance of governance). In my research, four case studies on environmental protection showed that public managers tackled the same policy topic in different countries with a different (but dynamic) governance framework. A fifth case study on inner security (community policing) suggested that metagovernance also happens in other policy areas.

One decade later I think that one of the added values of *Public management and the metagovernance of hierarchies, networks and markets* (Meuleman, 2008) has been the discovery that metagovernance is not only a theoretical concept. In fact, it is a still relatively ill-researched good practice of public management, and beyond this: also private sector and civil society organizations practice metagovernance. The challenge that springs out of this insight is that this 'good practice' needs further development into a (public) management method for combining

4 *What and for what is governance?*

different governance styles into tailor-made governance frameworks. Since 2008, I have presented and discussed metagovernance theory and practice at dozens of lectures and workshops, and practiced this approach in my daily work as line, project and process manager of national and European policy processes. At the same time, academics have published hundreds of research articles on the use of metagovernance as analytical, design and/or management tools for complex issues. Metagovernance has become accepted as a relevant political science concept (Bevir, 2011, Jessop, 2011, Rayner, 2015, Larsson, 2017) and scholarly literature utilizing the term has become abundant. The first reason to write this book was therefore to review recent literature in the light of the evolving theory and practice of metagovernance.

The second reason to write this follow-up book was to seek answers to the reverse of the central question of the first book, for which I had interviewed successful public managers: is it plausible that unsuccessful public managers are, generally spoken, not using a metagovernance lens to prevent and mitigate governance failures? What can we learn from the failures we can identify through metagovernance glasses?

The third and most important reason to write this book was that I was curious to see how the holistic concept of metagovernance could be applied to the most holistic of contemporary policies: the quest for sustainable development, and in particular for implementation by 2030 of the 2015 UN Agenda 2030 with its Sustainable Development Goals (SDGs). After several decades of working with the concept of sustainable development, the Rio+20 conference in 2012 and its outcome, the 2015 SDGs, can only be considered as a huge achievement of the UN and of the international community of member states, civil society, private sector and academia. The governance of the process to formulate strong but still commonly accepted Goals and targets through the 'Open Working Group' (OWG) was innovative at this scale, as the inside account of the process reveals:

> The OWG began its work at a time of frustration with multilateral sustainable development negotiations and long-standing mistrust between North and South. The experience of the OWG helped to rebuild this trust and improve multilateral negotiations in the UN system. Although not all of the techniques used by the OWG are replicable, such as the troika system, the notion that UN negotiations can experiment with new procedures and do not have to be wedded to the past is a major contribution of the OWG.
> (Kamau et al., 2018)

The joint work resulted in Goals and targets which are designed in a very smart way, in order to make them 'indivisible'.

Implementation of the SDGs requires in the first place readiness of public sector organizations and political leaders to design and manage effective governance. This requires rethinking of institutions, instruments and governance processes on at least the same scale as the rethinking of energy, economy, food and other systems that are essential for living well within the limits of the planet.

Sustainable development is about creating balances between the environmental, economic and social dimensions of life, also with a view to future generations. The three dimensions should be seen as communicating vessels: it is impossible to take them apart without facing consequences. Sustainable development is therefore a 'meta-policy' (Meadowcroft, 2011). The parallel is that metagovernance has also three dimensions: it is about creating balanced combinations of tools from three different governance styles, notably hierarchical, network and market governance. In isolation, these styles are weaker than in carefully designed combinations. Jordan (2008) argued that sustainable development and governance may be "the most essentially contested terms in the entire social sciences" – hardly a good foundation, one might think, for solid and insightful scholarship. With metagovernance the complexity is even greater.

Figure 1.1 combines the triangle of metagovernance and the triangle of the sustainable development dimensions, which immediately brings forth the question what the affinity is of each of the three sustainability dimensions with each of the three governance styles. Is it a coincidence that both holistic concepts are often presented as having three dimensions, modes or styles? How closely are the three pairs – environmental governance and hierarchy, economic governance and market governance, and social governance and network governance – related? This is something we need to find out. Of course, each of the sustainability dimensions can be dealt with using all governance styles but on first sight, there is this underlying preference: environmental policies traditionally rely on legislation; economic governance favours market-style instruments; and the social dimension shares key values with network governance.

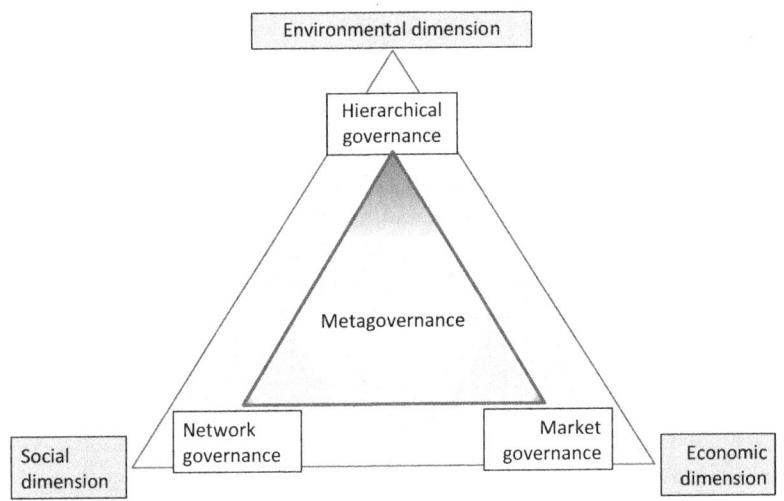

Figure 1.1 Sustainable development and metagovernance: holistic goals meet holistic governance

6 *What and for what is governance?*

The holistic nature of both parts of the focus of this book – metagovernance and sustainability, respectively the Sustainable Development Goals – necessitates an approach which is essentially multi-, inter- and transdisciplinary. Concepts and examples will be used from various social science disciplines and from various practices. In addition, how people argue, think, feel and dream and what drives them to action or non-action is an essential but often neglected part of the discussion on governance. Governmental organizations are built with and upon people – which is one of the reasons maybe why they can behave rationally one day and irrationally the next day. The reason to address this human dimension of governance and metagovernance of sustainable development is a very pragmatic one: it is because it happens to be influential in practice.

To conclude, what this book aims to achieve is bringing together new insights and practical solutions as regards the use of metagovernance to improve sustainability policies and their implementation, with the SDGs as main application case for the medium term (i.e. until 2030).

1.2 Sustainability governance: key terms

When the First Vice-President of the European Union's executive, the European Commission, spoke at the United Nations General Assembly on 27 September 2015, he emphasized the governance dimension of sustainable development, and in particular of the Sustainable Development Goals (SDGs). Also from the academic perspective it has been argued that sustainable development is above all about getting the governance right (Meadowcroft, 2011, p. 536):

> Sustainable development is really all about governance. The idea was formulated because of dissatisfaction with existing ('unsustainable') development patterns. And the assumption was that conscious and collective – i.e. political – intervention would be required to shift the societal development trajectory on to more sustainable lines.

This sounds simple as a starting point, but the difficulty is that the term governance has many meanings. Different people – scholars and practitioners alike – have defined governance in various and, quite often, mutually excluding ways. I use a broad definition of governance, which is in short that governance is how societal challenges are tackled and opportunities are created; in Sections 1.3 and 2.1, I will explain why. This definition focuses on the question of *'how to get things done'*, not on *'what should be done'*. This goes back to the classical distinction in political science between polity, politics and policy. Governance is about polity (institutions) and politics (processes), and not about policy (the substance). For me as a public administration practitioner, this distinction makes sense. Others argue that policy should be part of governance. Governance then becomes "a social function centred on steering individuals or group toward desired outcomes" (Young, 2017, p. 31). This view is relatively often seen in academic discourse on environmental governance.

Sustainable development is a laboratory for governance innovation (how will goals be achieved?), but also for policy innovation (which concrete goals need to be set in a specific situation?). Policy and governance are two sides of the same coin – the 'coin' being how to tackle effectively societal challenges. Within this line of thinking, governance is not an end, but a means to an end. Governance without carefully crafted policy is useless but policy without adequate governance cannot be implemented and is therefore without practical meaning, or put differently: governance needs policy and policy needs governance.

Although effective governance is essential for effective policy, the latter is usually politically more attractive because it is more visible for citizens. This makes sense: people who excel in a craft, be it a painter, carpenter, teacher or a medical doctor, will not become famous because of how they do it, with which tools etcetera, but because of what they do or achieve. Governance, however we define it, is often perceived as boring, talking shop, abstract and/or bureaucratic: citizens, politicians or the media are not interested in the craft of governance, which I am not sure is a problem. There are exceptions to this rule, when a perceived policy failure turns out to be in fact a governance failure: when bureaucratic procedures, lack of transparency, lack of efficiency or corruption prevent a policy to deliver the expected output or outcome. This dual complexity as regards policy and concerning governance is one of the reasons why sustainability governance is so difficult: complexity is a good predictor for trouble, as Chapter 3 on governance failures will illustrate.

The implementation of the SDGs is about defining, deciding, implementing and evaluating policies to support the common Goals. Usually public authorities – at different levels – will be the drivers of SDG implementation, with civil society and business as partners, clients or subordinates (depending on the governance approach applied by the public authority). The chapters in this book will develop a narrative on why and how to use metagovernance for sustainable development, in particular for implementation of the SDGs, while keeping the story close to the policies that need to be realized through governance. First, we should define the key terms.

Sustainable development

Sustainable development is one of the best examples of a complex policy field that is multi-level, multi-scalar, multi-sector and multi-actor at the same time, and requires governance frameworks, which are up to dealing with this complexity. The most prominent definition is that "**sustainable development** is development that meets the needs of the present without compromising the ability of future generations to meet their own needs" (Brundtland et al., 1987; emphasis added).

The SDGs strengthen the environmental dimension in development policy. This is important because environmental problems have become more and more intertwined with economic and social issues. Climate change, water scarcity, flooding, hunger, access to energy and to sanitation, and urban development are happening, force people to migrate and create political conflicts between countries. Political conflicts on water and other natural resources are on the rise and are socially and economically very disruptive.

Sustainable development has three dimensions (or 'pillars' in a more static view): the environmental, economic and social dimensions, sometimes also called people-planet-profit/prosperity. There is rather little dispute about this definition, although there have been proposals to add a fourth dimension, such as 'governance' (Niksic Radic, 2015) or 'good governance' (Pisano et al., 2015), 'culture' (Nurse, 2006) or 'peace and security'(United Nations, 2012). An analysis of 24 national sustainable development strategies in Europe showed that five countries (Estonia, Lithuania, Norway, Slovakia and Slovenia) included culture as fourth dimension of sustainability and two (Czechia and the Netherlands) chose governance (Steurer, 2012, p. 90).

Spangenberg et al. (2002) proposed taking 'institutions' as the fourth dimension. In a later article it was explained that this fourth dimension would have

> a slightly different status from the other three. While the established dimensions of sustainable development constitute broad areas of social endeavour, which may be conceptualized as social subsystems or components of the development trajectory (and in relation to which one can identify specific objectives and assess performance), the institutional vertex refers to structures and practices that frame activity within each of the other dimensions.
> (Meadowcroft et al., 2005)

All in all, sustainability remains a rather open-ended concept. This allows framing it in different ways in different contexts. In Dutch infrastructure policy, it was found that sustainability was used as a brokerage concept, as a symbolic concept and as a pull-push concept (Frantzeskaki et al., 2016).

The Agenda 2030 with its 17 Sustainable Development Goals represents a concrete political goal framework towards 2030. The SDGs are more ambitious than their predecessors the Millennium Development Goals (MDGs). Unlike the MDGs, they are universally applicable in all 193 UN member states, and connect the big challenges of our time with each other, such as hunger and poverty, biodiversity loss and climate change, health in an urbanized environment, and sustainable energy, mobility and economic development. Although the SDGs are not directly legally binding treaty obligations, it can be argued that they have an indirect legal status, as states are required to consider them as a commitment made by the international community in good faith (Atapattu, 2006).

The fact that UN member states in September 2015 agreed on 17 universal Goals for the transition to sustainable societies is a huge achievement. Some argue that it succeeded because it was a bottom-up process involving many member states and stakeholders, and commitment had been growing long before the final adoption. Considering the differences between countries as regard their starting positions, physical, economic, social and environmental conditions, the

SDGs will continue to be inspiring and generate many win-win results, but it will not be a walk in the park. Baskets full of performance indicators have already been proposed to monitor progress, in addition to the one good composite indicator, which is already there and allows timelines since 1980, which can serve as a good proxy for progress and direction: the combination of Human Development Index and Ecological Footprint.

Sustainability governance

What 'sustainability governance' means depends on how we define the term governance. The point of departure of an in-depth discussion about governance should be that governance is never neutral. It is not purely technical, even if benchmarks are used as 'evidence'. It is about steering (stemming from the old Greek word *kubernare*, the act of controlling a ship), but not in the mechanical way suggested by the term cybernetics. It is always based – albeit often implicit – on underlying principles and values (Meuleman, 2010, In 't Veld et al., 2011).

The history of the term governance started with a focus on a perceived transition from 'government' to 'governance', i.e. from state as the ruler of society to a process of interaction between state and other actors. For some scholars, governance became synonymous with 'network governance', and part of this school of thinking then became even anti-statist: governance is about solving societal problems without involvement of the state.

I define governance as addressing the whole of organizing problem-setting, solution-finding and implementation of societal challenges, partially because I am a practitioner of governance – and as a practitioner, I want a fully equipped toolbox if I can get it, full stop. Section 2.1 gives a short history of the term governance, showing that there are many definitions, most of which cover only part of the broad scope needed to address sustainable development challenges. At this place, it suffices to say that we need a very broad but still focused definition of governance if we want to use it for sustainable development, a holistic field of application.

An online glossary of the United Nations (retrieved on 11.02.2018) goes into this direction but the definition it gives (based on a UNDP source from 1997) is relatively state-centric – maybe not surprising, of course, for an association of 193 states. Governance is

> the exercise of economic, political and administrative authority to manage a country's affairs at all levels. It comprises the mechanisms, processes and institutions, through which citizens and groups articulate their interests, exercise their legal rights, meet their obligations and mediate their differences.

'Good governance' is then the exercise of this authority "with the participation, interest and livelihood of the governed as the driving force. In this sense, governance ceases to be a matter of government only" (United Nations, online source).

10 What and for what is governance?

I quite like a very early definition by (Kooiman, 1993) that governance is about "the patterns that emerge from the governing activities of social, political and administrative actors". Taking a later interpretation by (Kooiman, 2003, p. 4) as basis, I find the following definition the most practicable. It covers the three basic governance styles hierarchical, market and network governance (see Section 2.2) and emphasizes the relational dimension of the concept: "**Governance** is the totality of interactions in which government, other public bodies, private sector and civil society participate (in one way or another), aimed at solving public challenges or creating public opportunities" (Meuleman, 2008).

In practice, governance often appears in the form of a specific constellation of institutions, instruments, processes and actor roles on, for example, energy policy or tackling corruption, in a certain political and societal context. A **governance framework** is the totality of instruments, procedures, processes and roles of actors designed to tackle a societal problem. They should be adapted to legal, cultural and physical conditions of the problem environment, and be internally consistent; the normative assumptions (values, hypotheses) should be clear (after Meuleman, 2014).

Governance in the SDGs

The UN member states have mainstreamed governance in the SDGs in quite a smart way. There are two dedicated governance Goals (SDG 16 and 17) (United Nations, 2015):

- Goal 16: Promote peaceful and inclusive societies for sustainable development, provide access to justice for all and build effective, accountable and inclusive institutions at all levels.
- Goal 17: Strengthen the means of implementation and revitalize the Global Partnership for Sustainable Development.

In addition, there are concrete governance targets as enabler under each of the policy Goals 1–15. For example, under Goal 1 (Hunger), governance Target 1.a is about "ensuring significant mobilization of resources from a variety of sources . . . to implement programmes and policies to end poverty in all its dimensions". Target 1.b states that we "need to create sound policy frameworks at the national, regional and international levels . . . to support accelerated investment in poverty eradication actions".

There are 72 governance targets under the SDGs, which is 43% of all 169 targets. This includes all targets under the governance SDGs 16 and 17. Chapter 8 identifies examples of governance challenges for each of the 17 SDGs. At this place, it is fair to say that the mainstreaming of governance in the SDGs is much clearer than it was in the MDGs and is in some of the (national) sustainable development strategies. This is a good starting point for implementation of the SDGs, depending on the definition of governance, of course, and how it matches with the implementation challenges.

Governance: no golden bullet, no one-size-fits-all

A concrete governance framework for e.g. renewable energy, education or eco-innovation is always in interaction with its surroundings, in predictable and unpredictable ways. It is important to have an understanding of this 'environment', also because it may determine the feasibility of specific solutions. A *governance environment* can be defined as the whole of rules and laws, institutional setting, policy instruments, division of tasks and roles of governments and societal stakeholders, as well as the relevant culture, history and geographical conditions within a particular country or other boundary or transboundary context (adapted from Meuleman, 2015). Chapter 6 introduces the cultural and geographical conditions of governance, and Chapter 7 discusses mind-sets and mental silos, with a number of recipes from 'New Public Management' as examples.

The existence of a governance environment, which co-determines what works well and what not, and what is a preferred style of policy preparation, of policy instruments and of involvement of stakeholders, is sometimes ignored. The field of sustainability governance is no exception to the general observation that governance is a minefield of competing claims to universally applicable "best practices". As we will see, this competition misses the point: there is no golden bullet, no single blueprint for designing the governance needed for implementation of the SDGs, as the Committee of Experts on Public Administration (CEPA), a subsidiary UN body coordinated by UNDESA, concluded in 2017:

> There is no universal blueprint for implementing the Goals. Policy making and policy implementation will have to reflect specific national and local conditions, including the political environment. Ensuring participatory decision-making and consulting people is the way to arrive at the kind of policies that will succeed in implementing the Goals in a given society. It means finding arrangements to discuss policies in a very inclusive, yet specific and sector-focused way.
>
> (United Nations, 2017)

This is also the spirit of the World Public Sector Report 2018 (United Nations, 2018). We probably need something that at first looks like a paradox: 'common but differentiated governance' (Meuleman and Niestroy, 2015). Moving towards sustainable energy systems will be different in South Africa from in Switzerland, and sustainable mobility requires different methods and tools in Bhutan than in Brazil. History plays a role, among others because it has a huge influence on the system of democracy – or other systems of governing. National and sub-national values and traditions – in society in general and within public administration organizations – determine the default approaches that are first tried and the fall-back options when they fail (Meuleman, 2008).

A good example of how value-laden governance can be is the popular 'New Public Management' (NPM) approach to governance which is driven by the rational logic of business organizations and demands that governments should be small and first and foremost efficient (Osborne and Gaebler, 1993). As a counter reaction, 'New Public Governance' (NPG) follows the logic of pluralism and networks (Osborne, 2006). The World Bank's 'Good Governance' started with a focus on institutionalizing rule of law, accountability and transparency (World Bank, 1994) and has gradually developed an extensive list of indicators which includes elements of network and market-style governance.

Standardized governance approaches such as New Public Management and New Public Governance have been successful in some cases and places and failed elsewhere. They all have their merits and weaknesses. None of these prescriptions has turned out to be a *panacea* – i.e. universally applicable – although that had been claimed by some of their proponents. Mantras like "break down the silos" and "less is more" have brought about more efficiency, but often to the detriment of effectiveness. Therefore, I will not present ready-made recipes but arguments pro and contra about various ways of designing, organizing, managing and evaluating (sustainability) governance processes in specific situations, and with taking into account the dynamics of governance systems in relation to their environments.

There are many examples of successful governance and governance theory has contributed to novel approaches to emerging and old problems. However, especially in the field of sustainable development, governance failures seem to be everywhere. In Chapter 3, we will take a closer look at governance failures and discuss if and how metagovernance can help prevent or mitigate them. Before this, we should analyse the state of play of sustainable development governance, and in particular of governance of SDG implementation.

Sustainability governance

Governance for sustainable development has followed (and in some cases fuelled) the overall discussion about what governance is and should be. I use 'governance for sustainable development' and 'sustainability governance' as synonyms for pragmatic reasons. It is a trade-off: Actually, I do not like 'sustainability' very much because it sounds like a steady-state thing whereas 'sustainable development' should be a dynamic process (indeed, with three 'dimensions', instead of 'pillars'). However, in some countries, politicians and policy makers hear in 'sustainable development' the connotation of (sustainable) 'development cooperation'. With the SDGs we have passed that phase: they are adopted as universal goals for all countries.

Taking a similar broad definition of governance as I am using, as point of departure, sustainability governance was defined in 2005 as "the deliberate adjustment of practices of governance in order to ensure that society eventually proceeds along a sustainable trajectory". It implies "the deliberate adjustment of practices of governance in order to ensure that social development proceeds along a sustainable trajectory" (Meadowcroft et al., 2005, p. 5).

The preparation of the SDGs was a novel governance process itself, and it is very laudable that three of the most closely involved people have published an inside story about the whole preparation and negotiation process (Kamau et al., 2018). The 'Open Working Group' coordinating the work was indeed a very open group. A key result was the proposal to establish a High-Level Political Forum (HLPF) at the UN level as the main steering and monitoring mechanism for the implementation of the SDGs. Since the adoption of the 2030 Agenda and its 17 SDGs and 169 targets in September 2015, many countries have taken measures to create the framework for their implementation. This includes establishment or reinvigoration of high-level coordination groups and sustainable development councils at the national level. It covers policy instruments such as sustainable development strategies, monitoring reports and, notably, the Voluntary National Reviews to the HLPF (22 countries in 2016, 44 in 2017, and 47 in 2018).

In addition, new, more inclusive policy making and implementation processes are established which should guide effective SDG implementation. Finally, collaboration between the public sector and civil society and business is taking new shapes with new types of partnerships.

All UN member countries have agreed to turn these common Goals (Figure 1.2) into effective frameworks in their different realities. In doing so they also need to install mechanisms to prevent and mitigate governance failures, as this should prevent huge economic, social and environmental costs. This requires that those who are responsible for sustainability governance (the 'governors') are aware of the incompatibilities of governance styles, and of the fact that governance failure is not only contingent on unpredictable events but often results from a mismatch between a chosen governance approach and the problem in its societal context.

The governance challenge posed by sustainable development and in particular the SDGs has been a popular topic among scholars as well as public practitioners and stakeholders. In the years up to 2015 most publications focused on the selection and framing of the SDGs. For example, the need to have a separate SDG on urbanized areas – now SDG 11 – was subject to a broad lobby. Since around 2015, many surveys, studies and opinion articles have been published about the governance for the implementation of the SDGs. Current approaches to sustainability governance can be clustered in four types in which different principles dominate (see Section 2.4):

- Strong central leadership is necessary
- Strong bottom-up and partnership action are needed
- Focus on transition management
- Multi-perspective approach: apply situational governance.

1.3 Sustainability metagovernance

The acknowledgement within the Agenda 2030 that implementing the SDGs requires strong and tailor-made governance is a good start but not enough to make it an easy task. In this book I argue that we need metagovernance (Jessop, 1995, 2011, Kooiman, 2003, Meuleman, 2008, Sørensen, 2006, Larsson, 2015) or

Figure 1.2 The UN Sustainable Development Goals
(UN, 2015)

Introduction   15

'governance of governance'. This concept has become a theoretically sound and practically feasible approach to deal with very different governance challenges, ranging from complex and 'wicked' to simple and routine problems. The most relevant independent variables are the cultural, social and economic characteristics of the society under consideration, and the (framing of the) problems concerned.

The term metagovernance was proposed independently by three scholars from different theoretical traditions in the mid-1990s: the state theorist Bob Jessop, the systems theorist Jan Kooiman, and the political scientist Andrew Dunsire. Jessop (1997) defined metagovernance as "coordinating different forms of governance and ensuring a minimal coherence among them". Later he specified this description as "the organization of the conditions for governance", which involves "the judicious mixing of market, hierarchy and networks to achieve the best possible outcomes from the viewpoint of those engaged in metagovernance" (2003). It is about rebalancing market, hierarchy and networks (Jessop, 2005).

Metagovernance can be defined as "a means by which to produce some degree of coordinated governance, by designing and managing sound combinations of hierarchical, market and network governance, to achieve the best possible outcomes" (adapted from Meuleman, 2008). In my original definition (Meuleman, 2008), which combined elements from Jessop (2003), Kooiman (2003) and Sørensen (2006), I had added "from the viewpoint of those responsible for the performance of public-sector organizations: public managers as 'metagovernors'", as the research was on how public managers were using metagovernance. Meanwhile it has become clear that metagovernance takes place in the public as well as in the private domain, although with different scopes and objectives (Meuleman, 2008, Steurer, 2012, Derkx and Glasbergen, 2014). Private sector companies and/or civil society organizations, or their umbrella organizations are facing similar challenges as governments to reach their objectives, with regard to the complexity of the constantly changing environments, and with a need for a long-term perspective. This book concentrates on *public* metagoverance, but a short introduction on *private* metagovernance will be given in Section 4.4. Public and private metagovernance meet each other in partnerships of public, private and civil society organizations, such as the ones promoted by the SDGs (see e.g. Beisheim and Simon, 2015).

Applying metagovernance is no guarantee for success. It is not ballistics where you can calculate where and when a bullet will hit the target. We will even never be sure if the right bullet was chosen, not even after a successful hit of a policy target: there could be other reasons for a success, beyond the workings of the governance framework. Besides ambiguity and complexity, sheer luck or trouble is part of public governance in daily life. Jessop (2009) therefore suggested that metagovernance requires irony: participants in governance should recognize the likelihood of failure but still proceed as if success were possible. Nevertheless, metagovernance can help designing and managing governance frameworks, taking into account specificities of the actual 'governance environment', which helps preventing and mitigating governance failures to some extent.

In Chapter 4, a broad approach to metagovernance will be explained in more detail, including about how it differs from more narrow approaches using the same term. Chapter 8 formulates some metagovernance challenges for the SDGs, while Chapters 9–11 will discuss the practical use of metagovernance with a view to, respectively, realizing the conditions under which it can be used to develop situational governance frameworks, a seven-step method to apply metagovernance in a concrete situation, metagovernance and public sector reform and coherence promotion, and metagovernance capacity building.

1.4 Structure of the book

This first chapter presented key factors influencing the results of governance and metagovernance in general and in particular for sustainable development. In the next chapters I will analyse success and failure with a view to draw conclusions (in the last chapter) on three questions:

1. How have metagovernance theory and practice developed during the last decade (2008–2018), based on academic and other publications?
2. Is it plausible that unsuccessful public managers are, generally speaking, not using the metagovernance lens to prevent and mitigate governance failures?
3. How can the holistic concept of metagovernance be applied to the most holistic of contemporary policies: striving for sustainable development, and in particular for implementation of the 2015 UN Sustainable Development Goals (SDGs)?

In addition, this book takes a broad lens, taking concepts and conclusions not only from governance theory but also from other social sciences. It is in essence a *transdisciplinary* book as it integrates concepts and research results from various academic disciplines with observations, insights and experiences from practice inside public administration organizations.

The book has three parts.

Part 1. What and what for is governance?

Chapter 1 discussed why sustainability governance generally does not live up to its promises. The term 'governance' was defined and we saw how metagovernance has emerged as a method to design and manage (sustainability) governance frameworks. Chapter 2 shows that three basic governance styles tend to occur in combinations and that there are several proposals to consider a fourth style. Chapter 3 gives an overview and typology of governance failures, which can be dealt with through metagovernance.

Metagovernance will be introduced in Chapter 4 as the 'art' of combining the different available approaches into a feasible way to reach the desired results, which could help preventing and mitigating the effects of governance failure and open new avenues for breakthroughs. It will be argued that metagovernance can help to achieve progress much faster, and more effectively and efficiently than through traditional approaches like muddling through, top-down steering or leaving everything to market mechanisms, because it takes into consideration the specificities of each (national, subnational, local) context. I will also show that metagovernance is not without failure.

Part 2. Features of metagovernance

In order to apply metagovernance, good knowledge is required of the many differences between three basic governance styles. Chapter 5 presents fifty ways in which the three governance styles differ from each other. The governance styles and their combinations are linked to the vision and strategy of policy makers, which institutions (formal, informal) are thought to be effective, which policy tools and instruments are believed to be appropriate, how to organize decision-making processes, and which roles should be given to societal actors (administrations, business, civil society as well as science, independent think tanks and the media) and their partnerships.

How values, traditions and geography influence governance in different ways in different countries, and how this can lead to a preference for a certain governance style with related tools, while blocking other solutions, is the theme of Chapter 6. Chapter 7 addresses a special case of governance failure, namely when specific 'mind-sets' and not mindful deliberation seem to determine what happens. Why do we tend to think that 'best practices' are the best, and that 'less is more'? This chapter addresses a number of 'mental silos' originating from New Public Management that have resulted in governance failure, and suggests several strategies to tackle this problem.

Part 3. Metagovernance for sustainability

Chapter 8 focuses on application and applicability of metagovernance to the 17 Sustainable Development Goals. What can be said about key governance challenges for each Goal and what are typical constraints? Chapter 9 presents a sketch of a concrete seven-step method to apply metagovernance, with examples and suggested tools. Chapter 10 discusses why public sector modernization and reform need focusing on the SDGs, on policy coherence and on the capacity to apply metagovernance. Chapter 11 discusses the three questions asked in Section 1.5, the feasibility of metagovernance, how this is relevant for tackling systemic challenges and for effective partnerships, and wraps up the whole book.

The structure of the book requires that I treat several issues in two or more places, but from a different angle. For example, partnerships may encounter governance failures (Chapter 3), are a feature of governance (Chapter 5), are a central part of SDG 17 (Chapter 8), and will again be discussed in Chapter 11 on the use of metagovernance for implementation of the SDGs.

1.5 Conclusions

This chapter explained why sustainability governance is so difficult in practice, and presented definitions of key terms in this book. It concluded that:

- Metagovernance (a means) and sustainable development (a goal) are two holistic concepts with complex interrelations.
- Metagovernance or 'governance of governance' could be an appropriate way to deal with sustainability challenges and in particular with the SDGs.
- Three questions should be dealt with in this book, about (1) how has metagovernance theory and practice has developed during the last decade (2008–2018); (2) if it is plausible that unsuccessful public managers are not using the metagovernance method; and (3) how metagovernance can be applied for implementation of the 2015 UN Sustainable Development Goals.

Before I will elaborate the term metagovernance and its practical use more in-depth in Chapter 4, the next two chapters will pave the ground for metagovernance as method to deal with governance problems. Chapter 2 introduces three basic governance styles – hierarchical, network and market governance and their hybrids – which sometimes undermine each other because they are based on contrasting values and principles. Chapter 3 discusses examples of various kinds of governance failures.

References

Atapattu, S. 2006. *Emerging Principles of International Environmental Law*. Leiden: Brill-Nijhoff.

Beisheim, M. & Simon, N. 2015. Meta-governance of partnerships for sustainable development actors' perspectives on how the UN could improve partnerships' governance services in areas of limited statehood. SFB-Governance Working Papers Series. Berlin: Collaborative Research Center (SFB).

Brundtland, G., Khalid, M., Agnelli, S., et al. 1987. *Our Common Future ('Brundtland report')*. Oxford: Oxford University Press and United Nations World Commission on Environment and Development.

Derkx, B. & Glasbergen, P. 2014. Elaborating global private meta-governance: An inventory in the realm of voluntary sustainability standards. *Global Environmental Change*, 27, 41–50.

Frantzeskaki, N., Jhagroe, S. & Howlett, M. 2016. Greening the state? The framing of sustainability in Dutch infrastructure governance. *Environmental Science & Policy*, 58, 123–130.

Hersey, P. & Blanchard, K. H. 1988. *Management of Organizational Behavior: Utilizing Human Resources*. New York: Prentice Hall.

In 't Veld, R. J., Töpfer, K., Meuleman, L., et al. 2011. *Transgovernance: The Quest for Governance of Sustainable Development*. Potsdam: Institute for Advanced Sustainability Studies (IASS).

Jessop, B. 1995. The regulation approach, governance, and post-fordism: Alternative perspectives on economic and political change? *Economy and Society*, 24, 307–333.

Jessop, B. 1997. Capitalism and its future: Remarks on regulation, government and governance. *Review of International Political Economy*, 4, 561–581.
Jessop, B. 2003. Governance and metagovernance: On reflexivity, requisite variety, and requisite irony. In: Bang, H. (ed.) *Governance as Social and Political Communication*. Manchester: Manchester University Press.
Jessop, B. 2005. The political economy of scale and European governance. *Tijdschrift voor economische en sociale geografie*, 96, 225–230.
Jessop, B. 2009. From governance to governance failure and from multi-level governance to multi-scalar meta-governance. In: Arts, B. & Al, E. (eds.) *The Disoriented State: Shifts in Governmentality, Territoriality and Governance*. Heidelberg: Springer.
Jessop, B. 2011. Metagovernance. In: Bevir, M. (ed.) *The Sage Handbook of Governance*. London: Sage.
Jordan, A. 2008. The governance of sustainable development: Taking stock and looking forwards. *Environment and Planning C: Government and Policy*, 26, 17–33.
Kamau, M., Chasek, P. & O'connor, D. 2018. *Transforming Multilateral Diplomacy: The Inside Story of the Sustainable Development Goals*. London: Routledge.
Kooiman, J. 1993. *Governance and Governability: Using Complexity, Dynamics and Diversity*. London: Sage.
Kooiman, J. 2003. *Governing as Governance*. London: Sage.
Larsson, O. 2015. *The Governmentality of Meta-Governance. Identifying Theoretical and Empirical Challenges of Network Governance in the Political Field of Security and Beyond*. Skrifter utgivna av Statsvetenskapliga föreningen i Uppsala 193. 204 pp. Uppsala: Acta Universitatis Upsaliensis. ISBN 978-91-554-9296-0.
Meadowcroft, J. 2011. Sustainable development. In: Bevir, M. (ed.). London: Sage.
Meadowcroft, J., Farrell, K. & Spangenberg, J. 2005. Developing a Framework for Sustainability Governance in the European Union. *International Journal of Sustainable Development* 8.1–2, 3–11.
Meuleman, L. 2008. *Public Management and the Metagoverance of Hierarchies, Networks, and Markets*. Heidelberg: Springer.
Meuleman, L. 2010. The cultural dimension of metagovernance: Why governance doctrines may fail. *Public Organization Review*, 10, 49–70.
Meuleman, L. 2014. Governance frameworks. In B. Freedman (ed.) *Global Environmental Change*. Dordrecht: Springer.
Meuleman, L. 2015. Owl meets beehive: How impact assessment and governance relate. *Impact Assessment and Project Appraisal*, 33, 4–15.
Meuleman, L. & Niestroy, I. 2015. Common but differentiated governance: A metagovernance approach to make the SDGs work. *Sustainability*, 12295–12321.
Niksic Radic, M. 2015. FDI and good governance dimension of sustainability: Empirical evidence from Croatia. *3rd International Scientific Conference Tourism in Southern and Eastern Europe 2015*.
Nurse, K. 2006. Culture as the Fourth Pillar of Sustainable Development. In: *Small States: Economic Review and Basic Statistics (Vol. 16)*. London: Commonwealth Secretariat.
Osborne, S. 2006. The new public governance? *Public Management Review*, 8, 337–387.
Osborne, D. & Gaebler, T. 1993. Reinventing government: How the entrepreneurial spirit is transforming the public sector. *Resenhas*, 33, 97–99.
Pisano, U., Lange, L. & Berger, G. 2015. The 2030 agenda for sustainable development. Governance for SD principles, approaches and examples in Europe. *ESDN Quarterly Report*. Institute for Managing Sustainability Vienna University of Economics and Business.

Rayner, J. 2015. The past and future of governance studies: From governance to metagovernance? *In:* Capano, G., Howlett, M. & Ramesh, M. (eds.) *Varieties of Governance: Studies in the Political Economy of Public Policy*. Basingstoke: Palgrave Macmillan.

Sørensen, E. 2006. Metagovernance: The changing role of politicians in processes of democratic governance. *The American Review of Public Administration*, 36, 98–114.

Spangenberg, J. H., Pfahl, S. & Deller, K. 2002. Towards indicators for institutional sustainability: Lessons from an analysis of Agenda 21. *Ecological Indicators*, 2, 61–77.

Steurer, R. 2012. *The Governance of Sustainable Development: Putting the Pieces of Regulation Together*. Vienna: University of Natural Resources and Life Sciences.

United Nations. 2012. Secretary-general's message on the International day for preventing the exploitation of the environment in war and armed conflict. *United Nations Press Release* [Online]. Available: www.un.org/press/en/2012/sgsm14615.doc.htm.

United Nations. 2015. Transforming our world: The 2030 agenda for sustainable development. New York: United Nations.

United Nations. 2017. Report on the Sixteenth Session (24–28 April 2017) of the United Nations Committee of Experts on Public Administration (CEPA). New York: United Nations.

United Nations. 2018. *World Public Sector Report 2018. Working Together: Integration, Institutions and the Sustainable Development Goals*. New York: United Nations.

United Nations. *United Nations governance and public administration online glossary*. [Online] Available: https://publicadministration.un.org/en/About-Us/UN-Glossary

World Bank. 1994. *Governance – The World Bank's Experience: Development in Practice*. Washington, DC: World Bank.

Young, O. R. 2017. Conceptualization: Goal setting as a strategy for earth system governance. *In:* Kanie, N. & Biermann, F. (eds.) *Governing Through Goals: Sustainable Development Goals as Governance Innovation*. Cambridge, MA: MIT Press.

2 Three governance styles and their hybrids

> One chord is fine. Two chords are pushing it. Three chords and you're into jazz.
> – Quote attributed to Lou Reed

I have something with the number three, just like others like to see things in four versions. Sustainable development has three dimensions. Check. There are three seasons (hot, cool and wet – at least when you live in Thailand). Check. And then there are three classical governance styles. Check.

This chapter elaborates the broad governance definition presented in Chapter 1 (Section 2.1) and then distinguishes three basic governance styles: hierarchical, network and market governance (Section 2.2). They are proxies for quite different values and traditions that are driving governance. The next section presents hybrid forms of the three basic styles and addresses the search for a fourth style. In Section 2.4, four schools of thinking about sustainability governance are distinguished. Section 2.5 formulates three problems arising from interactions between governance styles and from their normative character.

2.1 What is meant by governance?

The term 'governance' emerged around 1980 in academic debate to address a shift from government as the main or even only actor responsible for and capable of solving societal problems. Government was considered generally as too much focused on hierarchical steering, command and control and too little on mechanisms and solutions emerging from outside government. In the 1980s, the first new governance style to become influential was market governance, in its hybrid form with hierarchical governance: New Public Management (NPM). NPM originated in Anglo-Saxon countries, was embraced in many (Western) countries, and became part of the World Bank philosophy – which made it mandatory for less developed countries.

In the 1990s, hierarchical and market governance became accompanied by a third style. Civil society and private sector began to demand more than just information: they wanted to be heard, wanted to interact with government, and sometimes even to co-decide. What we now find a normality, stakeholder involvement, is one of the main characteristics of this shift (Driessen et al., 2012). This focus on governance as steering through networks has remained important, especially in countries with a consensus culture. Early (network) governance literature tended to uncritically adopt a state declinist model (Bell and Park, 2006), but the current school of governance focusing on networks has become less anti-state and has accepted that the state has a role in governance.

The 2000s and 2010s have seen a continued dissemination of market governance ideas around the world, although in Western European countries the weaknesses of this approach – in particular the focus on efficiency to the detriment of effectiveness – became gradually more exposed and market governance lost some of its attractiveness. The network approach has become more popular than in the past in international policy making and implementation, and its partnership approach figures strongly in the SDGs.

We are now close to the 2020s and all three basic governance styles are 'alive and kicking'. Another important feature is the increasing insight that there is no one-size-fits-all in governance. It is therefore a good moment to revisit the definition of the term 'governance'.

As was argued in Chapter 1, governance is not about the content of policies (*what* to do?) or about the vision behind policies (*why* do it?), but concentrates on *how* to achieve objectives. Governance therefore is not about policy: the decisions of whether a 50% or 70% waste recycling rate will be demanded, or whether 10% or 15% youth unemployment is still acceptable are policy decisions. The choice of the policy tool (legal, financial, voluntary) is also political, albeit not in the realm of policy but in that of governance. Governance therefore includes polity (the institutions and instruments) and politics (the processes). These dimensions of governance are strongly interrelated:

- A sophisticated institutional framework is a necessary basis but not sufficient when dedicated policy instruments, implementation processes and involvement of relevant actors are absent. When one or more dimensions are weak, governance failure may emerge (see Chapter 3).
- While polity used to be a stable factor in the past, Hajer (2003) already stated that "policy making now often takes place in an 'institutional void' where there are no generally accepted rules and norms according to which politics is to be conducted and policy measures are to be agreed upon". He argued that "more than before, solutions for pressing problems transgress the sovereignty of specific polities". This is an important insight about the institutional dimension for SDG implementation: there is no single blueprint for polity, but polity has become a dynamic variable.

In his seminal *Handbook of Governance*, Bevir (2011) provides a very broad definition: governance is about "issues of social coordination and the nature of all patterns of life". Also Rayner (2015) defines governance in a broad way as a "heuristic lens through which the contextual realities of the coordination of multiple actors and institutions in the policy system can be reconstructed in detail".

I agree that the term governance should cover a lot. It should not only refer to structures and procedures, but also to the relations with those who are governed, regardless if they are considered as subordinates, partners or clients: governance is a relational concept. Governance is also intentional, based on assumptions about what would work best to achieve a certain policy objective.

There is nothing against the existence of many governance definitions: "All visions of governance are legitimate within their own parameters" (Osborne, 2006). Legitimacy does not say much, however, about effectiveness – a characteristic I am mostly interested in, being a practitioner of governance myself. What is important to understand is that governance definitions and frameworks are normative. They are carriers of values. Some embrace consensus and empathy, others entrepreneurship and competition, and again others authority and control. Therefore, "We can define governance as a collection of normative insights into the organization of influence, steering, power, checks and balances in human societies" (In 't Veld et al., 2011).

These values, although they are the drivers for governance practice, often play a rather obscure role and remain implicit (Kooiman and Jentoft, 2009). In other cases, the underlying values are explicit and actively used to market the 'brand'. An early definition of governance by the World Bank is that governance is "the exercise of political authority and the use of institutional resources to manage society's problems and affairs" (World Bank, 1994). With 'Good Governance', the World Bank wanted to promote the values of effectiveness, transparency, control and efficiency. Such values, even if the word 'good' is part of the brand name, are not automatically universal. Already in 2004, it was criticized that the list of indicators identified by the World Bank cannot realistically be applied everywhere. As a reaction, it was suggested to moving towards 'good enough governance' (Grindle, 2004):

> Working toward good enough governance means accepting a more nuanced understanding of the evolution of institutions and government capabilities; being explicit about trade-offs and priorities in a world in which all good things cannot be pursued at once; learning about what's working rather than focusing solely on governance gaps; taking the role of government in poverty alleviation seriously; and grounding action in the contextual realities of each country.

The World Bank has periodically adapted its list of indicators, which meanwhile also promotes democracy, rule of law, anti-corruption measures. In 2017, the World Bank defined governance more as a relational concept, namely as "the process through which state and non-state actors interact to design and implement policies within a given set of formal and informal rules that shape and are shaped by power" (World Bank, 2017).

Many other governance definitions (mainly variations of network governance, such as reflexive governance and adaptive governance) share the view that actors value empathy, trust and cooperation more than, for example, hierarchical governance values such as legitimacy, authority or accountability. Scholars from the Netherlands (Klijn, 2008, Hajer and Wagenaar, 2004), Denmark (Sørensen, 2006, Torfing and Triantafillou, 2011) and Sweden (Larsson, 2015) tend to appreciate such definitions, which match with the values of their national cultures. Also in the UK, governance has been seen as an interactive process between government and stakeholders: governance is

> a process whereby formal governing structures are no longer focused primarily on the political realms of public sector government (parliament, town/city hall, civil servants), but are increasingly incorporating a range of interests drawn also from the private sector and civil society.
>
> (Whitehead, 2003)

How we define governance is important because what works well in Finland may not do the trick in France, and Canadian successes may not work in China. The same problem could require a law in Germany, financial incentives in the USA or the UK, a partnership agreement in Denmark or the Netherlands, or naming and shaming in China. Although evidence shows that 'best practices' are rare and 'successful practices' are abundant, many practitioners of governance are still thinking in terms of 'best' – linking this appreciation to what they believe works best in their own country.

I consider the narrow approach to governance ('governance is network governance') not helpful for the holistic approach of the pathway to sustainability and for in particular for the implementation of the SDGs, because of its built-in conviction that network governance is the best approach in any situation in any country, at any time. It distracts from real-life power games for which we need to understand and be prepared to use hierarchical governance. Market governance and network governance can similarly be a blessing and a threat to effective governance, depending on what, where, who and why. Research and experiments based on network governance principles may produce useful results also in countries where the network approach is not the dominant style.

Putting the above in a broader context, globalization has brought about the idea that everything is global. This is not the case. The counter reaction, localization, is at least as important, and we are living in times where seemingly contradictory concepts exist together and may even need each other (Beck, 1992, In 't Veld et al., 2011): globalization and localization are together framed as 'glocalisation'.

Governance styles are not only normative because of the 'ways of life' (Thompson et al., 1990) they represent, but also because there are normative assumptions behind institutions and policy (implementation) instruments. Changing behaviour can be achieved with a variety of measures and tools, e.g. with penalties, financial incentives or voluntary collaboration – respectively tools from hierarchical, market and network governance. When political parties negotiate a coalition agreement to form a multi-party government, the fights are often not about the goals but about the means to reach the goals. It is sometimes easier to agree on the goals than on the tools that are required to achieve them. This suggests that the implementation of the SDGs may be an even more difficult quest than the adoption of them as common Goals.

Chapter 6 will show that relational values that define how people value other people's values can be linked to governance styles. For example, hierarchical governance is linked to the relational value of hegemony, network governance to pluralism and tolerance, and market governance to indifference (In 't Veld, 2010, Meuleman, 2013). A final common feature of the governance approaches discussed in this chapter is that their focus lies on *public* problems and opportunities. Of course, what is considered as public and what not, is subjective; it is a political choice.

The broad definition given in Section 1.2 covers the above-mentioned common characteristics: "**Governance** is the totality of interactions in which government, other public bodies, private sector and civil society participate (in one way or another), aimed at solving public challenges or creating public opportunities".

This definition covers more than what public actors actually *do*: public management is governance, but not all governance is public management (Kickert et al., 1997). It is a dynamic definition, in contrast to many other definitions, which are static. A static governance definition works like a prescription; a typical formula is 'TINA': 'there is no alternative' (Pollitt and Bouckaert, 2011). A dynamic governance definition allows many possible alternatives; one could say that 'there is always a solution' ('TIAS').

A recurring question during my discussions with students is if there are common values across all three basic governance styles. If any, would 'rule of law' or at least legitimacy not be the common value? Indeed, rule of law – if seen as a set of values – fits more in the logic of hierarchy than in the logics of markets or networks. This does not mean that cultures with a strong network or market governance preference would not agree to accepting rule of law. Only a '100% pure' network or market style might reject this, but this is very rare. The most common examples of countries without rule of law are not those with a network or market governance culture, but with an (extreme) hierarchical culture. There are exceptions: productive communities have emerged without common rules, bosses or formal accountability. Internet communities are one example, but there are also 'offline' examples. In Delft (the Netherlands), a commercially very successful consultancy had only one rule, namely that there were no rules. The basic approach was self-governance and the consultancy operated from two workplaces designed as 'grand cafés'. It became a fast-growing (500 staff in five years), fashionable company which was later integrated in a larger one (Derix, 2000).

Governance and metagovernance: revisiting the conceptual crowd

Figure 2.1 is an update after 10 years of the one I published in 2008, which showed that a 'conceptual crowd' had developed of variations or hybrids of 'new modes of governance', and particularly based on the network governance style. Besides the crowd around network governance, a clear 'conceptual void' was noticeable around hierarchical governance. In 2008, it was still more popular to promote

26 *What and for what is governance?*

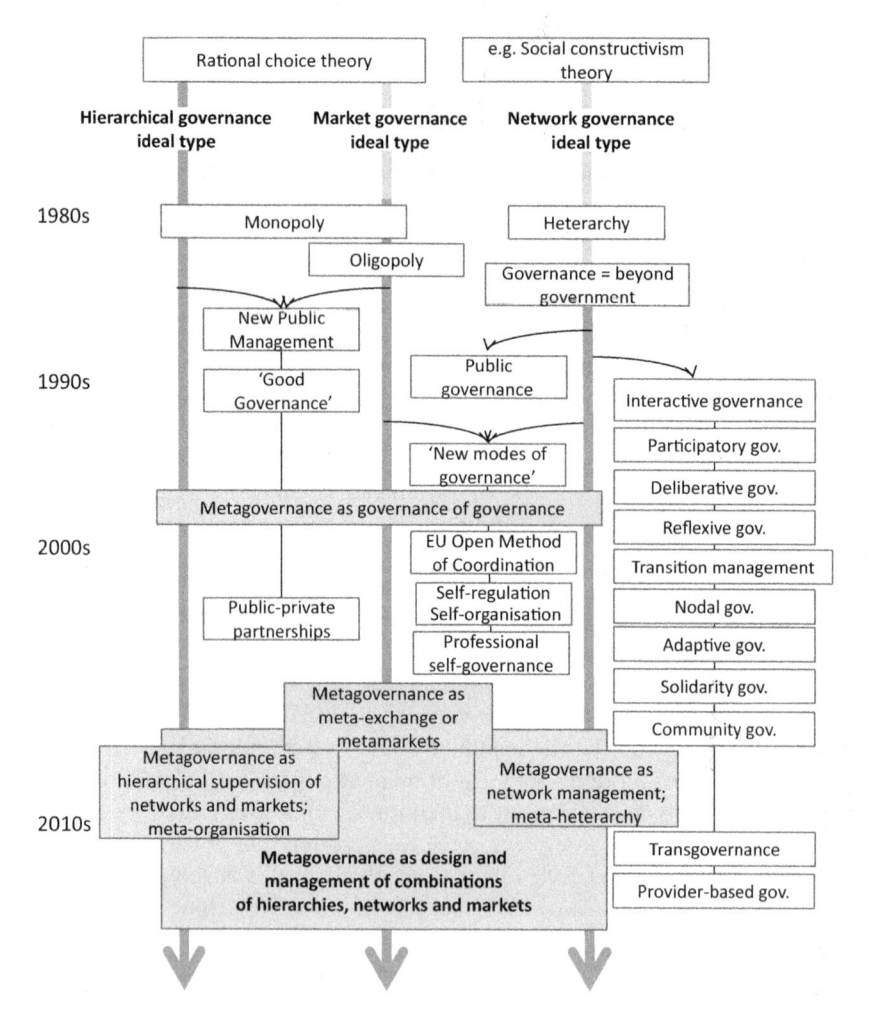

Figure 2.1 Governance and metagovernance: the conceptual crowd revisited
(After Meuleman, 2008, p. 73)

and research network-style governance than to take hierarchical governance seriously as a style that has something good to offer. Market governance was still going strong but already over its peak, as the disadvantages of pure efficiency thinking and using market-based instruments for all thinkable problems were slowly becoming clear.

Maybe the most striking thing about Figure 2.1 is that during the past 10 years, governance theory has not produced many new concepts that succeeded to enter the premier league of governance concepts. In addition, the governance-is-all-about-networks approach seems to have grown but also increased its isolation.

Much of the research publications are self-referential within the network paradigm, and I rarely notice that advocates of network governance relate to the broader governance and metagovernance literature in their publications.

2.2 Governance styles: hierarchical, network and market governance

In governance literature, the terms governance *styles* and *modes* are both used. I have been thinking why I prefer 'styles' to 'modes'. The term 'modes' is more popular: counting all hits on both terms on Google Scholar together gives 71% of hits on governance mode and only 29% on governance style.[1] In the plural form, the difference is even bigger: only 18% of the hits are about governance styles. The words mode and style mean more or less the same when linked to the word governance, but the slight difference is still telling: words *do* matter. The word 'mode' suggests prefabricated, given arrangements between which can be switched (like on my bicycle computer, or in an airplane between autopilot and manual), but which are themselves not dynamic. A pilot cannot select a mode between autopilot and manual flying. The word 'style' is less rigid, less prescribed. People working at a ministry will say "this is how we do things here: this is our *style*". A style has underlying values, is more personal and relational and this is precisely how I approach governance, and consequently metagovernance, in this book.

Governance styles can be defined as "the processes of decision making and implementation, including the manner in which the organizations involved relate to each other" (Van Kersbergen and Van Waarden, 2004, p. 143). They "define the roles and lines of responsibility of public sector and societal players in different ways: hierarchical, network and market governance" (Meuleman, 2008).

Most governance literature distinguishes three basic governance styles – hierarchical, network and market governance (Frances et al., 1991, Jessop, 1997, Kaufman, 1986, Kickert, 2003, Kooiman, 2003, Lowndes and Skelcher, 1998, Considine and Lewis, 2003, Meuleman, 2008, Peters, 1998, Powell, 1990, Schout and Jordan, 2003, Thorelli, 1986, Pollitt and Bouckaert, 2011, Jordan, 2008). These basic styles mirror three modes of social order, respectively state, community and market (Streeck and Schmitter, 1985). The styles are always mixed, can undermine each other, and can become belief systems. In the end, it is all about finding the right mixture. Hierarchical governance is still the most practiced style and remains important as solution-provider for certain issues and in distinct contexts. Network governance has gained momentum for sustainable development as includes ways to deal with complexity. Market governance has shown that it can lead to efficient, but at the same time ineffective solutions.

In addition to the three 'pure' styles, many hybrid forms can be distinguished. Since my metagovernance study of 2008 I have tried to find other basic styles but I failed. The three are 'ideal types' in the sense of Weber: logical constructions, which can be used to aid in the understanding of reality, and therefore they are rarely found in a pure form.² Successful hybrid forms have become the object of research and grew fashionable among politicians, sometimes to the extent that they became known as 'best practices'. However, networks, markets and (hierarchical) bureaucracies are rivalling ways of allocating resources and coordinating policy and its implementation (Rhodes, 2000, p. 345).

To understand why hierarchical, market and network governance styles conflict, contradict, undermine or support each other it is important to know about the potential tensions, trade-offs and synergies between them. Each of them "has its own postulated integrity, autonomy and tendency towards equilibrium and reproductivity. Each has distinctive properties and processes, and each corresponds to distinctive aspects of the human condition" (Streeck and Schmitter, 1985).

The three basic styles will be introduced below in a concise way. The next section addresses the permanent quest for a fourth style.

Hierarchical governance

Historically, hierarchical governance was the main approach of governments, at least since Napoleonic and Prussian times. The sociologist Max Weber, who formulated the key principles of hierarchical governance, believed that governmental tasks should be divided into sub-tasks and that decision making should be organized through top-down, rational processes based on legal instruments and without involvement of what we nowadays call "stakeholders". Hierarchical governance has a range of characteristics, which together compose a very strong logic. Hierarchical values and principles typically include rationality, reliability, stability, legitimacy, justice, accountability, risk averseness, government-centeredness, centralization, strategy as planning and design, authority, managing by instructions, one-way communication, dependency, subordinates, obedience, rules and command and control. A useful definition elaborated from (Steenkamp and Geyskens, 2012) is that **hierarchical governance is a governance style based on enforcement by means of legitimate authority, either through an employment relation (i.e. vertical integration) or a detailed contractual arrangement that provides decision-making authority in certain areas.**

With the emergence of network and market governance, the "new modes of governance" (Héritier, 2002), hierarchical governance instruments have become less fashionable in many countries. It is increasingly difficult to establish global legally binding agreements, as we have seen with climate change governance. However, Pierre and Peters' observation from 2000 is still correct: "the growing interest in governance should not be confused with a decline of the state but is rather proof of the state's ability to develop new strategies to maintain some degree of control" (Pierre and Peters, 2000, p. 136).

Network governance

In the 1980s and 1990s, public participation in the context of policy preparation emerged as a logical expression of the changing times in at least most Western countries. The emancipation of the public, higher education levels, and the establishment of many large and small environmental advocacy groups put pressure on governments to listen better and in an earlier stage, to the interests of "stakeholders". This was reflected, for example, in the fact that the Environmental Impact Assessment Directive of the EU (1985) contained provisions for public involvement in an early stage of project permitting procedures. In 1998, the United Nations Economic Commission for Europe (UNECE) established the Aarhus Convention on public participation, access to environmental information, and access to justice, to which 45 European nations have now subscribed. As a governance style, network governance is characterized more by cooperation rather than coercion or competition, by trust rather than authority or price, and by interdependency rather than dependency or independency. **Network governance** can be defined as "the 'management' of complex networks, consisting of many different actors from the national, regional and local government, from political groups and from societal groups (pressure, action and interest groups, societal institutions, private and business organisations" (Kickert et al., 1997, p. 735). It resembles the "muddling through" Lindblom observed (1959, 1979): it is not revolution but evolution, and more incrementalism than taking big steps.

Network governance favours dialogue and partnerships, which is quite different from the top-down approach of hierarchical governance and the individualist attitude of market governance in which there are only buyers and sellers. Typical terms connected to network governance are partnerships, collaborative learning, co-creation for innovation, informal arrangements, trust-based, harmony, communication as dialogue, process management, diplomacy, mutual dependence, mutual gains approach, consensus, voluntary agreements and covenants. Voluntary agreements and covenants can also be seen as part of market governance: they are a kind of social contract based on individual decisions, but often prepared in a network type of informal process.

Market governance

In the 1980s, inspired by the popularity of neo-liberal economics in countries like the UK, Australia, New Zealand and the USA, an influential movement came up in many Western nations, which advocated running governmental organizations as if they were businesses. The idea was to integrate market-oriented thinking into the organization and *modus operandi* of the public sector (Kay and Boxall, 2015). The basic assumption was that private companies were more efficient and better organized and that governments should copy the management mechanisms used inside companies and the competition within markets. This approach became popular as 'market governance' or similar terms such as self-governance (Kooiman and Jentoft, 2009) or market exchange (Jessop, 2009). **Market governance** can be defined as a governance style which is driven by market and business ideas, treats government organizations as if they were business units, prefers using market-based instruments like taxes, and focuses on principles such as efficiency, competition, devolution and empowerment.

Market governance rapidly became popular in a range of countries and in international organizations like the World Bank and resulted in the establishment of many more or less independent government agencies, in results-oriented management (in fact, it introduced the term 'management' as a novelty in public sector organizations), and in cost-benefit analysis as a crucial method to measure success. In countries with a strong hierarchical culture like Germany or France, this style has never become as popular as in the UK and the Netherlands, for example.

This governance style driven by business ideas resulted in outsourcing and privatization of traditionally public tasks like energy production, health care and the delivery of drinking water. It is, however, much more than that: it is a belief about how governments should work. Whereas hierarchical governance departs from a view of government "ruling" society, market governance considers governments as service providers for their "clients", i.e. citizens and businesses.

Like hierarchical governance, market governance comprises a set of characteristics that together form a typical logic. This includes rationality, cost-driven, flexibility, competition as driver for innovation, price, marketing, decentralized, bottom-up, individualist, autonomy, self-determination, empowering, services, contracts, incentives, awards and other market-based instruments, performance agreements, metrics and payments.

The best-known governance style hybrid, New Public Management, was born from market governance thinking but merged with hierarchical principles. It was (and still is) about devolvement of tasks (market style) combined with sophisticated control and reporting mechanisms (hierarchical style). In Chapter 7, some of the consequences – positive but also very much negative – will be shown of the NPM movement.

Governance style convergence or co-habitation?

Many Western countries have shown an evolution as regards their most popular governance approach. Starting with the classical Weberian hierarchical-bureaucratic style until the 1970s, in the 1980s, it became *en vogue* to consider governments a special type of business for which all business management models and strategies could be used. In the 1990s, this *market governance* approach received competition from the interactive, *network* model of *governance*, as public administrations gradually begun to understand that they could not solve their challenges on their own and needed the knowledge and support of other societal actors. It looked as if a succession took place from more simple to more complex governance approaches. This resulted, for example, in different views about how citizens see government and the other way around. From the viewpoint of politicians, citizens have evolved from 'subordinates' via 'clients' to 'partners'.

However, the market- and networks-inspired 'new modes of governance' (Héritier, 2002) have never pushed hierarchical governance out of business (Hill and Lynn, 2005), and there has never been a convergence to one overall approach (Wollmann, 2004). The three styles appear together in many different combinations. They coexist as layered realities for politicians and managers, with particular circumstances or contexts calling forth behaviours and decisions related to one or the other conception of governance and service delivery, . . . (and) each has both strengths and weaknesses for society.

(Hartley, 2005)

The three styles coexist at all levels of government. A case study in 96 departments in South Korea showed that all three governance modes were present, while hierarchy remained the dominant governance. The three governance styles were situationally mixed contingent upon the distinct departmental tasks (Yoo and Kim, 2012).

2.3 Governance hybrids and the quest for a fourth style

Some governance style hybrids have earned their own name; some of them are variations of network governance such as adaptive governance, participatory governance or reflexive governance. A special case is 'bazaar governance' (Demil and Lecocq, 2006), which does not build on trust: community members may not even know each other and may enter or leave the network unnoticed.

Other hybrids have no specific name but have in common that they have emerged in order to correct or reverse problems, which were the result of governance failure after a switch from hierarchical to market governance. Areas such as health and education have either returned to older governance frameworks or feature 'hybrid' forms of governance containing elements of hierarchical, market and network approaches (Howlett and Ramesh, 2014). Departing from this classical typology (Driessen et al., 2012) preferred to distinguish five governance modes (for environmental governance) which are in my view variations of hierarchy, networks and markets. 'Centralized governance' and 'decentralized governance' are both forms of hierarchical governance; 'public-private governance' and 'self-governance' are two variations of market governance, and 'interactive governance' is network governance.

One of the similarities between the holistic concepts metagovernance and sustainable development is that they work with three basic dimensions. Another similarity is that there is a never-ending stream of proposals to add a fourth dimension. In my 2008 book on metagovernance I needed around 65 pages to conclude that there is no need for a fourth governance style (Meuleman, 2008). Ten years later I still hold this position. However, proposals to add a fourth style often do make sense, because they point at the general neglect of certain themes in governance.

Private interest government

Already in 1985, Streeck and Schmitter (1985, p. 122) proposed regulated self-regulation by organized interests as a fourth governance style. Such *private interest government* creates 'corporative-associative order' in certain policy areas for which this would result in more socially adjusted and normatively acceptable results than 'pure' network, market or hierarchical governance. This approach can be seen as a hybrid of all three basic styles, with already some elements of metagovernance added. The authors claim that this is more than a mixture of the three other styles because it adds another layer: 'organizational concertation'. In conclusion, I would argue that private interest government is not a distinctive governance style, but rather a precursor of 'private metagovernance' (see Section 4.4).

Knowledge governance

Knowledge governance has been named as a fourth style. It is about "developing new insights, competencies and ideas via public investments in knowledge development and dissemination, which contribute to the emergence of new pathways for collective action" (Van Buuren and Eshuis, 2010). Knowledge governance "appeals to the self-organizing capacity of actors to adjust their behaviour based upon new insights and ideas". Another view is that knowledge is one of the around 50 dimensions of governance, for which hierarchical, network and market governance have different features (Chapter 5); this I find an appropriate option. Knowledge is important (see also Section 5.3) but not at the same level as the three ideal types which are rooted in cultural theory (Thompson et al., 1990).

Transition management

Drawing on the experience of Dutch environmental policy in the 1990s, transition management emerged as an approach to bring about systemic change to unsustainable practices (Meadowcroft, 2011).

> Although transitions cannot be managed, one can work towards them. This is what transition management attempts to do. . . . [Transition management] offers a promising alternative for a planning and control approach and the use of economic incentives that both suffer from serious problems.
> (Kemp and Rotmans, 2005, p. 54)

This approach, to be welcomed for its long-term orientation and vision, has been mentioned by some authors as a different type of governance (Kemp and Rotmans, 2005, Loorbach, 2007). Meadowcroft criticizes the technological focus of this approach, and that, as in a related approach ('adaptive management/resilience'), experts are central and the political and the political system are seen as exogenous (Meadowcroft, 2011, p. 547). I agree with this critical view: transition management is a hybrid of network and market governance with a strong disappreciation of hierarchical governance. I would not call it a distinct governance style.

Provider-based governance and community governance

Tenbensel (2005) advocated two different categories of alternatives to hierarchical and market governance, as he found network governance too broad a cluster of governance approaches: provider-based governance and community governance. The first is about utilizing the knowledge of providers, professionals and street-level officials. It is professional control or autonomy, and "requires devices for distinguishing between members and non-members, although the boundaries between insiders and outsiders cannot be too rigid". This variation of network governance looks like expert network governance, a form we also see in transition management (see above).

The second alternative Tenbensel proposes is "governance based on the collective values of communities". It is characterized by "commitment that can be mobilized to support or thwart governance attempt", and would be manifest in populist movements and grass-roots community activism (Tenbensel, 2005, p. 283). Because I take a perspective on governance in which normally government or public sector organizations are not excluded from governance, to say the least, I am not convinced that this community governance is very different from other variations of network governance which are rather anti-state, such as deliberative governance.

Professional self-regulation

In the area of medical performance, professional self-regulation was coined as a special style of governance, based on expert authority, with codes of practice and clinical guidelines set by professional bodies, and monitoring through peer review (Burau, 2007). This style mixes with the other three styles, mainly with hierarchical and network governance, less with market governance, although I would consider professional self-governance as a hybrid form of the three basic styles, with the self-determination value of market governance as a key ingredient.

Solidarity governance

One suggestion deserves special attention, and not only because the expert who coined the term (Jessop, 2009, Jessop, 2011, p. 113) is one of the most prominent governance theorists. Jessop distinguishes *solidarity* as fourth mode of governance. This style "involves a relatively unreflexive, value-oriented rationality" based on "unconditional reciprocal commitment and loyalty". According to Jessop, in its strongest form it is usually confined to small units (e.g. family), and it changes with larger units towards unilateral forms of trust in providers of goods and services. Solidarity, according to Jessop, matters particularly for sustainable development in communities and is evident in the practices of "communing". 'Spontaneous solidarity' was seen as the key principle of the community (i.e. network governance) mode of coordination that Streeck and Schmitter distinguished (Streeck,

1985, p. 335). It also reminds me of the phenomenon of collective empathy which emerged in the form of the collective mourning of millions of people after Britain's Princess Diana died in 1997 and has since then become part of the repertoire of showing solidarity in many countries (Walter, 1999).

Solidarity governance is related to what Ostrom (1990) identified as self-organization and self-governing the commons: a new mode of governance besides the state-centralism hierarchy and market-centralism, which she argued to be necessary to solve 'common-pool resource problems': valuable resources for which there is no mechanism to protect them from accessing and utilizing. It goes beyond trust as a key value of network governance and adds especially empathy and a concern for common (natural) resources. Ostrom defined this approach, which she illustrated with a large number of case studies, at the time when in (Western) democracies the call for public participation and co-decision was rapidly becoming louder and identified as what we meanwhile usually call network governance (Powell, 1990).

Solidarity is closest to network governance. Typical failures are betrayal/mistrust and co-dependency/asymmetry (Jessop, 2011, p. 114), which are also failures of network governance. I have concluded that the advantage of adding solidarity governance as a fourth basic governance style (namely more variation) is less than the disadvantage: it makes the ideal types fuzzier. Moreover, solidarity governance can be explained as a hybrid between (mainly) network governance and market governance: the autonomy dimension is part of the value set of market governance. In addition, solidarity governance is mostly seen as focusing on governing the commons: the protection of natural resources. Not only in Ostrom (1990) but also in more recent publications on local self-organization, this approach to governance distinguishes itself from other styles by its specific *objective*. Hierarchical, network and market governance are tool sets, which can be used, in principle, for any (public) objective. However, contrary to what the term 'the commons' suggests, people have more common objectives than their interest to protect natural resources. Examples which are covered by SDGs include quality of education and health, quality of governance, maintaining human rights. There are specific SDGs on protecting the commons, such as on biodiversity, land, oceans and resource efficiency. However, the successes of environmental policy since the 1970s have to an important extent relied on establishing legal requirements and standards: governing the commons cannot be done by self-organization alone.

The discussion about solidarity governance in the context of network governance is very useful because it illustrates the importance of (local) self-organization to help tackle complex problems around protection and management of natural resources. It has also shown that institutions are not only necessary in public administration but also in non-public organizations. The eight design principles Ostrom defined for such 'common pool resource' institutions (Ostrom, 1990, p. 90) are useful in any area where network governance is being used. They include mechanisms to deal with some of the inherent failures of networks, such as exclusion, non-transparency, time-consuming deliberations and conflict resolution, and have been integrated in work on the management of complex public policy processes (see e.g. De Bruijn et al., 2010).

2.4 Four schools of thinking about sustainability governance

Four different ways to define the purpose of sustainability governance can be distinguished.

Approach 1: strong central leadership is necessary

One line of argumentation is that current governance approaches are too fragmented, not well coordinated, lacking legally binding agreements. Strong central, hierarchical leadership is needed for the SDGs. At the global level, such central leadership has been institutionalized with the establishment of the High-Level Political Forum on Sustainable Development (HLPF). The HLPF has no legislative power and only limited direct authority but is still functioning as central 'orchestrator' of the implementation of the SDGs at all levels (Bernstein, 2017, p. 214). Central leadership at the national level has since long been seen as one of the crucial success factors of sustainable development, as has been shown in a variety of countries (e.g. Bhutan, Germany and Finland). A similar approach promoting a central role of government is advocated by e.g. Biermann et al. (2017, p. 75), who define governance as "the purposeful and authoritative steering of societal processes by political actors". This approach focuses on shared responsibility and shared action of governmental and non-governmental actors, but these actions should have "claim to authority, have some legitimacy, and are designed to steer behaviour". This implies, however, that a public-private partnership to build infrastructure to protect against floods, where the private actor has the lead, would fall outside the scope of governance. In addition, this approach does not distinguish between policy and governance. This approach has SDG 16 at its core and emphasizes institutional readiness and rule of law.

Approach 2: strong bottom-up partnership action

The second argument claims that the real problem is that sustainability governance is too much top-down and too little inclusive. In 2017, the UN Committee of Experts on Public Administration concluded that the Goals "require another type of governance that is multi-level and multi-stakeholder and based on the engagement of all actors and partnerships" (United Nations, 2017, p. 18). According to Hajer et al. (2015) the SDGs need in the first place an approach beyond 'cockpit-ism'. Cockpit-ism is "the illusion that top-down steering by governments and intergovernmental organizations alone can address global problems". Although the UN and its member states already in 2012 in the 'Rio+20 Outcome Document' (United Nations, 2012) concluded that all societal stakeholders should be involved in sustainable development, these authors observe a risk that civil society, cities and business will be left out. The argument for more bottom-up governance while keeping the state at a distance is strongly advocated by the Australian scholar Rhodes, who considers this an approach beyond governance and metagovernance which he calls 'decentred governance'. This theory

> rejects the notion of the state as a material object and governance as an emergent structure. It is a 'stateless' theory in the sense that it rejects the idea of the state as a pre-existing causal structure that can be understood as having an autonomous existence and causal effects over and apart from people's beliefs and actions.
>
> (Rhodes, 2017)

This approach concentrates on SDG 17, the 'partnership SDG', and is very popular (even dominating) among academics in consensus democracies such as the Netherlands, Denmark and Sweden. A counter argument to the strong focus on networks would be that at least two-thirds of the world's nations are currently governed by a central 'cockpit' and many of them are considered to be weak or failed states (see also Section 8.1). Secondly, there is no evidence that having a strong government with SDG leadership at the very top ('*Chefsache*' in Germany, where the Chancellor indeed has taken the lead), hampers effectiveness of governance.

The first two approaches reflect the classical divide between proponents of hierarchical governance and of the 'new modes of governance' (Héritier, 2002) which use network or market-style tools and mechanisms. Both arguments take insufficiently into account the complexity and other characteristics of human societies across the globe, lack a long-term orientation, deny the normative dimension of governance, and generally disapprove of plurality.

Approach 3: focus on transition management

A third line of thinking concentrates on one important aspect of the SDGs, namely the systemic transformation of our societies, through a combination of 'system jumps' and incremental change. Transition management has been seen as a distinct governance model (Kemp and Rotmans, 2005, p. 54) and has its origin in the pioneering environmental policy of the late 1990s in the Netherlands (Meadowcroft, 2011, p. 542). This approach acknowledges complexity and is characterized by a bottom-up approach through which innovations emerge in 'niches', then develop into 'islands' with, later, 'regimes' (Grin et al., 2010). I would categorize it as essentially a market governance (entrepreneurial, creative competition, individual initiatives) approach with an additional role for networking and a rather marginal role for hierarchy (standardization, 'regimes'). Just as the adaptive management and resilience approach to sustainability governance, the transition management approach relies on expert management (Meadowcroft, 2011, p. 547). Transition management does not claim to be the whole solution; indeed, it appears to operate in an apolitical void and concentrates on typical systemic problems, whereas not all sustainability governance challenges are of that type. Eco-innovation can result in much cleaner products and services without being part of a societal transformation, for example.

Approach 4: dynamic multi-perspective approach (metagovernance)

In response, the fourth approach claims that sustainability governance should combine different governance models and apply such combinations while taking into account the specificities of the problems to tackle and of the (economic, ecologic and social as well as administrative) environment in which governance should function. This links to a survey of governance trends for sustainable development which concluded that there is a need to combine governance for compliance of (international) agreements with expanding collaborative governance (Olsen et al., 2015, p. 43). Governments implementing the SDGs would need to contemplate how expanding collaboration and partnerships with stakeholders can complement and enhance the effectiveness of conventional top-down planning and implementation. This approach supports the multiple perspective and situational approach, for which metagovernance offers a model, which combines various governance principles and tools. Governance frameworks cannot be effective in situations where social, economic and political contexts impede successful implementation (Gulbrandsen, 2005).

Maybe it is no coincidence that this argument finds its proponents and supporters among public governance practitioners (Christopoulos et al., 2012, Olsen, 2006, Meuleman, 2008, In 't Veld et al., 2011, European Commission, 2017). The SDGs are in this view not themselves transitions, or a stable future state or a fixed target to be reached, but 'transformational guidance' using a range of policy instruments (Kuenkel, 2017), with the need to continuously crosscheck interconnections, interdependencies and impacts, and to apply common principles such as reflexivity, flexibility and long-term thinking (Meuleman and Niestroy, 2015). In addition, holistic thinking, inclusiveness and mindfulness (context-sensitivity) need to be applied, and the multi-sector, multi-level and multi-actor situations of each specific case are to be considered. The absence of these principles in SDG governance frameworks, or their inherent conflicts with hierarchical principles, is in this view considered as an obstacle for SDG implementation.

The four 'schools' of thinking about SDG governance depicted in Table 2.1 can be seen as complementary, but the fourth – metagovernance – school offers the broadest, multi-perspective view and therefore seems from a practitioner's

Table 2.1 Current approaches to SDG governance

1. Strong leadership (mainly H governance); Global disaster management	2. Strong partnerships (mainly N governance); Beyond 'cockpit-ism'	3. Bottom-up transitions (mainly M governance); Niches-islands-regimes
	4. Dynamic multi-perspective approach (metagovernance) No one-size-fits-all but situational governance	

perspective therefore the most pragmatic. The next step is to analyse why some governance means for the SDGs are effective, and why, where and when not. This requires an analysis of the deeper, 'root causes' of obstacles to implementation of the SDGs, for which I will use the analytical lens of metagovernance (Section 1.3 and more in-depth Chapter 4) and the three basic, ideal-typical governance styles (Section 2.2).

2.5 Governance problems and the need for governance of governance

Most of the existing studies and research papers published about the governance needed for the implementation of the SDGs are of a stocktaking (e.g. policy gap analysis) or prescriptive kind (e.g. which institutions are needed?). Thorough reports have described the 'readiness' of certain countries (Kroll, 2015) and the EU (Niestroy, 2016), the institutional challenges and observed good practices (Ziekow and Bethel, 2017), and different types of national sustainable development strategies (Scholz 2017), to give just a few examples. The long-term dimension of sustainable development (Meuleman and in 't Veld, 2010) is not yet addressed as a central issue in most of the recent literature.

Having goals is no guarantee for effective implementation. There is a growing concern that a strong institutional framework and sophisticated planning systems are not sufficient to ensure successful implementation of the SDGs. In the vast majority of countries, public sector organizations are organized according to Weberian principles of hierarchical governance, such as clear division of tasks ('silos'), clear chains of command, and a focus on stability and legality. These principles make sense, but are not sufficient and they can be counter-productive. They can lead to huge bureaucracies who may excel in doing business as usual and preventing change. When a country's national government has more than 50 ministries and almost 250 agencies – an existing practice – one should think that efficiency gains are low-hanging fruit.

When a governance framework has been established, for a policy challenge, implementation approach or an organizational reform project, this framework may become a kind of prison: it creates a lock-in effect and becomes a reality separated from other realities. Changes in the governance environment may not be identified – or too late. To use a railway metaphor: when the chosen governance approach is 'on track', any derailment is a disaster that has to be prevented with all means. The result may be that the governance gradually (or even suddenly) becomes ineffective. When there is a strong focus on efficiency this may not be noticed for quite some time – there are many examples of public sector organizations which have become efficient but not effective anymore.

Governance styles and their hybrids can lead to three types of problems:

1 *Governance styles have typical, built-in failures, weaknesses, or even perversions* (Jessop, 1998) (Kooiman, 1993, Jessop, 1998, Jessop, 2002, Meuleman, 2008). Hierarchical governance and market governance cannot cope

Three governance styles and their hybrids 39

with complexity, while network governance is too time-consuming for dealing with disasters and with routine problems. Hierarchical governance can lead to overkill of procedures, reporting requirements, detailed instructions. Network governance can lead to endless talks without conclusions – and if there is a conclusion, this may not be secured in an agreement. Market governance may lead to similar problems real markets have when they are not regulated. Typical perversions of the three styles are abuse of power (hierarchy), financial mismanagement (market governance) and manipulation (network governance).

2   **Governance styles may undermine each other.** Trust, which is a key feature of network governance, can easily be destroyed by top-down measures (hierarchy) or by individualist maximizing benefits for one's own cause (market governance). Maximizing flexibility (market governance) and reliability (hierarchical governance) at the same time is impossible – this is a dilemma calling for a trade-off.

3   **Governance styles may become belief systems.** Each of them may be seen as a solution for any problem (a panacea), because they offer an attractive logical set of values and principles to which people can subscribe – consciously or intuitively. However, Maslov's aphorism "If you only have a hammer, you tend to see every problem as a nail", which applies to hierarchical governance, can also be rephrased for the other styles. "If you only have trust, you tend to see every problem as a relational problem" points at a weakness of focusing too much on network governance, and "If you only have money, you tend to see every problem as a financial/monetary problem" explains a weakness of market governance.

An example from the early 2000s is the strong disagreement on the best governance style between the European Commission's Directorate General for the Environment and the then Directorate General for Enterprize. "While DG Enterprise considers self-regulation by industry desirable, DG Environment would have preferred legislation to reach desired objectives" (Héritier and Lehmkuhl, 2008, p. 13). This internal disagreement undermined the external credibility of the threat of having legislation.

Therefore, we need a mechanism to take a step back, or up – a bird's eye view – to monitor governance frameworks and decide on interventions when needed. This is what in the next chapter will be elaborated as metagovernance.

2.6 Conclusions

In this chapter it was concluded that:

- Governance is not so much about the content of policies (what to do?) or about the vision behind policies (why do it?), but concentrates on how to achieve objectives.
- Hierarchical, network and market governance are still the main three ideal-typical governance styles, usually present in combinations. Other suggested

styles should be identified as hybrids or variants of the three main governance styles.
- Four schools of thought about sustainability governance can be distinguished.
- Combinations of the three classical governance styles can result in synergy but also in governance failure. We need an approach which can deal with such problems: a governance of governance – metagovernance – approach.

Before we discuss metagovernance as a practical tool to design and manage governance frameworks in Chapter 4, Chapter 3 will introduce a typology of governance failure and their causes.

Notes

1 Source: Google Scholar, retrieved on 11 October 2017.
2 See for elaborate discussion Meuleman (2008, pp. 9–69).

References

Beck, U. 1992. *Risk Society: Towards a New Modernity*. London: Sage.
Bell, S. & Park, A. 2006. The problematic metagovernance of networks: Water reform in New South Wales. *Journal of Public Policy*, 26, 63–63.
Bernstein, S. 2017. The United Nations and the governance of sustainable development goals. In: Kanie, N. & Biermann, F. (eds.) *Governing Through Goals: Sustainable Development Goals as Governance Innovation*. Cambridge, MA: MIT Press.
Bevir, M. 2011. Governance as theory, practice, and dilemma. In: Bevir, M. (ed.) *The SAGE Handbook of Governance*. London: Sage.
Biermann, F., Stevens, C., Bernstein, S., et al. 2017. Global goal setting for improving national governance and policy. *Governing Through Goals: Sustainable Development Goals as Governance Innovation*. Cambridge, MA: MIT Press.
Burau, V. 2007. The complexity of governance change: Reforming the governance of medical performance in Germany. *Health Economics, Policy and Law*, 2(4), 391–407.
Christopoulos, S., Horvath, B. & Kull, M. 2012. Advancing the governance of cross-sectoral policies for sustainable development: A metagovernance perspective. *Public Administration and Development*, 32, 305–323.
Considine, M. & Lewis, J. M. 2003. Bureaucracy, network, or enterprise? Comparing models of governance in Australia, Britain, the Netherlands, and New Zealand. *Public Administration Review*, 63, 131–140.
De Bruijn, H., Ten Heuvelhof, E. & In 'T Veld, R. 2010. *Process Management: Why Project Management Fails in Complex Decision Making Processes*. Berlin, Heidelberg: Springer.
Demil, B. & Lecocq, X. 2006. Neither market nor hierarchy nor network: The emergence of bazaar governance. *Organization Studies*, 27, 1447–1466.
Derix, G. 2000. *The vision web*. Schiedam: Scriptum.
Driessen, P. P. J., Dieperink, C., Van Laerhoven, F., et al. 2012. Towards a conceptual framework for the study of shifts in modes of environmental governance – experiences from The Netherlands. *Environmental Policy and Governance*, 22, 143–160.
European Commission. 2017. Quality of public administration. European Semester Thematic Factsheet. [Online]. Available: https://ec.europa.eu/info/sites/info/files/

european-semester_thematic-factsheet_public-procurement_en.pdf. Brussels: European Commission.
Frances, J., Levacic, R., Mitchell, J. & Thompson, G. 1991. Introduction. In: Thompson, G., Frances, J., Levačić, R., & Mitchell, J. (eds.), *Markets, Hierarchies and Networks the Coordination of Social life*. London: Sage.
Grin, J., Rotmans, J., Schot, J., et al. 2010. *Transitions to Sustainable Development: New Directions in the Study of Long Term Transformative Change*. New York: Routledge.
Grindle, M. S. 2004. Good enough governance: Poverty reduction and reform in developing countries. *Governance*, 17, 525–548.
Gulbrandsen, L. H. 2005. Sustainable forestry in Sweden: The effect of competition among private certification schemes. *The Journal of Environment & Development*, 14, 338–355.
Hajer, M. 2003. Policy without polity? Policy analysis and the institutional void. *Policy Sciences*, 36, 175–195.
Hajer, M., Nilsson, M., Raworth, K., et al. 2015. Beyond cockpit-ism: Four insights to enhance the transformative potential of the sustainable development goals. *Sustainability*, 7, 1651–1660.
Hajer, M. A. & Wagenaar, H. 2004. *Deliberative Policy Analysis: Understanding Governance in the Network Society*. Cambridge: Cambridge University Press.
Hartley, J. 2005. Innovation in governance and public services: Past and present. *Public Money & Management*, 25, 27–34.
Héritier, A. 2002. New modes of governance in Europe: Policy making without legislating? IHS Political Science Series – No. 81. [Working Paper]. Vienna: Institute for Advanced Studies.
Héritier, A. & Lehmkuhl, D. 2008. The shadow of hierarchy and new modes of governance. *Journal of Public Policy*, 28, 1–17.
Hill, C. J. & Lynn, L. E. 2005. Is hierarchical governance in decline? Evidence from empirical research. *Journal of Public Administration Research and Theory*, 15(2), 173–195.
Howlett, M. & Ramesh, M. 2014. The two orders of governance failure: Design mismatches and policy capacity issues in modern governance. *Policy and Society*, 33, 317–327.
In 't Veld, R. 2010. *Knowledge Democracy: Consequences for Science, Politics, and Media*. Heidelberg: Springer.
In 't Veld, R. J., Töpfer, K., Meuleman, L., et al. 2011. *Transgovernance. The Quest for Governance of Sustainable Development*. Potsdam: Institute for Advanced Sustainability Studies (IASS).
Jessop, B. 1997. Capitalism and its future: Remarks on regulation, government and governance. *Review of International Political Economy*, 4, 561–581.
Jessop, B. 1998. The rise of governance and the risks of failure: The case of economic development. *International Social Science Journal*, 50, 29–45.
Jessop, B. 2002. Governance and meta-governance in the face of complexity: On the roles of requisite variety, reflexive observation, and romantic irony in participatory governance. In: Heinelt, H., Getimis, P., Kafkalas, G., Smith, R. & Swyngedouw, E. (eds.) *Participatory Governance in Multi-Level Context*. Wiesbaden: VS Verlag für Sozialwissenschaften.
Jessop, B. 2009. From governance to governance failure and from multi-level governance to multi-scalar meta-governance. In: Arts, B. & Al, E. (eds.) *The Disoriented State: Shifts in Governmentality, Territoriality and Governance*. Heidelberg: Springer.
Jessop, B. 2011. Metagovernance. In: Bevir, M. (ed.). *The Sage Handbook of Governance*. London: Sage.

Jordan, A. 2008. The governance of sustainable development: Taking stock and looking forwards. *Environment and Planning C: Government and Policy*, 26, 17–33.

Kaufman, F.-X. 1986. The relationship between guidance, control and evaluation. In: Kaufman, F.-X., Majone G. & Ostrom V. (eds.). *Guidance, Control, And Evaluation in the Public Sector: The Bielefeld Interdisciplinary Project*. De Gruyter studies in organization; 4. Berlin: de Gruyter.

Kay, A. & Boxall, A. M. 2015. Success and failure in public policy: Twin imposters or avenues for reform? Selected evidence from 40 years of health-care reform in Australia. *Australian Journal of Public Administration*, 74, 33–41.

Kemp, R. & Rotmans, J. 2005. The management of the co-evolution of technical, environmental and social systems. In: Weber, M. & Hemmelskamp, J. (eds.) *Towards Environmental Innovation Systems*. Berlin, Heidelberg: Springer.

Kickert, W. J. M. 2003. Beneath consensual corporatism: traditions of governance in the Netherlands. *Public Administration*, 81, 119–140.

Kickert, W. J. M., Klijn, E. H. & Koppenjan, J. F. M. 1997. *Managing Complex Networks: Strategies for the Public Sector*. London: Sage.

Klijn, E.-H. 2008. Governance and governance networks in Europe. *Public Management Review*, 10, 505–525.

Kooiman, J. 1993. *Governance and Governability: Using Complexity, Dynamics and Diversity*. London: Sage.

Kooiman, J. 2003. *Governing as Governance*. London: Sage.

Kooiman, J. a. N. & Jentoft, S. 2009. Meta-governance: Values, norms and principles, and the making of hard choices. *Public Administration*, 87, 818–836.

Kroll, C. 2015. *Sustainable Development Goals: Are the Rich Countries Ready?* Gütersloh, Germany: Bertelsmann Stiftung.

Kuenkel, P. 2017. A pattern approach to leading sustainability transformation – How the 17 SDGs can become a starting point for systemic change. *Collective Leadership Series*, Working Paper No 4. Potsdam/Germany: The Collective Leadership Institute.

Larsson, O. 2015. *The Governmentality of Meta-governance. Identifying Theoretical and Empirical Challenges of Network Governance in the Political Field of Security and Beyond*. Skrifter utgivna av Statsvetenskapliga föreningen i Uppsala 193. 204 pp. Uppsala: Acta Universitatis Upsaliensis. ISBN 978-91-554-9296-0.

Lindblom, C. E. 1959. The science of 'Muddling Through'. *Public Administration Review*, 19, 79–88.

Lindblom, C. E. 1979. Still muddling, not yet through. *Public Administration Review*, 39, 517–526.

Loorbach, D. 2007. *Transition Management: New Mode of Governance for Sustainable Development*. PhD, Erasmus University, Rotterdam, The Netherlands.

Lowndes, V. & Skelcher, C. 1998. The dynamics of multi-organizational partnerships: an analysis of changing modes of governance. *Public Administration*, 76, 313–333.

Meadowcroft, J. 2011. Sustainable development. In: Bevir, M. (ed.). *The SAGE Handbook of Governance*. London: Sage.

Meuleman, L. 2008. *Public Management and the Metagovernance of Hierarchies, Networks, and Markets*. Heidelberg: Springer.

Meuleman, L. 2013. Cultural diversity and sustainability metagovernance. In: Meuleman, L. (ed.). *Transgovernance – Advancing Sustainability Governance*. Berlin/Heidelberg: Springer Verlag.

Meuleman, L. & In 't Veld, R. J. 2010. Sustainable Development and the Governance of Long-Term Decisions. In: In 't Veld, R.J. (ed.). *Knowledge Democracy*. Berlin/Heidelberg: Springer.

Meuleman, L. & Niestroy, I. 2015. Common but differentiated governance: A metagovernance approach to make the SDGs work. *Sustainability*, 12295–12321.

Niestroy, I. 2016. *How Are We Getting Ready? The 2030 Agenda for Sustainable Development in the EU and Its Member States: Analysis and Action So Far*. Bonn: Deutsches Institut für Entwicklungspolitik.

Olsen, J. P. 2006. Maybe it is time to rediscover bureaucracy. *Journal of Public Administration Research and Theory*, 16, 1–24.

Olsen, S. H., Zusman, E. & Cadman, T. 2015. Trends in the international sustainable development policy discourse: Compliance, collaboration or both? In: Bengtsson, M., Olsen, S. H. & Zusman, E. (eds.) *Achieving the Sustainable Development Goals: From Agenda to Action*. Kamiyamaguchi, Hayama, Kanagawa, Japan: Institute for Global Environmental Strategies.

Osborne, S. 2006. The new public governance? *Public Management Review*, 8, 337–387.

Ostrom, E. 1990. *Governing the Commons: The Evolution of Institutions for Collective Action*. Cambridge: Cambridge University Press.

Peters, B. G. 1998. Managing horizontal government: The politics of co-ordination. *Public Administration*, 76, 295–311.

Pierre, J., & Peters, B.G. 2000. *Governance, Politics, and the State*. Basingstoke: Macmillan.

Pollitt, C. & Bouckaert, G. 2011. *Public Management Reform: A Comparative Analysis: New Public Management, Governance, and the Neo-Weberian State*. Third Edition. Oxford: Oxford University Press.

Powell, W. 1990. Neither market nor hierarchy: Network forms of organization. *Research in Organizational Behavior*, 12, 295–336.

Rayner, J. 2015. The past and future of governance studies: From governance to metagovernance? In: Capano, G., Howlett, M. & Ramesh, M. (eds.). *Varieties of Governance. Studies in the Political Economy of Public Policy*. Basingstoke: Palgrave Macmillan.

Rhodes, R. A. W. 2000. The governance narrative: Key findings and lessons from the ESR's Whitehall programme. *Public Administration*, 78, 345–363.

Rhodes, R. A. W. 2017. Understanding governance: 20 years on. *Organization Studies*, 28(8), 1243–1264.

Scholz, I. 2017. National strategies for sustainable development between Rio 1992 and New York 2015. In: Von Hauff, M. & Kuhnke, C. (eds.) *Sustainable Development Policy: A European Perspective*. London: Routledge.

Schout, A. & Jordan, A. 2003. Coordinated European governance: Self organising or centrally steered? *Working Paper – Centre for Social and Economic Research on the Global Environment*, 1–28.

Sørensen, E. 2006. Metagovernance: The changing role of politicians in processes of democratic governance. *The American Review of Public Administration*, 36, 98–114.

Steenkamp, J.-B. E. M. & Geyskens, I. 2012. Transaction cost economics and the roles of national culture: A test of hypotheses based on Inglehart and Hofstede. *Journal of the Academy of Marketing Science*, 40, 252–270.

Streeck, W. & Schmitter, P. C. 1985. Community, market, the prospective contribution of interest governance to social order. *European Sociological Review*, 1, 119–138.

Tenbensel, T. 2005. Multiple modes of governance: Disentangling the alternatives to hierarchies and markets. *Public Management Review*, 7, 267–288.

Thompson, M., Ellis, R. & Wildavsky, A. 1990. *Cultural Theory*. Boulder, CO: Westview Press.

Thorelli, H. B. 1986. Networks: Between markets and hierarchies. *Strategic Management Journal*, 7, 37–51.

Torfing, J. & Triantafillou, P. 2011. Introduction: Interactive policy making, metagovernance and democracy. In: Torfing, J. & Triantafillou, P. (eds.). *Interactive Policy Making, Metagovernance and Democracy*. Colchester, UK: ECPR Press.

United Nations. 2012. The future we want. Resolution adopted by the General Assembly on 27 July 2012.

United Nations. 2017. Report on the sixteenth session (24–28 April 2017) of the United Nations Committee of Experts on Public Administration (CEPA). New York: United Nations.

Van Buuren, A. & Eshuis, J. 2010. *Knowledge Governance: Complementing Hierarchies, Networks and Markets?* In: In 't Veld, R.J. (ed.). *Knowledge Democracy*. Berlin/Heidelberg: Springer.

Van Kersbergen, K. & Van Waarden, F. 2004. 'Governance' as a bridge between disciplines: Cross-disciplinary inspiration regarding shifts in governance and problems of governability, accountability and legitimacy. *European Journal of Political Research*, 43, 143–171.

Walter, T. 1999. *The Mourning for Diana*. Oxford: Berg.

Whitehead, M. 2003. 'In the shadow of hierarchy': Meta-governance, policy reform and urban regeneration in the West Midlands. *Area*, 35, 6–14.

Wollmann, H. 2004. Policy change in public sector reforms in crosscountry perspective: Between convergence and divergence. In: Abraham, B. P. & Munsh, S. (eds.). *Good Governance, Democratic Societies and Globalisation*. New Delhi: Sage.

World Bank. 1994. *Governance – The World Bank's Experience: Development in Practice*. Washington, DC: World Bank.

World Bank. 2017. *World Development Report 2017: Governance and the Law*. Washington, DC: World Bank.

Yoo, J. W. & Kim, S. E. 2012. Understanding the mixture of governance modes in korean local governments: An empirical analysis. *Public Administration*, 90, 816–828.

Ziekow, J. & Bethel, R. 2017. *Institutional Arrangements for the Sustainable Development Goals*. New York: UN Committee of Experts on Public Administration (CEPA).

3 Governance failures and their causes

> NASA has this phrase that they like, 'Failure is not an option'. But failure has to be an option. In art and exploration, failure has to be an option. Because it is a leap of faith. And no important endeavour that required innovation was done without risk. You have to be willing to take those risks. . . . In whatever you are doing, failure is an option. But fear is not.
> – Film director James Cameron, address to the 2010 TED conference (13 February 2010)[1]

3.1 Failure is normal but comes with a cost

In Chapter 2, we saw that each governance style has specific strengths and stands for typical successes in terms of output and outcome, but also that governance styles are characterized by typical weaknesses in different problem contexts, which results in different types of *governance failure*. In addition, governance styles may undermine each other, and each of them has believers who think that a certain style can be the solution for all problems.

Failure is normality just like success. However, not all failure is equal – some failure is more predictable, or better manageable, than other failure. Moreover, different forms of governance fail in different ways (Jessop, 2011:113). Some failure can be linked to specific governance styles, which makes them to some extent predictable. This is good news for governance practitioners. Failure can be the result of unsuccessful design or weak management, but external factors can be at least as determining. Failure may also occur because of a mismatch between a governance approach and the governance context in which and for which purpose it is designed and applied. When failures are identified, this usually signals the start of a search for a 'culprit'. In the context of this book, it is more relevant to find out what the systemic causes are of such failures, and how manageable they are in the context of different governance styles.

Although governance failure cannot be defined in an objective way, its impact results in objective as well as subjective (i.e. perceived) economic, social and environmental costs. For the Sustainable Development Goals, governance failure brings about risks that the 2030 targets will not be met. The cost of governance failure in the form of non-action was illustrated convincingly in the seminal study 'Late lessons from early warnings' (European Environment Agency, 2013). The European Commission's Environmental Implementation Review states that implementation gaps in the member states regarding the EU's environmental policy and law are responsible for around €50 billion per year in terms of societal costs (European Commission, 2017a), which equals around one-third of the annual EU budget. Most of these implementation gaps are the result of governance failures. Fifty billion Euros per year seems quite a lot, but maybe it is even more. After all, the height of the estimated costs of non-implementation depends to a large extend on how they are calculated, and mainly: what discount rate was chosen, as the 'Late Lessons' report states. A discount rate of 1.3% results in a 3.5 times higher costs of inaction than a discount rate of 2.8%. The EEA rightly argues,

> To many non-economists, the implied shrinking of future values with the discounting technique is at odds with the core idea of intergenerational equity that is central to sustainability. A related problem is that many economists are applying discount rates that are typically used in the corporate sector – rising up to as much as 10 % – without reflecting on the specific context of environmental challenges.
>
> (Andersen and Clubb, 2013)

3.2 Governance failure or policy failure?

The first questions to address are what the difference is between governance failure and policy failure, and how the two concepts are interlinked. *Policy failure* has many root causes (Peters, 2015) and can be defined as the failure to reach a policy objective. Howlett (2012) defines six dimensions of policy failures (extent (size), avoidability, visibility, intentionality, duration and intensity), but these could as well be used as dimensions of governance failure. Howlett, Ramesh and Wu (2015) argue that policy failures can be persistent across time and space because governments may lack the capacity to identify or tackle the root causes of policy failures and/or to draw lessons from identified failure, which is again similar to governance failure. Finally, they argue that *governance failure* has emerged in literature as a 'dependent variable' of *policy failure*. Indeed, governance does not make much sense without policy, but a policy without governance is doomed to fail.

It is difficult to define policy failure unambiguously because agreement on the success or failure of a *policy* is rare (European Commission, 2017b). Instead of trying to determine whether a policy is a success or failure, it might be more useful to try to identify which aspects of a policy have failed and to explain why these aspects ought to be considered to have failed; different forms of failure should not be confused (Newman and Head, 2015). The same can be said about success or failure of *governance*. In addition, policy challenges and governance challenges are social constructs: they are selected and constructed around a mixture of world view, ambition, sense of opportunity and knowledge.

In Chapter 2, I defined governance broadly as covering institutions (polity) as well as processes (politics), leaving strategic choices about goals and how they should be reached (policy) outside the scope. Policy *instruments*, however, belong to the domain of governance (Le Galès, 2011). Governance failure has been defined as the "perceived ineffectiveness of a governance process" (Bovens et al., 2001). This triggers the question of who is the perceiver of failure. Different actors will always perceive effectiveness of governance differently, if only because they have different values (see Section 6.1). I agree with McConnell (2016), that a reflection on the interpretation and construction of failures by different actors is important.

On the other hand, the definition of governance failure should not be confined to governance processes but be seen in a broader context, covering the whole composition of governance: governance frameworks. Jessop's definition of governance failure, namely "failure amongst a group of agents to renegotiate shared objectives and negotiate collective action towards them" (Jessop, 2000) is relatively value-neutral. However, maybe a definition that acknowledges the subjective and normative character of what governance failure is works better in governance practice, as practice is normally far from value-neutral. A very short and simple definition of governance failure could therefore be "the ineffectiveness of a governance framework to achieve its policy goals".

There is another point to take into account, however. Governance failure not only occurs when an administration does not succeed in implementation of policy goals, such as to build road infrastructure or to reduce inequality. Also a lack of implementation of key governance goals – such as on rule of law and on corruption – is a governance failure. The best illustration of this broad approach to governance is the smart way governance is integrated in the Agenda 2030 and its SDGs. Whereas *policy failure* is the failure to achieve the objectives of the Goals 1–15 and their numbered (1, 2, 3, . . .) targets, *governance failure* is failure to achieve the objectives of the specific governance Goals 16 and 17 and of the enabling governance targets under Goals 1–15 which are labelled with letters (a, b, c, . . .). In addition, failure to design and manage the governance frameworks for implementation of all Goals and targets should be considered governance failure. Therefore, a more complete definition of **governance failure** is "the ineffectiveness of governance goals, a governance framework or the management thereof, to achieve policy goals".

Policy and governance failures meet in the concept of *government* failure. Keech and Munger (2015) argue that many so-called market failures are 'government failures' because government defines the institutions in which markets succeed or fail. Government failure has two faces: "government may fail to do things it should do, or government may do things it should not do" (ibidem).

Table 3.1 illustrates that there is a certain overlap between policy and governance failure. Failure linked to the choice and roles of institutions (i.e. polity) is usually about governance. Failure linked to the choice or mixture of policy instruments and tools and the level and type of participation can be part of policy design, but also of the design and management of a governance framework. An

Table 3.1 Policy failure and governance failure

	Locus of failure	Type of failure
Polity	Design/roles/choice/management of institutions	Governance failure
Politics	Design/choice/management of instruments/tools/processes/ involvement	Governance or policy failure (or hybrid)
Policy	Policy design	Policy failure

in principle well-designed policy may lead to governance failure if the policy is not implementable in the existing governance context. For example, support by other relevant authorities/ministries may be essential for successful policy implementation by a national ministry, but the related governance conditions to ensure policy cross-sectoral coherence and horizontal coordination may not be fulfilled. Finally yet importantly, policy failures can be the result of weak policy design. An example is the introduction of a ground water extraction charge in a water-scarce area, which is so low that it fails to incentivize reducing ground water use. No governance framework will be able to 'repair' this failure.

Peters (2015) distinguishes the categories of systemic failure when both policy and governance fail, and systemic success when both succeed. Policy failure without clear governance failure he names 'business as usual', and governance failure without policy failure he calls 'micro success'. He rightly argues that

> policy failures, while certainly common and regrettable, are the easiest of these forms of failure within the public sector to resolve. If a policy does not work, it is not linked to the organic structure of the state but is only an instrument that can be manipulated in order to produce better results.
>
> (Peters, 2015, p. 264)

European Union waste management provides a good example of the relation between policy and governance failure. The policy failure is that (in 2017) 22 out of 28 EU member states are struggling to reach the legal municipal waste recycling target of 50% (European Commission, 2017a). Waste recycling is usually a task devolved to local authorities. The governance failure is that in many EU countries, municipalities are too small to have the resources to hire the experts and to purchase the waste management infrastructure they need to fulfil the obligation. In 15 member states, the average population per municipality is below 10,000; in six member states it is even below 5,000 (Thijs et al., 2018). This is part of a broader problem: In many countries, local governments (especially those of small and rural municipalities) lack the necessary human resources and technical capacities to fulfil their increasing number of tasks as a result of the process of decentralization (Tacoli, 2014).

An interesting example of how policy failure can gradually lead to undermining governance capacity comes from an analysis of 40 years of reforms in the Australian health sector (Kay and Boxall, 2015). The authors conclude that:

the informal tends to subvert the formal, as policy failure undermines institutions. Formal or law like institutions that are codified and enforced by third parties are by their nature 'stickier' than informal institutions. Policy failure as a negative feedback affects informal institutional practice more immediately, absent of any grand reform, as things are done slightly differently in response to failure inquiries. This may be unconscious but can steadily accumulate over time to the point where related formal institutions are undermined. This is beginning to happen in health care with the stark divide between the public and private systems gradually eroding at the service delivery level even though there have been no major changes made to our mixed public-private health insurance arrangements for over 30 years.

For an extensive analysis of governance failure related to Agenda 2030 and the SDGs adopted in 2015 it is still too early, but what we can do is look at the Agenda 21 – a programme of action for sustainable development worldwide – adopted at the Earth Summit in 1992. The Rio Declaration on Environment and Development, adopted by 178 UN member states, consisted of 27 principles, and was adopted together with Agenda. The Rio Principles represented agreements on policy goals rather than on governance, but Agenda 21 itself was more governance-oriented. The Agenda had the ambition of being "a comprehensive blueprint for action to be taken globally, from now into the twenty-first century". Meanwhile we are 25 years later and the term 'blueprint' does not appear anymore in the SDGs and documents guiding their implementation: Blueprinting, i.e. copying a successful practice from one to another situation without adaptation, is a good predictor of governance failure, as it does not take into account the local/national governance environment with its values, traditions and history.

A review on the implementation of Agenda 21 and the Rio Principles mentions the following governance failures (Stakeholder Forum, 2012):

- Lack of information on progress and gaps in the implementation of sustainable development commitments, and where it exists, information is often scattered.
- Lack of policy integration, leading to segmentation and silo-ization, both at the international level and at the national level.
- Lack of a dynamic approach, as Agenda 21 did not address the interconnectedness of the various goals, nor did it explore the fundamental drivers of sectoral and inter-country outcomes.
- Lack of funding: with additional funding coordination and implementation in the UN agencies and programmes could have achieved much greater levels of implementation.
- Lack of public participation in the decision making as well as implementation process, especially at the national and local levels.
- Lack of implementation of key policy principles: access to justice, the precautionary principle and the polluter pays principle.

50 What and for what is governance?

In the 'Environmental Implementation Review' (EIR) (European Commission, 2017a), the European Commission argues that the root causes of failures to implement waste management, water quality, nature protection or air quality requirements are among others insufficient capacity at local level or weak policy coordination and integration, which are *governance* failures. In addition to its focus on governance failures, the EIR process has a built-in feedback loop for *policy* failures. If implementation dialogues with(in) EU countries show that a national implementation gap can be linked back to the design of a policy or law at EU level (e.g. a legal requirement which is very difficult to implement in a certain type of member states), this could lead to reviewing the respective legal requirement at EU level, or the transposition of it in national law.

The political document summarizing the common challenges observed in the 800 pages of analysis of implementation gaps in the 28 EU countries (European Commission, 2017a, pp. 12–13) mentions five root causes of weak environmental implementation which largely overlap with the main governance failures mentioned above regarding the implementation of the Rio 1992 Principles and Agenda 21:

- Ineffective *coordination* among local, other sub-national and national authorities
- Lack of administrative *capacity* and insufficient *financing*
- Lack of *knowledge and data*
- Insufficient *compliance assurance* mechanisms
- Lack of *integration and policy coherence*.

The last point is both a policy failure and a governance failure. A related governance failure is the lack of incentives and arrangements to support policy integration and coherence.

To conclude: it should be kept in mind that policy failure and governance failure are often connected and may be two sides of the same coin. In the next section, we will focus on *governance* failures and in particular on what constitutes their root causes.

3.3 Root causes of governance failure: a typology

Concrete governance frameworks for a policy goal are always normative. They are based on certain values, principles and assumptions which need to be taken into account in order to assess success and failure. The design, management and implementation of a governance framework are complex and multi-layered processes. Failure can happen in one crucial or many less crucial parts of the framework. Diving deep into what we could call the 'root causes' of governance failure may be more useful than sticking to the surface of failure.

Bovens, 't Hart and Peters (Bovens et al., 2001) have suggested two logics of evaluating policy failures: a more 'objective' programmatic evaluation based on costs and benefits, and a political evaluation which is fundamentally subjective and constructed. Both types are not highly correlated: measurable programmatic success may not be recognized at all in the political arena. They label three types of failure as, respectively, *farce* (weak results but political success), *tragedy* (strong results but no political acknowledgement) and *fiasco* (weak results and weak political credits) (Bovens and 't Hart, 2016). In conclusion, they state that the topic is too broad and messy to develop a grand theory of all types of policy failure, and that it is more useful to analyse specific policy failures or ranges of policy failures. Because of the similarity of policy failure and governance failure, I would argue that also for governance failure an ultimate typology is not something we should strive for. However, some clustering will be useful to guide towards common solutions, good practices and lessons learned.

Peters (2015) distinguishes two types of governance failure. The first type is caused by failures of political parties and their leaders to cooperate. The second type concerns failures of organizations within the bureaucracy to cooperate and coordinate. Jessop (2009) defines two forms of governance failure in relation to the territorial dimension: an overemphasis on the local territorial grounding and/or genesis of the problems that need to be governed, and an overemphasis on the 'global' context of local problems. The first leads to neglect of supra-local aspects of failure, the second to a neglect of local solutions to a failure. McConnell (2016) identified two different views on what constitutes failure, which resemble the above-mentioned two logics of Bovens and 't Hart. The first assumes – in a rational scientific tradition – that failure is an objective fact. In terms of governance styles, this links best to the two rational styles, hierarchical and market governance. The second view assumes that the world is contingent on individual perceptions, which typically vary, depending on who is 'perceiving'; this is a typical constructivist and discursive view, matching with network governance. Both views are plausible – but they are at the same time difficult to combine and this can result in heated debates.

Although the above-mentioned typologies are useful in helping the analysis of failure, as a practitioner and pragmatist – not necessarily synonyms, indeed – I would like to work with a typology of governance failure, which links the categories to action perspectives. Closer to this objective is the framework that groups governance failures in two categories (European Commission, 2017b): governance *design* failures link to the politics of preparing and deciding on governance frameworks, and governance *capacity* failures link to the characteristics of the governance environment. **Governance design failure** results from the mismatch of problem context and governance style. In this case, the governance style (combination) is incapable to address successfully a specific problem type. **Governance capacity failure** results from the mismatch of governance style and governance capacity. In this case, the chosen governance style may be suitable to address a specific problem type, but governmental actors do not possess the necessary capacity (i.e. competences, skills, capabilities) to bring about results. I believe that we need a third category to cover another important cause of failure: failure resulting from ineffective management: **Governance management failure**: failure resulting from ineffective management of governance frameworks. This includes what I would call governance 'savviness' issues such as the mismatch of the ambition, the level of information, and ability to make good judgement beyond ideological assumptions, related to the governance challenge to be addressed.

Capacity, design and management would constitute my proposed typology of areas from which governance failure emerges. Design and management failures partially overlap with metagovernance failures (Chapter 4).

In Sections 3.3–3.5 this typology is used to present a non-exhaustive list of concrete examples of governance failure causes, based on my own experience and other sources (Bovens et al., 2001, Meuleman, 2008, Jessop, 2009, Howlett and Ramesh, 2014, European Commission, 2017a, European Commission, 2017b). The examples are, where relevant, tentatively linked to governance styles. The typical governance style weaknesses of each governance style result in governance failure in all three areas: capacity, design and management. Therefore, Section 3.4 presents such failures before other examples will be discussed in Sections 3.5–3.7.

3.4 Hierarchical, network and market governance failure

Many governance failures can be directly linked to the mind-set or the repertoire of one of the three governance styles: they are the result of, or part of the logic of each style.

Inherent hierarchical governance weaknesses

Hierarchical governance has several weaknesses, such as the tendency to create "red tape" (too much bureaucracy), to not being able to deal with complexity and uncertainty (for example because the idea is based on a clear and fixed division of tasks), and to create opposition (for example because actors who have something important at stake are not listened to). The hierarchical style of communication has one direction, from sender to receiver, which is also not useful when the success of a policy depends on understanding and acceptance by affected groups. Dixon and Dogan (2002) observed that, when confronted with governance failure, the typical hierarchical reaction is to strengthen hierarchical controls. This approach "discounts any need for the governors to engage with the governed in the governance process". Weaknesses observed in a case study on partnerships in Wales included imposed objectives, unbending rules and unhelpful departments (Entwistle et al., 2007).

Climate change governance is an example from sustainable development, which for a long time has been primarily hierarchical. This may have been an important reason for the unsatisfactory results of the 2009 Copenhagen climate summit (Meuleman, 2010). The focus on global, legally binding agreements and central organization was in line with the most common framing of the problem in those days: climate change had been successfully framed as a global disaster that could only be solved with centralized measures and global norms. The risk of this frame was that among the main governors there was an addiction to command, and seeing only emergencies instead of complexity (Grint, 2010). However, problems are socially constructed and climate change is no exception. After the failure of the 2009 UN climate summit in Copenhagen, another frame became more popular among scholars and policy makers: climate change as a "wicked problem" (Rittel and Webber, 1973). Wicked problems are situations with a complex structure and where there is consensus neither on values nor on knowledge; they cannot be solved but they still need to be dealt with. The success of the 2015 Paris Climate agreement followed many lessons learned from Copenhagen.

Pure hierarchical governance usually has no answer to the challenges posed by 'wicked problems'. Exceptions apply, though: in 2004, the then president of the small Caucasian country Georgia tackled the in-grown corruption in the traffic police organization in a top-down way by firing all (30,000) traffic police officers at the same day (Centre for Public Impact, 2016) (see also Section 8.1).

In many less developed countries, hierarchical governance is the main style applied by government, at least by central government. Donor countries could build on this culture of hierarchy by supporting institutional reform, which has led to well-established planning ministries and agencies in many countries. The downside is that the governance structure now has a 'heavy head' at the top and nothing, not even softer governance mechanisms in place, at the level of implementation where citizens will benefit of more sustainability. Corruption is one of the darker phenomena of abuse of power in a hierarchical governance context. The reflex of donor countries has often been to combat corruption by adding specific conditionalities in aid programmes – a hierarchical reflex – but recipient countries have argued in reaction that corruption should be seen in the wider context of poverty and capacity constraints (Pomerantz, 2011), which needs a broader governance reaction than just 'more hierarchy'. SDG 16 promotes the positive side of hierarchy such as ensuring rule of law, having effective institutions overall but in particular also in the judiciary.

Inherent network governance weaknesses

Network governance also has its typical deficiencies. The dialogue on improving soil governance in the Netherlands in the late 1990s resulted in endless talks with no results, other than the shared conclusion of most parties that the Environment Ministry was their common enemy (Meuleman, 2008). Another common problem is the lack of clear lines of responsibility: in a multi-party process, everyone can blame the others if it fails. The key value of trust is also a risk: as a Dutch saying goes, "Trust arrives by foot but leaves by horse", meaning that trust can be lost quickly and takes time and effort to be restored. A key weakness of network governance is therefore that it is sensitive to manipulation. A typical reaction when confronted with governance failure is changing the composition of the network until the problem is solved. This can be done through

> the removal from governance structures of those who are not fully committed to co-determination, and the empowerment of different and more committed actors to engage in interactive governance processes, with a view to

developing strategies that would not only correct the governance failure, but also evoke a moral commitment to interactive governance success, thereby encouraging compliance.

(Dixon and Dogan, 2002)

Another network dysfunction is that it may lead to excessive demands of a relatively small group of people (Entwistle et al., 2007).

Many publications have mentioned the democratic weakness of network governance, (e.g. Sørensen, 2006). In a recent Swedish urban case study, Larsson (2017) observed that networks "tend to promote collaboration among elites while leaving weak groups and policy recipients without a voice". Howlett and Ramesh (2014) call these type of governance failures 'network failures' or, when about private sector networks, 'corporatist network failures'. Another study, on the national implementation of the EU Bird and Habitat directives, showed that, in the Netherlands the introduction of network governance only addressed some of the problems and created new problems. In the absence of hierarchy, "power struggles became more overt, discursive conflicts occurred between the state department and the lower administrative levels as well as between advocates and adversaries, and rules of the game faced similar contestations" (Arnouts and Arts, 2009). A research project in Germany analysed the central assumption of previous research that collective actor constellations and collaborative network-based governance are crucial success factors in regional energy governance in. It was found that the massive development and the high share of electricity generated from renewables couldn't be explained by collaborative governance alone: other factors such as power and geographical conditions were also important (Gailing and Röhring, 2016).

Others have commented that the return from a network approach to a more regulated approach may be difficult: Because power is by design diffused in network governance arrangements, such arrangements can be much harder than traditional hierarchical forms to reform or dismantle when problems arise (Howlett and Ramesh, 2014).

The implementation of SDG 17, which is about partnerships, depends on effective network governance, while taking into account the weaknesses of this governance style.

Inherent market governance weaknesses

Market governance, on the other hand, can produce many of the problems, which are usually termed as "market failure", because markets are 'imperfect' (Dommett and Flinders, 2015). NPM-fueled reforms aimed at creating a smaller and smarter state, but resulted in many countries in a highly complex and fragmented public sector which operates beyond the direct control of elected politicians (Dommett and Flinders, 2015). The use of market mechanisms in public policies has resulted in a loss of democratic control over independent agencies, in a priority on efficiency rather than on effectiveness, and in monopolies. Dommett and Flinders observe five 'pathologies' of delegated governance, which are also causes of governance failures. They are mainly about insufficiencies in the relation between 'sponsor' departments and newly established arm's-length bodies: lack of clarity of roles and responsibilities, lack of mechanisms to maintain productive relationships, and lack of skills development at the devolved agencies. Others observed that market forms of coordination required endless applications for relatively small grants (Entwistle et al., 2007). The creation of 'markets' for CO_2 emissions trading has resulted in the giving away of many emission permits for free by the same governments which established the system.

Multi-stakeholder partnerships which are promoted in SDG 17 not only require effective network governance, but also the entrepreneurial spirit of market governance, and accountability rules and other 'rules of the game' (hierarchical governance).

3.5 Governance capacity failure

Governance capacity can be defined as "the resources and skills a government requires to steer a governance mode so as to make sound policy choices and implement them effectively" (Howlett and Ramesh, 2014). Without sufficient governance capacity, design and management are bound to fail. Therefore, I will treat capacity failures first.

Governance capacity is almost synonymous with the term **governability**, which Kooiman defined as "the overall capacity for governance of any societal entity or system" (Kooiman, 2008). It is a dynamic concept: "Governability as a whole, or any of its components, can be influenced by acts of governance. However, many external factors influence governability as well, some of which can only be poorly handled in governance, or not at all".

Table 3.2 presents six examples which will be introduced.

C1. Lack of motivation ('political will')

In the 1990s, I was involved in cumbersome debates between the Ministries for Agriculture and Planning & Environment in the Netherlands about what should be done – which governance arrangements – about the increasing soil and water pollution caused by applying more manure on the soil than it could take up. On the side of the powerful Agriculture Ministry, there was no political will to conclude on any structural measure, which was rationalized with their repeated statement that the problem was not yet sufficiently researched. The Environment Department was ambitious but less powerful. Their strategy was to frame the problem as an environmental disaster. I represented the Spatial Planning Department angle, which framed the problem as weak spatial allocation of conflicting functions (agriculture and nature). In the end, time was on our side: a change of government enabled new problem frames and enabled setting some first steps.

Although everybody has an idea of what *political* will means, it is a very diffuse concept. It has been defined as representing an actor's willingness to expend energy in pursuit of political goals, and is viewed as an essential precursor to engaging in political behaviour (Treadway et al., 2005, Painter, 2014). Lack of

56 *What and for what is governance?*

Table 3.2 Causes of governance failure linking to governance capacity

■ = primary type, □ = secondary type

Examples of causes of governance failure	Types of governance failure		
	Governance capacity failure	Governance design failure	Governance management failure
C1. Lack of motivation ('political will')	■		□
C2. Mismatch of governance framework and governance capacity	■	□	
C3. Lack of knowledge and experience on one or more governance styles	■		
C4. Lack of resources (human, financial)	■		
C5. Lack of skills to deal with governance challenges/failures	■		
C6. Lack of compliance capacity and mechanisms	■	□	

political will is the result of weighing pros and cons of action and non-action at many levels (personal, political, impact on voters/stakeholders, timing, etc.) by a political leader or college. Mintzberg (1983) argued that people should not only have the *political will* to enact political behaviours, but also possess the necessary *political skill*: the ability to execute these behaviours in politically astute and effective ways. Treadway et al. (2005) concluded from research that political skill is not only a feature of politicians but also a value for employees in political organizations such as ministries; it should therefore be part of the organization's selection processes and be enhanced through training programmes.

A political decision to act or not act is rational in the eye of its beholder, or based on intuition or on ideological considerations. As a remedy to non-action it may help to bring in new facts (e.g. about costs and benefits), or to organize countervailing powers. My personal view as a policy maker is that lack of political will to go ahead is better than no will or view at all. Resistance requires energy, which is a good basis to investigate where this energy could be used in a positive way (the *judo* principle). This could be done by reframing the issue (offering a related problem that is acceptable within the mind-set of the political decision maker), or proposing a pilot project as a start. A pilot project is often a good start, because as soon as anything moves, new doors will open and new, unforeseen (political) opportunities may emerge.

The factor 'lack of political will to act upon a challenge' has different operational forms. A politician who thinks hierarchically may express this in the form of a decision to postpone new regulation or to shift to soft instruments, which may not work well in a hierarchical culture. A market-governance way of showing political resistance could be to refuse using market-based policy instruments such as taxation, dedicated funds and public procurement. This is what happened when a Soil Action Plan for England was being prepared and the Treasury excluded the use of such instruments (Meuleman, 2008). Network governors may show a lack of political will by creating distrust among the partners in a network.

C2. Mismatch of governance framework and governance capacity

A cause of governance failure is when a governance framework is designed that requires certain skills and capabilities which are absent. If the governance framework is mainly about legislative tools, one would expect that the authorities responsible for implementation have a sufficient amount and quality of legal officers. In the past, in Germany, more than 70% of civil servants in the ministries had a legal background; in the neighbouring Netherlands this was less than 30% (Van der Meer et al., 1997). This means, for example, that even if political leaders decided to switch from a networked consensus approach to a focus on legal tools, the Dutch government would not be able to accommodate this, at least not on short notice.

Wicked problems are very difficult to deal with, with any governance approach, but for hierarchical governance this is virtually impossible. Surprisingly, the European Commission, which is at its core a hierarchical organization, has been reported to be relatively good in this respect (e.g. Candel et al., 2015).

C3. Lack of knowledge and experience on one or more governance styles

The capacity to apply metagovernance by switching between or combining the three basic governance styles and their respective toolboxes depends on good knowledge of and experience with each of the styles. The styles may also undermine each other (Hulst and Van Montfort 2007). The capacity for hierarchy, e.g. command and control, may be weakened by market governance measures to promote efficiency or to devolve responsibilities. The capacity for network governance, which includes time and energy to build up trust in a network, may be undermined by a sudden top-down (hierarchical) intervention. I experienced this when we had achieved consensus with sub-national authorities and civil society about a large water/environment infrastructure investment and my minister suddenly disagreed. The capacity for market governance may be undermined from within (e.g. because of lack of market regulation as a result of an efficiency operation inside a ministry) or because the paradigm of consensus (network governance) is paralyzing dynamics and innovation.

C4. Lack of resources (human, financial)

Where implementation of policy and legislation is the task of local government, human and financial resources are often mentioned as the main governance

58 *What and for what is governance?*

problem. The first solution to this problem is of course reallocation of budgets. However, smart measures to do more with the same budget may be unexploited; it might be worthwhile to find out what solutions have emerged elsewhere. In the field of environmental compliance assurance, risk-based allocation of resources for inspection visits is gaining momentum (European Commission, 2018b).

Another approach is pooling of resources with other (local or regional) authorities, creating economies of scale (Kelly, 2007). This makes purchasing sustainable products (including e.g. waste management installations) more feasible, makes applying for funds and grants less complicated and time-consuming, and makes management and maintenance more cost-effective. Inter Municipal Cooperation (IMC) is widely used in continental Europe but much less in England, where – not surprisingly considering the national preference for market governance – hiring private service providers was historically the preferred alternative to collaboration among (public) peers. A thorough analysis of different models of IMC in various sectors such as the environment or transport was put together by UNDP and the Council of Europe (2010). Four models emerged: informal IMC, weakly formalized IMC, IMC in functional "enterprises", and IMC as a model of integrated territorial cooperation. An additional model I learned from the mayor of Cape Town is when cities or metropolitan areas invite under-resourced municipalities to benefit from the purchasing power of the bigger city: "I meet the 30 mayors around Cape Town monthly and discuss where we can do things together – and where we can help them".[2]

C5. Lack of skills to deal with governance challenges/failures

Handling governance failure requires that public officials understand the root causes of failures. For this, deep understanding is needed of the governance environment (see Section 1.2) in which the failures have emerged. Dedicated training could be made mandatory – that is, if mandatory training is accepted in the dominant governance culture. Public officers in a network governance environment may want to co-decide which training is required. Under market governance, officials are encouraged to 'shop' in the training menu for the specific training they want.

In the early 1990s, the senior management of the Dutch (i.e. consensus oriented) Environment Ministry made a three-day training in the mutual gains approach to negotiation mandatory for all policy officers and their junior management. Heads of units were asked to become trainers themselves. Because everybody went through the same training, a common 'language' developed and it became a new normality to look for 'win-win' packages together with stakeholders. At a certain moment, however, the Ministry became known for always applying this approach as a solution for all problems, and a weekly magazine criticized the 'win-win dogma' of the Ministry in a cover article.

C6. Lack of compliance capacity and mechanisms

With regard to securing compliance with a policy or legal objective, Howlett and Ramesh (2014) present the example of incentives not functioning and suggest that finding another solution/tool within the same governance style (market governance) would be easier than when a style switch is necessary; the same applies when coproduction of a policy gets frustrated: another approach from the network governance toolbox might work, based on the same values and principles. Indeed, the most efficient reaction seems to be looking first for a solution close to the mind-set with which the original (not functioning) tool was conceived. When, however, such is not possible or effective, a switch to another toolbox/governance style should be considered.

The European Commission (2018a) promotes a wide range of (environmental) compliance assurance measures, inspired by different governance styles. The idea is that the Commission works

> hand in hand with member states and professionals such as inspectors, auditors, police officers and prosecutors in order to create a smart and collaborative culture of compliance with EU environmental rules on activities such as industrial production, waste disposal and agriculture.

In metagovernance style, the Commission offers in addition to classical (hierarchical) forms of compliance assurance (enforcement), support through network governance tools (networks for practitioners, evaluation of and information portals on national compliance systems to enable comparison and mutual learning, mechanisms to facilitate cooperation, information and guidance on good practices, and training programmes. In addition, financial support is offered (a market governance tool).

3.6 Governance design failures

A governance design failure occurs when there is a fundamental mismatch between an established governance style (or styles combination) and the type of problem it should address (Howlett and Ramesh, 2014). Table 3.3 presents some examples.

D1. Incompatibility of governance styles in a combination

The second problem regards the incompatibility of certain characteristics of the three main governance styles. A major reason why the conflict potential is high is that the three styles express different types of relations with other parties: dependency (hierarchy), interdependency (network) or independency/autonomy (market) (Kickert, 2003, p. 127). A hierarchical "command and control" style of leadership will seldom lead to a consensus (network style) – even if this was

Table 3.3 Causes of governance failure linking to governance design

■ = primary type, □ = secondary type

	Types of governance failure		
Examples of causes of governance failure	Governance capacity failure	Governance design failure	Governance management failure
D1. Incompatibility of governance styles in a combination	□	■	□
D2. Incompatibility of governance style and problem type	□	■	□
D3. Lack of integration of key policy principles		■	
D4. Lack of effective institutions		■	□
D5. Resistance to reframe the problem type		■	□
D6. Mismatch between policy goals and governance tools and settings		■	
D7. Lack of mechanisms for implementation		■	□
D8. Lack of mechanisms for policy and institutional coherence	□	■	□
D9. Lack of taking into account the cultural dimension of governance	□	■	□
D10. Blueprinting: using one governance style as cure-for-all		■	□
D11. Inadequate/disputed/scattered knowledge and data on the concrete policy challenges, leading to poor problem diagnosis	□	■	□
D12. Neglect of the long-term dimension		■	□
D13. Lack of built-in reflexivity and resilience (incl. bird's-eye perspective; not sensitive to 'frog in pan' situation)		■	□
D14. Lack of inclusion of relevant stakeholders/citizens at all relevant levels of government, in policy preparation, decision making and implementation		■	□

the only feasible outcome of a policy process which the government was not able to "steer" with legal instruments. Decentralization or outsourcing (a typical market governance strategy) makes actors more autonomous. They may become frustrated when detailed control mechanisms are introduced (hierarchical governance). The coexistence of "new modes of governance" with compulsory regulation or hierarchy is problematic.

Sometimes attempts at a network approach are undermined by hierarchical governance inside one and the same public sector organization. From the perspective of the classical hierarchical governance style, network governance is problematic because "governments, like the church, will find networks messy and carp at the mess" (Bakker et al., 2008). Internal competition with the traditional hierarchical governance style is one of the reasons that the introduction of network governance sometimes fails. This competition can lead to obstruction from other public sector organizations or other parts of the same organization and to unreliable behaviour (not keeping promises, or sudden withdrawal of a negotiation mandate).

D2. Incompatibility of governance style and problem type

As a rule of thumb, the relations between the three basic governance styles and different types of problems are as follows:

- hierarchical governance cannot deal with complexity and unpredictability;
- market governance and network governance dislike command and control – although they would benefit from a small portion of it – and may generate governance failures such as the breaking apart of a partnerships (in network governance) or entrepreneurial ways to escape control (market governance);
- network governance cannot deal with sudden disasters: there is no 'command' and no 'control', and there is no time to deliberatively decide who should put out the fire;
- market governance works best when the challenge is of a routine, not disputed type, or when innovation through competition is the issue to tackle. When complexity and 'wickedness' of a problem become too big, market governance is not able to produce the collaboration based on trust and empathy that may be needed – market governance 'partnerships' are contractual, not empathetic.

D3. Lack of integration of key policy principles

Sustainable development is a normative concept that expects that a number of key policy principles are taken into account – or better: well integrated – in policy and law within countries. Some of these principles have their origin in environmental protection agreements: access to justice (UNECE Aarhus Convention), the precautionary principle and the polluter pays principle. On each of them a whole literature has developed, discussing how they should be integrated and why this is mostly such a cumbersome process. The principles have in common that they are rather vague and thus can be interpreted in many ways. This does not have to be a problem: at least this leads to useful deliberations which would not be there without the principles.

In hierarchical countries, these principles need a clear definition and their use is part of many procedures at, for example, the Constitutional Court. In network-oriented countries, where sometimes (e.g. the Netherlands) there is no Constitutional Court, people tend to prefer a deliberation leading to a consensus on the meaning of such principles.

D4. Lack of effective institutions

What an effective institution is depends among others on the challenges and the context in which they need to function. Informal institutions may not work well in a hierarchical administrative culture, while too formal arrangement may cause irritation and non-compliance in consensus-oriented or market governance countries. Identifying governance failure because of the absence of the effective functioning of dedicated institutions – or their absence as such – therefore needs to be seen in its context. However, sometimes there is a built-in 'blindness' to see such relations; if you only have a hammer you tend to see every problem as a nail – while some relevant problems may be overlooked because they have a different shape.

D5. Resistance to reframe the problem type

Some work done by public sector organizations is like trying to hammer a nail in a concrete wall. It may be the wrong tool for the right problem, or the other way around – but it is normally a waste of resources. Unless, of course, it is a political priority to buy time because the political 'window' to take action may not yet be visible. Once I had to commission a three-month study on a subject we already knew enough about, for the only reason that my minister needed that time for her lobbying and 'massaging' behind the scene.

Reframing a problem from one that cannot be solved with the resources one has, to another that might be better manageable can meet (emotional) resistance when the reframing implies switching to another governance style. In the 1990s, the Netherlands government completely turned around its approach as regards flood protection, namely from 'working against water' (building ever higher dikes) to 'working with water'. The latter implied giving 'room for the river' in order to mitigate the impact of high water levels. This included even breaking down some of the existing dikes. It was a rational choice but it met with some resistance: for hundreds of years, the Dutch had protected their low-laying land with dikes. In the end, both problem frames meant the similar action (infrastructural work), but with a very different design and win-win with nature development.

D6. Mismatch between policy goals and governance tools and settings

An example of the tension between a policy goal and the available governance tools is the situation that resulted in the 'Better Regulation' policy of the European Commission (2015). The Commission, with its German/French internal culture, has a long tradition of solving policy problems by regulation. Non-regulatory tools became only more important after the 2001 White Paper on Governance, which stimulated the use of network and market governance tools (European Commission, 2001).

D7. Lack of mechanisms for implementation

Policy implementation is traditionally seen as subordinate to policy making, and it is usually not a political priority – until something goes terribly wrong. The divide between policy making and implementation is a classic in environmental policy and law, for example, where signals from inspectors doing field visits often do not reach the policy making, let alone the political level.

Supporting implementation begins with making sure that there are bodies (agencies or other) that have the skills and resources to fulfil the tasks. Secondly, an effective interface between implementation and policy making is needed.

Multi-level governance is a chain of policy and implementation: for example, when the European Union adopts a Directive, the member states have to implement this. They do that by developing a strategy and legislation, to be implemented by e.g. regional authorities. In some areas, regional authorities then make rules for local authorities to be implemented. Along this whole chain, governance failure may happen and needs to be identified as early as possible.

The UN Agenda 2030 is also an example of multi-level governance, but this depends much more on voluntary agreement and collaboration by its member states than in the EU. Of course, the many multilateral agreements on environment and other areas imply legal binding once they have been ratified, and they contribute to the implementation of the SDGs. However, as regards the nexus themes and overall, the transformations needed to achieve the Goals, the implementation tools are restricted to voluntary action, peer-to-peer exchange, peer pressure and multi-stakeholder partnerships.

D8. Lack of mechanisms for policy and institutional coherence

Governance failure caused by insufficient policy and/or institutional coherence can be the result of weak management of a governance framework, but often something already has gone wrong in the design phase. Implementation of an integrated policy on an agro-environmental issue such as eutrophication of surface water requires coherent policy preparation in the first place. However, it may not be necessary or even desirable to develop specific coherence promotion mechanisms for each governance challenge: a tool such as interdepartmental 'dossier teams' to which also decentralized authorities can be invited (like in the Netherlands), is usable for many purposes.

Promoting policy and institutional coherence is Target 14 of SDG 17. Section 10.2 presents a number of ways to deal with a lack of policy and institutional coherence.

D9. Lack of taking into account the cultural dimension of governance

Neglect of the cultural, historical and geographical context of governance can lead to a variety of governance failures. Chapter 6 addresses the cultural dimension of (sustainability) governance.

D10. Blueprinting: using one governance style as cure-for-all

One of the great challenges of governance is that there are people and organizations who consider using one of the three basic governance styles as a cure-for-all, for their own use as well as for 'export'. Others, on the receiving side, may want to try any tool designed within the context of their favourite style – or they are not aware of the cultural context of imported governance recipes and use 'best practices' without adaptation. Chapter 7 addresses 'best practices' and other popular governance solutions which are not as neutral as they seem to be at first sight.

D11. Inadequate/disputed/scattered knowledge and data on the concrete policy challenges, leading to poor problem diagnosis

Access to data has become an essential topic for the implementation systemic changes (see among others SDG 16 and 17).

D12. Neglect of the long-term dimension

See Section 4.9, where long-term aspects are discussed as part of metagovernance.

D13. Lack of built-in reflexivity and resilience (incl. bird's-eye perspective; not sensitive to 'frog in pan' situation)

Governments have a long tradition in establishing (and then again abolishing) 'watchdog' or early warning mechanisms. These arrangements may have the form of independent expert/stakeholder advisory bodies such as sustainable development councils, or internal bodies such as internal think tanks close to the Prime Minister, or, as currently in the European Commission, an advisory think tank directly reporting to the Commission's President. The British Sustainable Development Council explicitly had such a watchdog function but was abolished in 2009, like the Dutch Advisory Council for Research on Nature and Environment and several other independent advisory councils.

D14. Lack of inclusion of relevant stakeholders/citizens at all relevant levels of government, in policy preparation, decision making and implementation

In a hierarchical governance context, stakeholders could get an executive role or, the opposite, are considered as bystanders hindering the achievement of policy goals. Under network governance, the enthusiasm for stakeholder involvement may be so great that it becomes random in terms of selection of participants, unspecific in terms of topic, and unguided in terms of 'rules of the game'. The result can be an angry civil society, business representatives or general public. The latter I described elsewhere as example: 10,000 Dutch citizens gave input on national spatial planning strategy, at many after-work discussion meetings, who afterwards did not receive any feedback about what the national government had done with their ideas. Many of them will not have participated in such meetings again (Meuleman, 2003). In the same book I gave the example of a market governance failure. A Ministry wrapped a marketing campaign to sell the Minister's ideas about flood protection measures around a narrative about listening to citizens' needs. A campaign bus visited each and every small village along the river – and left their citizens behind in a mixture of confusion and anger, because they were not listened to.

Governance failure also occurs when decision-making structures and related institutions contain systematic biases against groups who are officially supported by a policy (in this case poor households), as observed in the case of urban water supply in Indonesia (Bakker et al., 2008).

3.7 Governance management failures

Governance failures based on weak management and leadership can occur in many ways. Three examples are presented below.

Table 3.4 Causes of governance failure linking to governance management

■ = primary type, □ = secondary type

Examples of causes of governance failure	Governance capacity failure	Governance design failure	Governance management failure
M1. Weak horizontal coordination across government departments; fragmentation; silo-thinking		□	■
M2. Weak vertical coordination among government levels		□	■
M3. Lack of interconnectedness of goals and problems (e.g. nexus approach)	□	□	■

M1. Weak horizontal coordination across government departments; fragmentation; silo-thinking

Both fragmentation and silo-thinking have a negative connotation. Fragmentation could also be framed positively, namely as 'specialization'. When environmental, social and/or economic policies are well integrated in sectoral policies, they may become at the same time less visible and 'fragmented'. In such a case it is important to 'regain' the lost influence through establishment of a coordination point at the central level, accompanied by a cross-cutting strategic document (i.e. the sustainable development strategy many countries have at the national level).

How to deal with silo-thinking is a similar case. When silos prevent policy integration or alignment, the first reaction may be breaking down the institutional silos, whereas creating bridges across mental silos might be a better first step. Section 7.7 addresses this challenge based on an earlier publication (Niestroy and Meuleman, 2016).

M2. Weak vertical coordination among government levels

Insufficient vertical coordination between government levels can lead to governance failure, for example when national government is not well aware of the constraints of sub-national authorities, or when different political parties are ruling at different levels. It can be impossible to combine strong multi-level coordination and strong horizontal coordination at one level. A 2015 research article concludes that a gradual development had taken place, towards more direct contacts between the European Commission and national agencies, thus counteracting stronger coordination at national level. This is a clear 'coordination dilemma' (Egeberg and Trondal, 2016) with a strong governance dimension: the European Commission and national agencies collaborate in a network governance environment, because of which the hierarchical control of these agencies by their national ministries decreases.

M3. Lack of interconnectedness of goals and problems (e.g. nexus approach)

A strong, hierarchical division of tasks can result in high-quality, specialized results, but at the same time to silo-thinking. Making linkages with other sectors may then become a (career) risk. In such a situation, the indivisibility of the SDGs becomes a problem: a nexus approach is impossible.

Formal and informal governance structures can be designed to address this problem. It is also a matter of organizational culture. Once, I experimented with a new approach at the start of an interdepartmental project that I would lead. At the first project team meeting, my first question was not who the participants represented, but what they could contribute to the project. This resulted in a more collaborative atmosphere than I expected. A similar example was the start of a project about sustainable sea management, in which several agencies and ministries involved had contrasting opinions and interests. A colleague of mine who was the project leader hired a meeting room on the beach (in The Hague) and asked the participants, before we started the meeting, to go for a walk on the beach in duos and ask each other what their very first memories of the sea were. This made the issue personal and participants more or less forgot what they were against: they had remembered a shared interest in protecting the quality of the sea.

3.8 Governance failures and policy phases

In Figure 3.1, I have plotted the key issues around governance failure against phases in the *policy 'cycle'*. The policy cycle model with its chronology of agenda-setting, policy formulation, decision making, implementation and evaluation has become the conventional way to describe a policy process (Wegrich and Jann, 2007). I will use it in Figure 3.1 for practical reasons (it is probably the best-known model across government officials). Besides, it acknowledges that policies are part of a continuum: there will always be new policy challenges, or the same problem will return in a different disguise. The downside of the model is that it suggests that the phases occur in a neat order (which is not always the case), and that it neglects contingency and the autonomous behaviour of actors.

Teisman (2000) distinguished besides the policy cycle model (which he calls the phase model) two other models, which have different assumptions on what decision making is: the *stream* model and the *rounds* model. These models are not completely neutral as regards governance styles: The policy cycle model aligns best with hierarchical governance: it assumes that there is a cockpit in

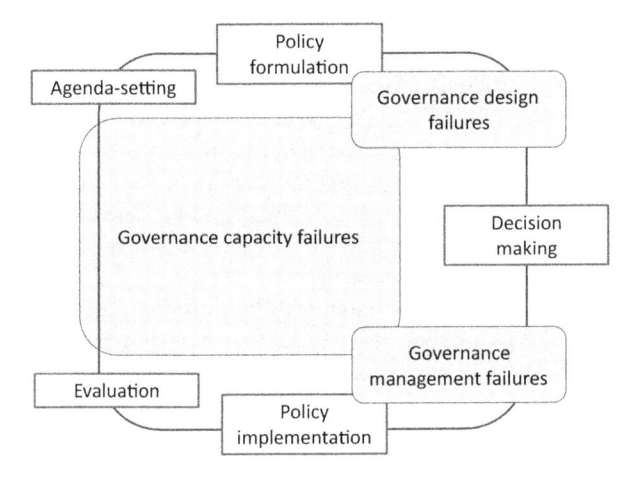

Figure 3.1 A model of governance failure occurrence in the policy cycle

which steering takes place. The stream model, in which streams of participants, problems and solutions connect, emphasizes the contingency of the process environment and the autonomy of actors. We could link this loosely to the logic of market governance. The rounds model focuses on the interaction between actors, while they introduce combinations of problems and solutions, and create progress. This model is related to network governance; it assumes interdependency of actors (Wegrich and Jann, 2007).

3.9 Conclusions

So far, we have discussed the challenges of sustainability governance (Chapter 1) with the lens of the three ideal-typical governance styles – hierarchy, network and market (Chapter 2), and governance failures and their causes, with transboundary governance as a special case (Chapter 3). The concept of governance failure overlaps partially with policy failure but should also be treated separately. Both failure types may influence each other and other features of a governance framework. Failure can be persistent and recurrent, and dealing with them requires understanding what several scholars call 'meta-causes' such as the capacity of policy actors and better designing policy processes and institutions (Jessop, 1997, Kooiman, 1993, Dunsire, 1996). These 'root causes' of failure can be of the capacity, design or management type.

The examples in this chapter show that analyzing and repairing a governance failure is not a purely rational exercise. Experience, intuition and training should support detecting failure in an early stage. For this, public managers/officers need to understand the full range of governance failures, and in particular their relation to specific governance styles: things one doesn't understand are much more difficult to see than those one expects as at least a possibility. Nevertheless, many governance failures cannot be foreseen, prevented or even repaired. What seems to be needed is, apart from experience and skills, that governance frameworks include built-in early warning mechanisms.

Taking a metagovernance perspective, with some distance from failure as it emerges, ensures that all governance type toolboxes are available to develop remedies. Moreover, curing governance failure itself can be a case of metagovernance, when, for example, the cause(s) of failure are tackled by switching to another governance style's repertoire. I would go even further and argue that dealing consciously with the (emergence, possibility, probability) of governance failure is a requisite of metagovernance.

In the next chapter metagovernance will be introduced from several angles. One of them is the point that also metagovernance brings forth failure, at a higher level of abstraction: *metagovernance failure*.

Notes

1 www.ted.com/talks/james_cameron_before_avatar_a_curious_boy.html/
2 Oral communication, June 2017, Brussels.

References

Andersen, M. S. & Clubb, D. O. 2013. Understanding and accounting for the costs of inaction. In: European Environment Agency (ed.) *Late Lessons from Early Warnings: Science, Precaution, Innovation.* Copenhagen: European Environment Agency.

Arnouts, R. & Arts, B. 2009. Environmental governance failure: The 'Dark Side' of an essentially optimistic concept. In: Arts, B., Lagendijk, A. & Van Houtum, H. (eds.) *The Disoriented State: Shifts in Governmentality, Territoriality and Governance.* Heidelberg: Springer.

Bakker, K., Kooy, M., Shofiani, N. E. & Martijn, E.-J. 2008. Governance failure: Rethinking the institutional dimensions of urban water supply to poor households. *World Development*, 36 (10), 1891–1915.

Bovens, M. & 't Hart, P. 2016. Revisiting the study of policy failures. *Journal of European Public Policy*, 23, 653–666.

Bovens, M., 't Hart, P. & Peters, B. G. 2001. Analysing governance success and failure in six European states. In: *Success and Failure in Public Governance: A Comparative Analysis.* Cheltenham, Glos: Edward Elgar Publishing.

Candel, J., Breeman, G. & Termeer, C. 2015. The European commission's ability to deal with wicked problems: An in-depth case study of the governance of food security. *Journal of European Public Policy*. 6, 789–813.

Centre for Public Impact. 2016. Seizing the moment: Rebuilding Georgia's police [Online]. Available: https://www.centreforpublicimpact.org/case-study/siezing-moment-rebuilding-georgias-police/#.

Council of Europe & UNDP. 2010. Toolkit manual inter-municipal cooperation. Strasbourg: Council of Europe.

Dixon, J. & Dogan, R. 2002. Hierarchies, networks and markets: Responses to societal governance failure. *Administrative Theory & Praxis*, 24, 175–196.

Dommett, K. & Flinders, M. 2015. The centre strikes back: Meta-governance, delegation, and the core executive in the united kingdom, 2010–14. *Public Administration*, 93(1): 1–16.

Dunsire, A. 1996. Tipping the balance: Autopoiesis and governance. *Administration & Society*, 28(3), 299–334.

Egeberg, M. & Trondal, J. 2016. Why strong coordination at one level of government is incompatible with strong coordination across levels (and how to live with it): The case of the European Union. *Public Administration*, 94, 579–592.

Entwistle, T., Bristow, G., Hines, F., et al. 2007. The dysfunctions of markets, hierarchies and networks in the meta-governance of partnership. *Urban Studies*, 44, 63–79.

European Commission. 2001. European governance: A white paper. COM *(2001)*, 428, 25.

European Commission. 2015. Better regulation for better results – An EU agenda. COM(2015) 215 final. Brussels.

European Commission. 2017a. The EU environmental implementation review: Common challenges and how to combine efforts to deliver better results. COM(2017) 63 final. Brussels.

European Commission. 2017b. *Quality of Public Administration – A Toolbox for Practitioners.* 2017 edition. Abridged version.

European Commission. 2018a. Environmental compliance assurance – Scope, concept and need for EU actions. SWD/2018/010 final. Brussels.

European Commission. 2018b. EU actions to improve environmental compliance and governance. COM/2018/010 final. Brussels.

European Environment Agency. 2013. Late lessons from early warnings II. Copenhagen.
Gailing, L. & Röhring, A. 2016. Is it all about collaborative governance? Alternative ways of understanding the success of energy regions. *Utilities Policy*, 41, 237–245.
Grint, K. 2010. The cuckoo clock syndrome: Addicted to command, allergic to leadership. *European Management Journal*, 28(4), 306–313.
Howlett, M. 2012. The lessons of failure: Learning and blame avoidance in public policymaking. *International Political Science Review*, 33, 539–555.
Howlett, M. & Ramesh, M. 2014. The two orders of governance failure: Design mismatches and policy capacity issues in modern governance. *Policy and Society*, 33, 317–327.
Howlett, M., Ramesh, M. & Wu, X. 2015. Understanding the persistence of policy failures: The role of politics, governance and uncertainty. *Public Policy and Administration*, 30, 209–220.
Hulst, R. & Van Montfort, A. 2007 (eds.). *Inter-Municipal Cooperation in Europe*. Dordrecht: Springer.
Jessop, B. 1997. Capitalism and its future: Remarks on regulation, government and governance. *Review of International Political Economy*, 4(3) (1997): 561–581.
Jessop, B. 2000. Governance failure. In: Stoker, G. (ed.) *The New Politics of British Local Governance*. Basingstoke: Palgrave Macmillan.
Jessop, B. 2009. From governance to governance failure and from multi-level governance to multi-scalar meta-governance. In: Arts, B. & Al, E. (eds.) *The Disoriented State: Shifts in Governmentality, Territoriality and Governance*. Heidelberg: Springer.
Jessop, B. 2011. Metagovernance. In: Bevir, M. (ed.). *The Sage Handbook of Governance*. London: Sage.Kay, A. & Boxall, A. M. 2015. Success and failure in public policy: Twin imposters or avenues for reform? Selected evidence from 40 years of health-care reform in Australia. *Australian Journal of Public Administration*, 74, 33–41.
Keech, W. & Munger, M. 2015. The anatomy of government failure. *Public Choice*, 164, 1–42.
Kelly, J. 2007. The curious absence of inter-municipal cooperation in England. *Public Policy and Administration*, 22(3), 319–334.
Kickert, W. J. M. 2003. Beneath consensual corporatism: Traditions of governance in the Netherlands. *Public Administration*, 81, 119–140.
Kooiman, J. 1993. *Governance and Governability: Using Complexity, Dynamics and Diversity*. London: Sage.
Kooiman, J. 2008. Exploring the concept of governability. *Journal of Comparative Policy Analysis: Research and Practice*, 10, 171–190.
Larsson, O. L. 2017. Meta-governance and the segregated city: Analyzing the turn to network governance, knowledge alliances, and democratic reforms in Malmo city, Sweden.
Le Galès, P. 2011. *Policy Instruments and Governance*. London: Sage.
McConnell, A. 2017. Policy success and failure. In: Thompson, W. R. (ed) *Oxford Research Encyclopedia of Politics*. Oxford: Oxford University Press.
Meuleman, L. 2003. *The Pegasus Principle – Reinventing a Credible Public Sector*. Utrecht: Lemma.
Meuleman, L. 2008. *Public Management and the Metagoverance of Hierarchies, Networks, and Markets*. Heidelberg: Springer.
Meuleman, L. 2010. Metagovernance of climate politics: Moving towards more variation. Paper presented at the Unitar/Yale conference 'Strengthening Institutions to Address Climate Change and Advance a Green Economy' (Yale University, New Haven, Connecticut, 17–19 September 2010).

Mintzberg, H. 1983. *Power in and Around Organizations*. Englewood Cliffs, NJ: Prentice-Hall.
Newman, J. & Head, B. W. 2015. Categories of failure in climate change mitigation policy in Australia. *Public Policy and Administration*, 30, 342–358.
Niestroy, I. & Meuleman, L. 2016. Teaching silos to dance: A condition to implement the SDGs. *IISD SD Policy & Practice. Guest Article, posted on 21 July 2016* [Online]. Available: http://sd.iisd.org/guest-articles/teaching-silos-to-dance-a-condition-to-implement-the-sdgs/.
Painter, M. 2014. Governance reforms in China and Vietnam: Marketisation, leapfrogging and retro-fitting. *Journal of Contemporary Asia*, 44, 204–220.
Peters, B. G. 2015. State failure, governance failure and policy failure: Exploring the linkages. *Public Policy and Administration*, 30, 261–276.
Pomerantz, P. R. 2011. Development theory. In: Bevir, M. (ed.) *The SAGE Handbook of Governance*. London: Sage.
Rittel, H. W. J. & Webber, M. M. 1973. Dilemmas in a general theory of planning. *Policy Sciences*, 4, 155–169.
Sørensen, E. 2006. Metagovernance: The changing role of politicians in processes of democratic governance. *The American Review of Public Administration*, 36, 98–114.
Stakeholder Forum. 2012. Review of implementation of agenda 21 and the Rio Principles. Synthesis. London: Stakeholder Forum.
Tacoli, C. 2014. The role of small and intermediate urban centres and market towns and the value of regional approaches to rural poverty reduction policy. *OECD DAC POVNET Agriculture and Pro-Poor Growth Task Team Helsinki Workshop, 17–18 June 2004*.
Teisman, G. 2000. Models for Research into Decision-Making Processes: On Phases, Streams and Decision-Making Rounds. *Public Administration*, 78(4), 937–956.
Thijs, N., Hammerschmid, G. & Palaric, E. 2018. *A Comparative Overview of Public Administration Characteristics and Performance in EU28*. Brussels: European Commission.
Treadway, D. C., Hochwarter, W. A., Kacmar, C. J., et al. 2005. Political will, political skill, and political behavior. *Journal of Organizational Behavior*, 26, 229–245.
Van der Meer, F. M., Dijkstra, G. S. A. & Roborgh, R. J. 1997. The Dutch civil service system. In: *Civil Service Systems in Comparative Perspective, School of Public and Environmental Affairs*. Bloomington, IN: Indiana University.
Wegrich, K. & Jann, W. 2007. Theories of the policy cycle. In: Fischer, F. & Miller, G. J. (eds.). *Handbook of Public Policy Analysis: Theory, Politics, and Methods*. London: Routledge.

4 Introducing metagovernance
Governance of governance

> We do not make an argument for bureaucracy, contracts or networks. We argue that each can work or fail. It depends on the context and the mix. At some point we have to accept that the structures mix like oil and water. . . . [T]he future will not lie with either markets, or hierarchies or networks but with all three. The trick will not be to manage contracts or steer networks but to mix the three systems effectively when they conflict with and undermine one another.
> – Fleming et al., 2005

After the general introduction on why sustainability governance is so difficult (Chapter 1), the concept of three basic governance styles (Chapter 2), and an overview of governance failures (Chapter 3), this chapter introduces the concept and the practice of metagovernance. Thereafter, further chapters give this practice more context, including in relation to the SDGs, and in Chapter 9 a metagovernance method will be sketched.

4.1 Metagovernance as governance of governance: a heuristic concept

In 2000, a new director of Soil Protection was appointed at the Ministry of the Environment of the Netherlands. Two things struck him from the beginning. Firstly, he observed that three units in his directorate who populated different floors in the building did not communicate much with each other. They had developed own distinct cultures matching with their tasks. The unit developing legal norms for soil protection thought hierarchically, the unit who dealt with stakeholders had a network governance attitude, and the unit dealing with remediation of contaminated soils had a business attitude: money was the language and market governance the approach. The new director brought the three units together on one floor and made them communicate better again. Secondly, he noticed that since more than four years, a dialogue with stakeholders was ongoing about modernization of the national soil policy. This network approach had stalled without concrete results and turned into a crisis in which all parties pointed at the Ministry as the cause of the problem. He concluded that it was time to switch to another governance style; he closed the doors for six months during which his staff wrote a complete new draft of the soil policy. This hierarchical approach benefited from the insights from the networking period and ended with a serious – and successful – attempt to restore trust. In both interventions, he switched between and combined governance styles and acted therefore as metagovernor.

At the same time, project leaders on soil policy in Germany and England were facing similar, but different problems. In Germany, the coordinator of the preparation of the German soil protection law noticed that the political support was weak – soil protection is difficult everywhere because it touches upon private land ownership. His intuitive reaction was to start flanking actions such as creating soil expert networks and initiating a public awareness campaign through postal stamps on the meaning of soil and a children's book on soil protection. In England, the project manager of the Soil Action plan for England had hoped to find support from the Treasury to be able to use financial instruments to fuel a new soil protection policy, but found out that this was not what the Treasury had in mind for such a 'minor' policy issue. Hence, he also switched to a softer, networked governance approach, with an action plan as output instead of a strategy or a legal proposal. The cases from the Netherlands, Germany and England, which were analysed more in detail in my PhD dissertation (Meuleman, 2008a) have in common that the responsible public manager intuitively applied what we now call metagovernance. For various reasons they had to switch from their 'default' governance style to another, and/or recombine the mixture. The soil policy cases illustrate that the question is not whether metagovernance occurs. It does. The question is how it works, and how it can be used as a practical method.

Moreover, I would argue – based on research and numerous workshops with public managers in several countries and from all kinds of ministries – that metagovernance is in the first place a heuristic concept that describes how public managers deal with complexity (Meuleman, 2008a). It was 'discovered' as practice and not only created as theory. It is an existing approach and method to address complexity and with the need to apply multiple perspectives. Metagovernance can be used as an analytical tool to help making a systematic comparison between a related set of case studies (Daugbjerg and Fawcett, 2015).

Secondly, it is a conceptual reply to the challenge of dealing with governance failures such as those caused by inherent weaknesses of certain governance styles or by incompatibility of governance styles. Applied to regulation, metagovernance is concerned with overseeing, reflecting on and orchestrating different types of regulation, including hybrid ones (Steurer, 2012). Concrete, metagovernance can be used to design and to manage governance style combinations.

Successful 'metagovernors' who didn't know the term while they were leading their team, project or programme, gave two different reactions when they were informed about metagovernance (Meuleman, 2008b). Some were pleased to hear that what they considered good practice was also theorized by academia. Others were somewhat irritated; what they did – designing composite governance frameworks and switching between governance styles when needed – did not

74 *What and for what is governance?*

need a fancy name: they considered it a matter of common sense, based on long experience. The following interview quote from a police commander in one of the larger cities in the Netherlands illustrates this:

> We are chameleons: We switch between styles depending on the situation at hand. People in our organisation have a sense for this. When an incident occurs, they know that there is no time for discussion. Nobody asks 'Why?' 'Shouldn't we involve other parties?' 'Isn't this too expensive?' After the incident, network and market governance elements reappear.
> (interview with a Dutch police commander, Meuleman, 2008a, p. 240)

A policy coordinator in the UK government had a similar view:

> We are basically doing meta-governance in some way . . . We create the conditions that make it easier for departments to govern. That might be signing-up the ministers, providing the guidance or facilitating a different mix of service providers. But we are basically 'hands off' – only sometimes 'hands on'.
> (interview with a UK Cabinet Office official, Dommett and Flinders, 2015)

Larsson (2015) is concerned that metagovernance is a "far more difficult practice than has been anticipated by existing theories and policy recommendations. Turning to metagovernance as a way to govern and control organizations may in fact lead to further fragmentation and distortion of public politics". The discussion of whether metagovernance is feasible under all conditions is useful, will continue, and then certainly lead to new insights, but meanwhile, out there, in the ambiguous and messy daily reality of the public sector, there is this quite useful approach, which public policy makers and managers – including me – are practicing on a daily basis. Metagovernance of hierarchies, networks and markets is a reality.

4.2 The genesis of the concept of metagovernance

Metagovernance was introduced in the 1990s in Western Europe mainly as a reaction to the complexity of governance and the failures of the existing governance styles – notably market governance, according to Jessop in his analysis of the genesis, early development, problems and success of metagovernance (Jessop, 2011, p. 106 ff.). The literature on complex governance challenges since the 2000s suggests that metagovernance is a promising reply to the challenge of dealing with governance failures and their root causes.

As defined in Section 1.4 and illustrated in Figure 4.1,

> **Metagovernance** is a means by which to produce some degree of coordinated governance, by designing and managing sound combinations of hierarchical, market and network governance, to achieve the best possible outcomes.

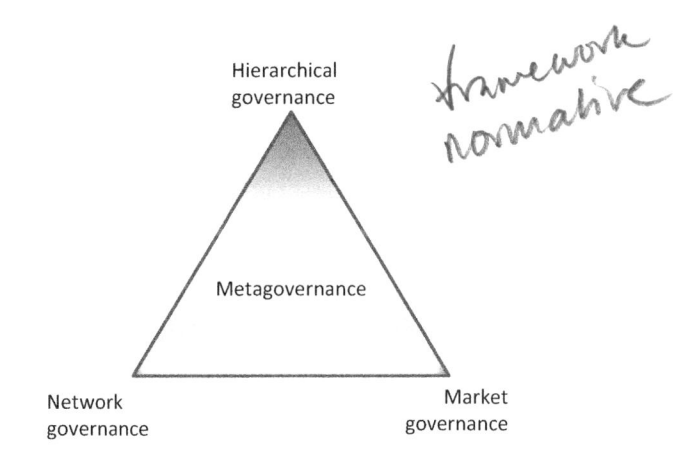

Figure 4.1 Metagovernance as coordination of hierarchical, network and market governance

As metaphysics is not physics, metagovernance is not governance and it is therefore not a governance style. It is about "organizing the conditions of governance" (Jessop, 2002) and at the same time a *method* to design practicable governance frameworks by developing dynamic combinations of elements from different governance styles, and to manage such combinations. It aims at a "judicious mixing of market, hierarchy and networks to achieve the best possible outcomes from the viewpoint of those engaged in metagovernance" (Jessop, 2009). It takes a bird's-eye perspective on governance in order to see how different approaches interact in synergetic or undermining ways. Metagovernance is a critical approach to the study of governance and a way of thinking about how governance is structured and applied (Amore and Hall, 2017). It is also a practical design and management method for governance challenges (Meuleman, 2008a). The 'art' of combining governance styles and switching between them is especially relevant for complex policy challenges such as the ones that run as a red thread through this book, the SDGs.

The basic assumptions of metagovernance that governance styles are normative, and that governance frameworks should be adapted to national or sub-national specific circumstances, make the concept universally applicable to all public governance challenges in all countries and adapted to their specific requirements and conditions. Therefore, it can be an adequate tool to solve the paradoxical imperative that we need "common but different governance" (Meuleman and Niestroy, 2015) to effectively implement the SDGs (see also Section 11.2).

Most of the analysis in the following chapters will be about *public* metagoverance, but many conceptual and practical topics are similar for non-state, i.e. *private* metagovernance, executed by private sector companies and/or civil society organizations or their umbrella organizations. I will present some specific examples of private metagovernance in Section 4.4.

Internal and external metagovernance

Kooiman (2003, p. 182) saw metagovernance as a third order of governance, after problem solving and opportunity creation (first-order governance) and care for institutions (second-order governance). How governance modes emerge and are organized *inside* public sector organizations (the institutional aspect: second-order governance, or the meso and micro levels) in relation to first-order governance, is still relatively ill-researched. I know from experience that internal and external governance can be very different. The organizational culture of ministries in one and the same country can be so different that they have problems communicating effectively with each other. For example, different ministries of one government may use the same terms with different meanings. Economic and financial departments tend using the term 'sustainable' in the meaning of 'long-lasting'. The term 'environment' is often used in concepts such as business environment or tax environment. Within a rather technical Ministry of Infrastructure a more rational culture can be expected than in a Ministry of the Interior, which is more political. This means that governance frameworks for policy goals which relate to more than one ministry may need a different (meta)governance within the national government, than with external parties.

Internal metagovernance is about dealing with potential conflicts and synergies between hierarchy and other governance styles within public administration organizations. Practitioners as well as scholars seem to recognize this: a conference paper on 'Internal metagovernance' (Meuleman, 2006) is one of my most-cited publications. Bell and Park (2006) noticed in a case on water governance in Australia the strong political dimension internal metagovernance can have, as it involves decisions on political steer, resource allocation, including information, identifying key stakeholders, setting agendas and structuring outcomes.

Achieving policy and institutional coherence requires that internal and external metagovernance are well connected – which does not mean that they have to be the same: a good interface is essential, however. This prevents situations I have experienced where a ministry externally practiced network governance, but I as participant in a multi-stakeholder team was not able to be a reliable partner because my minister – who was on a hierarchical track – was in no way committed to agreements reached in the network.

4.3 Metagovernance of hierarchies, networks *and* markets

What metagovernance means in practice depends on how one defines the term governance. My definition of metagovernance is not normative in the sense that it does not select one (normative) governance style *a priori* as the best. Hierarchical, network and market governance and their hybrids are very different governance styles, which strengthen or undermine each other when they are combined. Ten years ago I made an in-depth analysis of the debate on governance definitions (Meuleman, 2008a, pp. 10–66), which resulted in the conclusion that excluding one of the three ideal-typical styles – hierarchical, network or market governance – would be not logical because all three exist and interact with each other in practice. This broad approach to governance is meanwhile the mainstream in both academia and public administration practice. The Agenda 2030 (United Nations, 2015) also uses governance in a broad sense. It combines the hierarchical (rule of law, legal institutions) approach in SDG 16 and the network (partnerships, coherence) and market-style (e.g. investments) focus in SDG 17. It considers the whole set of tools derived from hierarchical, network and market styles of governance, as well as from combinations of these styles. This means that the approach is usable in very different administrative and societal traditions.

Metagovernance of hierarchies, networks and markets has been called second-order metagovernance (Jessop, 2011, Meuleman, 2011). It 'handles' more complexity than what is called first-order metagovernance: governance of one governance style. Most examples of first-order metagovernance or metagovernance 'light' have sprung from the tradition to define governance as managing networks. Besides metagovernance of networks, some examples exist of metagovernance of markets and of hierarchies.

Metagovernance of hierarchies or networks or markets

Countering the view that governance (and therefore also metagovernance) should be about all existing styles, another use of metagovernance has remained influential, where metagovernance is about strengthening one specific governance style by learning from its failures and introducing elements from other styles to remediate the weaknesses of this style. A substantial part of the metagovernance literature is centred on this approach, with network governance as the most popular focus.

Metagovernance of hierarchies

The broad approach to metagovernance introduced above is not per definition state-centric, although it has been characterized as such by some who define governance as something different from (and way from) government, and in particular something that works through network principles. 'Getting government in again' in governance is necessary because hierarchy is a real-life mechanism: hierarchical governance has not disappeared (Hill and Lynn, 2005). There is, however, also a specific state-centrist variation of metagovernance: this is a form of metagovernance, which has the objective to mitigate the inherent weaknesses of hierarchical governance:

> In this way metagovernance appears to provide a way of exploring how new articulations of state power become expressed in and through governance structures and the ways in which governance systems are in turn forged in the persistent 'shadow of hierarchical authority'.
>
> (Whitehead, 2003)

In this view, "metagovernance functions are the prime responsibility of the state" (Bell and Hindmoor, 2009, p. 46). Jessop (2009) calls such metagovernance of hierarchies 'meta-organisation', in as far it concerns the management and (re)organization of public sector organizations. It aims at "reflexive redesign of organizations, the creation of intermediating organizations, the reordering of interorganizational relations, and the management of organizational ecologies".

Metagovernance of networks

The idea that network governance is the new paradigm and that metagovernance should concentrate on managing the challenges of network governance has travelled all over the world and settled in the minds of decision makers. In this view, governance is an alternative to government. With network governance as focus, Sørensen et al. (2005) have described metagovernance as a way of enhancing coordination of governance in a fragmented political system, based on a high degree of autonomy for networks and institutions. Larsson (2015, p. 54) claims that the broad form of metagovernance which covers hierarchical, network and market governance styles is too broad because "it includes all types of actions taken by any authority, thereby rendering metagovernance both all and nothing". It is of course a matter of perception of what "all and nothing" means. In my view, metagovernance concerns polity and politics but not policies; metagovernance is therefore not 'all'.

Not surprisingly, many scholars defining metagovernance as 'enhanced network governance' work and live in consensus democracies and share similar values – which illustrates that governance is a normative issue. Authors who joined this approach tend to originate from Scandinavian countries (Sørensen, 2006, Torfing and Triantafillou, 2011, Larsson, 2015) or the Netherlands (Klijn and Koppenjan, 2000, Nooteboom and Teisman, 2003) and Belgium (Struyven and Van Parys, 2016).

Beisheim et al. (2017, p. 6) seem to follow Sørensen's approach and define metagovernance to support the Agenda 2030 as a set of over-arching principles and rules which should help stimulate, accompany or evaluate the work of partnerships. Also academics focusing on private metagovernance (Section 4.4) tend to use the narrow, network-focused definition of metagovernance (Glasbergen, 2011, Derkx, 2013).

Jessop (2009) names this form of metagovernance 'meta-heterarchy'. It involves

> the organization of the conditions of self-organisation by redefining the framework for reflexive self-organisation, for example through various measures to promote improved forms of interpersonal networking and interorganizational negotiation or through institutional innovations to promote more effective inter-systemic communication. The OMC [Open Method of Coordination] adopted in the Europeanisation of employment policy is a good illustration.

Without wanting to discredit this approach as a whole – although I find it too narrow to tackle contemporary governance challenges – there are several weaknesses. First, the fact that network governance has been welcomed broadly as an important addition to the principles and tools of public governance – including in the Agenda 2030 – should not lead to believe that it could (or should) become the dominating governance style. Network governance utilizes a set of values and norms that is not universal: in most countries, hierarchical or market-type values dominate how the public sector works. Rayner (2015, p. 247) observes the networked approach to metagovernance as a correction to the "misplaced enthusiasm for self-organizing networks as a desirable 'third way' that avoids a straight choice between markets and hierarchies". Such a correction is necessary because there is "an abundance of examples in which networks either did not self-organize in response to problems or provided manifestly sub-optimal solutions to them". Some proponents of network metagovernance claim that metagovernance has emerged in response to the great interest in how networks are steered (Daugbjerg and Fawcett, 2015), claiming therefore that metagovernance is only an academic idea, whereas I have argued that it is at least also a heuristic concept that works in practice and just had to be 'found'.

The second weakness is the assumption among network governance researchers that hierarchical governance is in decline. This is, as a general trend across the globe, empirically not true (Hill and Lynn, 2005), but is indeed happening in certain countries and situations. The public law scholar Schuppert has argued that with a narrow governance definition it is impossible to include the historically most successful form of governance, namely hierarchical-bureaucratic governing (Schuppert, 2007, p. 8). Capano et al. (2015) argue that if the fact that governments continue to play a "pivotal" role in policy making is not taken into account, it will look as if governance risk is "anchored to a merely normative or prescriptive view rather than an empirically robust one".

The third weakness is more conceptual. If it is accepted that governance, including the choice of instruments, is normative, then the focus on the network style implies that the values underlying hierarchical governance (such as legitimacy, accountability and justice) and market governance (such as flexibility, empowerment and innovation) are excluded from the governance framework and toolbox. If governance is not normative, there is no need to take into account the governance context: indeed, when the thinking is thus, one-size-fits all could be promoted again, just like when New Public Management was in the peak of its popularity.

Fourth, Brandtner et al. (2016) criticize the approach because "it mainly addresses emerging governmental tasks that come with high-complex network governance" which is not sufficient to understand urban governance challenges, for example. Indeed, not all government tasks are about managing highly complex networks. Now and then, governing is about managing disasters, and hence command and control is a more effective idea than deliberative argumentation in networks. Complexity can addressed with more measures than only those related to network governance. Nevertheless, the 'metagovernance as enhancing network governance' approach has clear merits from which this broader approach can benefit. This includes a better understanding of how weaknesses of network governance can be overcome by introducing elements such as rules and procedures from other governance styles, mainly from hierarchical governance.

The applicability of the meta-network-governance approach for SDG frameworks seems to be limited to cases where a network/partnership approach should 'do the trick' and is feasible. The latter is rather difficult to imagine for the implementation of the SDGs in countries with autocratic governments: we are talking about a majority of the UN member states. An important benefit of metagovernance of networks is that it can help 'anchor' democratic quality (Sørensen, 2006) where this risks being lost.

Ultimately, I use the broader kind of metagovernance because the issues I am working on as a practitioner such as environmental implementation and sustainable development require coordinated interaction of all governance styles, not only the network style. Moreover, institutional and policy coherence require linkages between departments with different preferred governance styles. Metagovernance should focus explicitly on the practices and procedures that secure governmental influence, command and control within governance regimes (Whitehead, 2003), but this does not exclude involvement of other actors. This is the case at the EU level, my current occupation, but also at national and sub-national levels where I have worked in a consensus country.

Also for his fourth governance style, solidarity (which I tend to see as a variation of network governance), Jessop (2009) has an accompanying form of metagovernance: 'meta-solidarity'. This involves "the promotion of opportunities for 'spontaneous sociability' and other measures to enhance solidarity at all levels, from neighbourhood and work place to cosmopolitan humanitarianism". He mentions policies to build 'social capital' as a good illustration of this form of meta-coordination.

Metagovernance of markets

The third sub-type of metagovernance could be defined as enhancing the management of market governance. It would be about ensuring that the weaknesses of market governance such as an inconsiderate focus on efficiency and cutting 'red tape' are being dealt with. In governance literature, this is a yet almost unwritten page.

Jessop (2009) distinguishes two forms. The first is 'meta-exchange', which is about "reflexive redesign of individual markets (e.g. for land, labour, money, commodities, knowledge – or parts or subdivisions thereof) and/or the reflexive reordering of relations among two or more markets by modifying their operation and articulation". The second is metagovernance oriented to 'metamarkets': securing the conditions for the operation of markets, though e.g. "deregulation; liberalization; privatization; introduction of market proxies into the residual state sector; internationalization; and direct tax cuts to widen the scope for market-oriented decisions on the part of market agents".

4.4 Metagovernance by private sector actors

In the 2012 Rio+20 outcome document and the commitments of the 2030 Agenda adopted in 2015, there is a clear call on the private sector to contribute to the implementation of the Sustainable Development Goals (SDGs) and mobilize their ideas and creativity. Business actors are often powerful players and may have even more resources than national governments, especially in regions such as Africa. Companies work in the real economy where investment decisions are being made and new business cases developed with a long-term view. Economists call this micro-economics. Government economists are often macro-economists who look at highly aggregated numbers such as GDP growth and unemployment rates. In governments, a little more micro and a little less macro might be a good idea, generally – in my perception. What does this imply in terms of challenges for the private sector and in particular for corporate social responsibility (CSR)?

In the first place it means that business has become a crucial partner in the implementation of the SDGs. SDG 17 calls for such partnerships, and this is supported at the global level by the UN Global Compact, an important voluntary initiative based on CEO commitments to implement universal sustainability principles and to take steps to support UN goals. Networks of business professionals help bridging the gap between global and local innovation and transformation.

Secondly, the adoption of the SDGs as universal goals for all UN member states means that we are now talking about creating a 'new normal'. Sustainability is not a fringe issue but part of core business strategy, and it should not remain in the boardrooms alone. It looks indeed that companies, who succeed to integrate SDG implementation in their regular ongoing activities, in their daily business, might be the future winners.

Thirdly, business' responsibility for sustainable development also includes a responsibility to support statehood where needed, both in terms of human rights and in terms of ensuring level playing fields and sufficient standards to stimulate innovation and transformation towards sustainability. Weak public administration is not only a nightmare for citizens but also for economic operators. A recent report gives examples of how companies from the SDG Fund's Private Sector Advisory Group are working to prevent corruption, including by instituting a zero-tolerance policy and implementing a reporting system for potential compliance violations. It reports, for example, that a number of companies in Colombia are collaborating to integrate small farmers in high-conflict areas into their supply chain in order to create jobs and support reconciliation efforts (SDG Fund, 2017). Therefore it is important that all 17 SDG have dedicated targets on governance and that Goals 16 and 17 can be considered as two governance Goals.

As I have already stressed earlier (Meuleman, 2008a), metagovernance can also be executed by private actors. Indeed, there is a case for "poly-centred metagovernance beyond governments" (Steurer, 2012, p. 23). Although Sørensen (2006, p. 103) emphasized that "metagovernance can potentially be exercised by any resourceful actor – public or private", the private side of the phenomenon

82 What and for what is governance?

has hardly been addressed so far, with some notable exceptions (e.g. Glasbergen, 2011, Fransen, 2018). Metagovernance by private actors is often seen in the field of standard-setting, guiding such partnerships' governance activities (Mundle et al., 2017, Lujie, 2017). Networks of business advisors have emerged which constitute metagovernance arrangements that are able to disrupt the path-dependent behaviours of organizations (Rachel and Damian, 2013).

Private governance is tackling social and ecological sustainability issues, but also financial and economic issues. The latter has led to a range of new global standards organizations, such as the Fair-Trade Mark, the Forest Stewardship Council and the Extractive Industries Transparency Initiative. The private governance landscape is today marked by the presence of what are called private metagovernance institutions, which include the International Organization for Standardization (ISO) and the World Trade Organization (WTO), which essentially provide standards and guidelines for the development of standards. The latter means that that private actors can indeed take on metagovernance roles with some success. Another example is the G-20's Financial Stability Board (FSB). The G-20-FSB framework points to the future of governance systems in which the state participates in a collaborative governance structure, but in which states share rulemaking power with public and private non-state actors.

4.5 Examples from different locations and topics

Over the past decade, metagovernance has been used to analyse governance practice and failure in different geographical locations and cultural settings, and on a wide range of topics. Table 4.1 gives some examples. Christopoulos et al. (2012) who were among the first to apply metagovernance to sustainable development see its use in a sustainability context as concerning the reflexive coordination and organization of the framework conditions under which governance takes place. In particular,

> It is the coordination potential of metagovernance that can make it valuable for addressing the complexity and the vagueness of SD, given the different socio-economic and environmental needs of different communities, countries and regions. In terms of solutions for SD, simply put, metagovernance has the potential to cast light on what is the right patchwork of governance elements in order to develop the right cross-sectoral mix of policies.

Almost 25 years after the term was coined, there is a broad agreement that governance for complex challenges such as sustainable development and especially the implementation of the SDGs should always be situational. The report summarizing the findings of the 2016 High-Level Political Forum (HLPF) on the SDGs stated that "creating ownership of the Sustainable Development Goals and their interlinkages will require building on national and local contexts, values and cultures, avoiding the use of blueprints" (United Nations, 2016). Follow-up UN conferences in Korea and the Bahamas respectively concluded that "there is no stereotype institutional arrangement for mainstreaming the 2030 Agenda and adapting SDGs to a national context, owing to uniqueness in history, values and culture" (UNOSD, 2016) and that "Institutions/ministries should work closely together to break silos" (UNPAN, 2017). Finally, also the UN Committee of Experts on Public Administration underscored in 2017

> that there is no single blueprint for implementing the Sustainable Development Goals and that identifying the most effective policies in a given context requires the participation and engagement of all stakeholders, inter alia local authorities, civil society and the private sector, on various policy options.
>
> (United Nations, 2017)

Academic publications provide meanwhile case studies on all topics covered by the SDGs (see Chapter 8). A good example of theoretical reflection and practical application is a study of three cases in Croatia (on energy efficiency of buildings), Nepal (on decentralized renewables) and Mongolia (inclusive finance and micro credits) which concluded that metagovernance had resulted in coordination, formal conventions and less formal agreements of what is acceptable and appropriate administratively and policy-wise, to solving complex sustainability challenges (Christopoulos et al., 2012). Table 4.1 brings together some more examples.

In addition, metagovernance plays an important role in multi-level governance processes of the UN (Derkx, 2013), in collaborative governance in the East African Community (EAC)[1] (Stark, 2015), and in particular in the European Union governance (Section 4.3).

The fast global dissemination of the use of metagovernance can be attributed to two factors. First, dissemination is accelerated through the wide availability of the Internet with its online (and more and more Open Access) research papers and articles, compared to 10 years ago. However, other concepts did not travel that fast. Schuppert (2007, p. 9) argues that Western governance approaches are not *Reisefähig* (able to travel) to many other countries because two-thirds of the states in the world have a weak statehood. This is plausible for network governance ideas, but we have seen an almost global spread of market governance based New Public Management ideas, supported by the apparently convincing financial arguments of World Bank and other donors. The second and probably most important reason why metagovernance is being used in so many different countries is that the concept is sensitive to cultural and other situational differences.

Increased use of metagovernance as a scholarly concept

The number of publications using the term 'metagovernance' (or the less-used orthography 'meta-governance') is rapidly growing (Figure 4.2), although it has never become a 'trending topic'. One of the reasons may be that the concept as such offers a more or less value-neutral method, but the field of application – governance – is so value-laden and normative that any putting-into-perspective

Table 4.1 Examples of metagovernance research: various locations and topics

Region	Country	Topic	Author(s)
Africa	Kenya	Partnerships	(Beisheim et al., 2017)
	Nigeria	Leadership	(Agu et al., 2014)
	South Africa	Road infrastructure	(Tsheola, 2013)
	South Africa	Fisheries and nature	(Sowman, 2015)
Asia	China	Local governance	(Li, 2015)
	China	Tourism	(Wan and Bramwell, 2015)
	China	Waste management	(Yanwei et al., 2016)
	Japan	Fuel cell technology	(Murakami et al., 2010)
	Mongolia	Finance and micro credits	(Christopoulos et al., 2012)
	Nepal	Renewable energy	(Christopoulos et al., 2012)
	South Korea	Local government	(Yoo and Kim, 2012)
	Vietnam	Fisheries	(Thang, 2018)
Australia	Australia	Water reform	(Bell and Park, 2006)
	Australia	Youth care	(Voets et al., 2015)
	New Zealand	Tourism	(Amore and Hall, 2017)
Europe	Belgium	Innovation policy	(Stevens and Verhoest, 2015)
	Croatia	Energy efficiency	(Christopoulos et al., 2012)
	Finland	Local government	(Haveri et al., 2009)
	Germany	Environmental policy	(Meuleman, 2008a)
	Germany	Health	(Burau, 2007)
	Greece	State reform	(Lampropoulou and Oikonomou, 2016)
	Netherlands	Community policing	(Meuleman, 2008a)
	Norway	Regional governance	(Higdem, 2015)
	Portugal	Education policy	(Magalhaes et al., 2013)
	Sweden	Crisis management;	(Larsson, 2017)
	United Kingdom	Agentification	(Dommett and Flinders, 2015)
North and Central America	Canada	Urban governance	(Doberstein, 2013)
	USA	Organic seed regulation	(Renaud et al., 2016)
	Mexico	Organic seed regulation	(Renaud et al., 2016)
Southern America	Brazil	Oil production	(de Aguiar and Freire, 2017)
	Brazil	Nature conservation	(Kull et al., 2017)
	Brazil	Social policy	(Kirschbaum, 2015)
	Colombia	Decentralization	(Bonivento, 2014)

of any governance style, which is what metagovernance does, meets suspicion or even animosity.

Among scholars who use metagovernance as analytical concept, the long delay or frequent rejection of their publications in peer-reviewed journals has been a popular discussion topic. Not only in public sector organizations but also in social sciences, silos seem to exist which make it difficult for scholars to step out of their comfort zones. This reminds of the struggles the sociologist Hofstede had to go through before an article was accepted in which he had dared to suggest

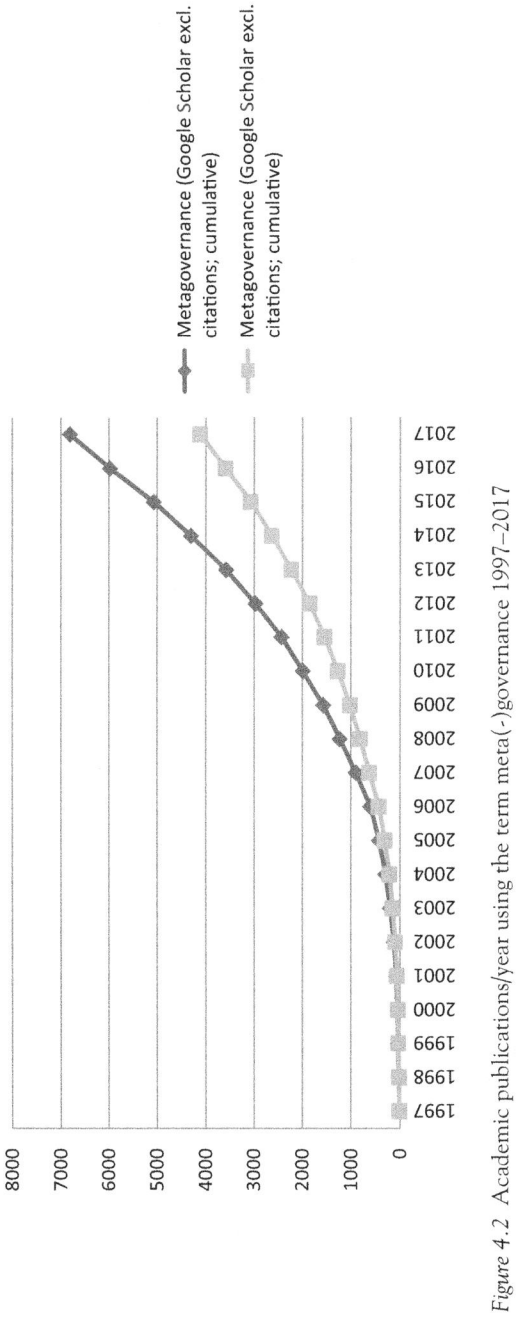

Figure 4.2 Academic publications/year using the term meta(-)governance 1997–2017 (Google Scholar, retrieved 31.03.2018)

that all theories and methods reflect the cultural environment in which they are conceived (Hofstede, 1980).

4.6 The European Commission as a centre of metagovernance

Many scholars have found that the EU level, and in particular the European Commission as its executive organization, is an excellent place to practice metagovernance (for example Grote and Gbikpi, 2002, Jessop, 2005, Meuleman, 2008a, Jessop, 2009, Cini, 2012, Porras-Gómez, 2013, Stark, 2015, Cini, 2015, Maas and Caviedes, 2016, Tömmel, 2016, Niestroy, 2016). In 2009, Jessop (2009, p. 89) noted that the EU had developed "a wide and deep array of both governance and metagovernance capacities that enable it to influence economic and social policy in most areas and on most scales". He also observed a division of metagovernance labour between the European Council, the specialized sectoral Councils of ministers and the European Commission, which I recognize:

> The European Council is the political metagovernance network of prime ministers that decide on the overall political dynamic around economic and social objectives. . . . The European Commission plays a key metagovernance role in organizing parallel power networks, providing expertise and recommendations, developing benchmarks, monitoring progress, exchanging best practice, promoting mutual learning, and ensuring continuity and coherence across Presidencies.
>
> (Jessop, 2008)

One of the first expressions of metagovernance at the level of the EU was the Open Method of Coordination (OMC), an approach established by the Lisbon European Council (23–24 March 2000), to implement the Lisbon Strategy (Radaelli, 2003). This has been an attempt to combine effectively the classical hierarchical governance of the EU with 'new modes of governance' (Héritier, 2002). The OMC is principally based on jointly identifying and defining objectives to be achieved (adopted by the Council); jointly established measuring instruments (statistics, indicators, guidelines); benchmarking, i.e. comparison of EU countries' performance and the exchange of best practices (monitored by the Commission).[2] According to Jessop (2009, p. 94), this had instituted permanent reflection about the nature of problems; monitoring of different approaches to coordination; a flexible repertoire of responses that can be deployed on different scales, with different scales for different sets of problems; and an orientation which includes intertemporal as well as inter-scalar aspects.

The EU's governance of governance system is tailored to the multi-level structure of the EU, the need to balance contradictory policy objectives, governance modes and implementation strategies at EU and national government levels, as well as the differences among the member states (Tömmel, 2016). European governance has evolved "from a simple state interventionist model to a complex system of 'governance of governance', employing both hierarchical and non-hierarchical governance modes combined in innovative ways" (Maas and Caviedes, 2016).

Being a multi-cultural public sector organization with policy officers from almost 30 countries, using metagovernance (although the term is seldom used) has become part of the governance 'software' of the European Commission's work:

> There is no one-size-fits-all approach to public sector organisation. There is a need for a differentiated approach that takes into account the costs and benefits of designing and managing different governance-style combinations for the achievement of desired outcomes and optimized resource allocation based on different needs.
> (European Commission, 2017b, p. 9)

As an 'in-house reflexive practitioner' inside the European Commission I have to admit that I agree. Also the recently updated Commission's Toolbox Quality of Public Administration recommends metagovernance of hierarchies, networks and markets as a way to promote public officials to be open-minded about how they interact with each other and external stakeholders:

> The term meta-governance encapsulates the concept of public servants being able and encouraged to select, blend or switch between the most suitable governance style, according to circumstance, each having pros and cons.
> (European Commission, 2017a, p. 214)

In a more general context about what defines a modern public administration, the European Commission recognized that governance-of-governance is needed:

> "There is no one-size-fits-all approach to public sector organisation. There is a need for a differentiated approach taking into account the costs and benefits of designing and managing different governance style combinations for the achievement of desired outcomes and optimized resource allocation according to different needs". Focus on one style alone such as market governance is risky, because Blindly breaking down institutional entities can make formal structures disappear, which are needed for effectiveness and accountability.
> (European Commission, 2017b)

Reasons why metagovernance seems to work relatively well in the EU include the multi-level organization of the EU, the permanent criticism of lack of democratic legitimacy of the European Commission, and the fact that decisions are collegial among the Commissioners – none of them can 'silo' him/herself into a not-agreed position. New policy proposals are usually 'packages' built from a whole range of different instruments, varying from legislation to financial incentives and partnerships.

In line with this, (Knill and Lenschow, 2005) observe that the EU uses three governance modes to steer relevant actors toward desired behaviour: compliance, competition and communication in networks – respectively hierarchical, market and network governance. Recent examples from different policy fields are the Energy Union package of 2015 (on the energy-climate nexus – SDGs 7 and 13), the Circular Economy packages of 2015 and 2018 (a nexus of responsible consumption and production (SDG 12), innovation (SDG 9) and decent work and economic growth (SDG 8). Also the Social Pillar package (2017)(a nexus linking reduced inequality (SDG 10), gender equality (SDG 5), decent work (SDG 8) and justice (SDG 16) should be mentioned. A more focused example is the Plastics Strategy, proposed in January 2018, which combines instruments from all three basic in a well-reflected way (see Box 9.3 in Chapter 9).

In order to keep the holistic view while implementing the SDGs, since several years the so-called 'nexus' approach is being promoted. Nexus is about integrating goals across sectors, taking into account the interdependencies between individual SDGs and targets. It is an analytical approach to derive optimized solutions based on an integrated assessment of the related challenges and opportunities (Venghaus and Hake, 2018). A recent survey of nexus thinking in recent EU policies, with a focus on the food-energy-water nexus, leads the authors to conclude that in the EU

> a broad spectrum of different governance measures has developed so that questions of metagovernance (i.e. the 'governance of governance') will likely play an important role in the strive for policy coherence among the nexus sectors.
>
> (Venghaus and Hake, 2018)

The European Commission's capacity to apply metagovernance may be linked to the fact, as suggested by Candel et al. (2015), that the Commission is also surprisingly effective in dealing with 'wicked' problems As the relatively small[3] executive organization supporting countries with 500 million citizens it has been and still can be considered a laboratory for governance innovation.

Metagovernance implies that the state assumes an active role in relation to other forces and modes of governance and develops the systems and structures needed to improve the performance of governance (Yu and He, 2012). Stark (2015) points at the phenomenon that the European Commission as a non-state actor is able to apply hierarchical means in its governance frameworks to influence, steer and even take member states who do not implement EU law to the European Court.

Consistent with their focus on network metagovernance, Sørensen et al. (2015) conclude in a case of EU employment policy that the EU is not able to fulfil its metagovernance role in an effective way because member states do not want to give up their sovereignty and independence in this policy area. Typical metagovernance (of networks) tools such as peer-review processes, reporting arrangements, guidelines that measure the degree to which member states work together, and workshops and events that inspire and facilitate such collaboration could not be put to work under these circumstances.

4.7 Guiding principles

When I say I agree to do something "in principle", this can be a strong commitment with a slight chance that it will not happen, or no commitment at all. Which of the two it is depends among others on the culture in which I grew up. It does have an impact when people rigidly adhere to principles (or hide behind them) – the German word *Prinzipienreiterei* points at this phenomenon and has no synonym in English. In many other countries, "in principle" merely points at an intention which can be overruled at any time by other intentions. Therefore, many governance principles are contextual. Indeed, governance principles need room for interpretation: "Just as there is no single model of democracy, there is no single way of organizing responsive and effective institutions, although such institutions may share common features" (United Nations, 2017, p. 5). Other principles are universal, e.g. human rights and local self-determination, and they may be in conflict with contextual principles (Kooiman and Jentoft, 2009).

In any case, in order to measure effectiveness of governance we need measuring rods – and 'principles' is the accepted term for this. In addition, if there is a need to compare effectiveness across countries, governance principles may be the right level of abstraction, because the concrete choice of governance tools is – or should be – related to the context (the governance environment).

Principles are more specific and imperative than values and norms, but less specific than concrete choices. Values, norms and principles are not always so clearly related to each other that such a 'hierarchy' can be seen. What is called a principle is partially a matter of subjective assessment. In governance practice, the focus is usually on choices, while norms, principles and values remain implicit. Moreover, "when they are made explicit, they are rarely absolute. They are not always deliberated, and when addressed they often lead to nothing concrete and practical", Kooiman and Jentoft (2009, p. 818) argued.

Governance principles differ from policy principles such as access to justice, the precautionary principle and the polluter pays principle. When governance principles are used too rigidly, they may undermine each other. Many principles require choices or trade-offs between them because they cannot be fully applied at the same time. For example, a governance framework cannot at the same time maximize reliability and flexibility. Flexibility may require changing the rules regularly to adapt them to changed circumstances, while reliability suggests unchanged rules over at least some years, offering a stable investment climate for businesses.

Kooiman and Jentoft suggested that metagovernance is the level where values, norms and principles are developed and promoted, based on which governance practices can be designed and evaluated. Not surprisingly, various sets of principles for governance, metagovernance and/or sustainability (meta) governance have been proposed. Table 4.1 offers an overview of key principles used for governance, metagovernance and sustainability governance, from various sources. Some of these proposed principles can be directly linked to one of the three basic governance styles. It is clear that *rule of law, equity and transparency* are hierarchical governance principles. *Reflexivity, inclusiveness* and *collaboration* are key network governance principles, and *flexibility* is key to market governance.

Other principles link to all three styles, but in different ways. *Accountability* can be organized top-down, or through networks, or through contracts. Such principles should be linked to metagovernance. The same applies to principles which are clearly over-arching: the need to always take into account the *multi-sector, multi-actor, multi-level, knowledge* and *long-term orientation* dimensions of governance also belong to the metagovernance principles. Niestroy has proposed to make these five principles central for the implementation of the SDGs (Niestroy, 2014) which seems indeed a good basis. (Gupta et al., 2017, p. 276) also propose the multi-level and multi-actor aspect as key principles for actions related to the SDGs, followed by *context specificity*. The latter is related to the principle of 'common but differentiated governance' (Meuleman and Niestroy, 2015) and the targets on horizontal and vertical coherence in SDG 17.

Jessop (2002) formulated three metagovernance principles, which should guide the attitude and qualifications of those who apply metagovernance:

- deliberate *cultivation of a flexible repertoire of responses* ('requisite variety') so that strategies and tactics can be combined in order to reduce the likelihood of failure;
- a *reflexive orientation* about what would be an acceptable outcome in the case, and regular re-assessment of the extent to which current actions are producing desired outcomes; *self-reflection* supports gaining more congruent belief systems in a multi-actor process and thus integration and process orientation (Niestroy, 2000);
- *self-reflexive 'irony'* such that participants in governance recognize the likelihood of failure but still proceed as if success were possible.

For the governance of complex, multi-actor, multi-sector and multi-level land use and other plans and programmes, a case study comparing planning regarding two large waterways in California and Germany, respectively, it was concluded that self-reflection, transparency and process orientation were central principles (Niestroy, 2000). Transparency is also a principle of Agenda 2030 (Goal 16.6). Process orientation is an important addition. It reduces the complexity of the problem respectively the plan and hence can increase efficiency of planning; it requires an "iterative planning procedure with subsequent steps of decisions to structure and 'tier' the process". This principle is also important to strike a balance between process management and project management. Project management is about managing resources (time, money, human resources) but process management is about managing uncertainties, disputes and other contingencies (see for an elaborate guidance on the principles and practice of process management De Bruijn et al., 2010).

I have suggested three additional principles for metagovernance (Meuleman, 2013), which should ensure that the governance context is taken on board:

- *problem-orientedness*: analyse the problem in its framing and in its context;
- *temporality and locality*: governance for sustainability is multi-level, -scalar and -actor;
- *culturality*: this principle refers to the understanding of cultural phenomena "based on dynamics, transformations, fusion and manipulations" (Abdallah-Pretceille, 2006, p. 497).

The UN Committee of Experts on Public Administration (United Nations, 2017, pp. 11–14) states that the Agenda 2030 contains three governance strands, focusing on, respectively, effective, accountable and inclusive institutions. They propose a 'provisional taxonomy' with 10 possible principles: competence, sound public policy, cooperation, integrity, transparency, independent oversight, non-discrimination, participation, subsidiarity and intergenerational equity. In 2018, a revised set of 11 principles on effectiveness, accountability and inclusiveness was proposed, with an overview of commonly used strategies (United Nations, 2018).

Principles are meant to be benevolent, but in practice they can turn into their opposites or malevolent versions ('perversions'). Flexibility can turn into arbitrariness, rule of law into abuse of power, and participation can turn into manipulation. Table 4.2 shows the relation between principles, each of the governance styles and metagovernance. In some cases, three forms of a principle exist. For example, accountability has its classical top-down control and reporting form (hierarchy), a bottom-up form with involvement of citizens or civil society groups (network), and a privatized form (market).

4.8 Metagovernance strategies

Metagovernors apply various strategies when they design and manage a governance framework. Earlier, I distinguished combining, switching and maintenance as the main strategies (Meuleman, 2008a). I think that framing should be added as a fourth strategy; it was already mentioned in this context by (Sørensen, 2006). The four strategies are then:

- *Framing* of problems to make the solvable in an existing governance environment;
- *Designing* a framework by combining elements of different governance styles into compatible arrangements;
- *Switching* from one to another (main) governance style;
- *Maintaining* of a chosen governance framework or arrangement, for example, protecting it against perverse/undermining influences in the governance environment. Maintenance complements the combining and switching strategies.

Table 4.2 (Meta)governance principles: an overview

Metagovernance principles	Hierarchical governance principles	Network governance principles	Market governance principles
Moral responsibility and integrity	Equity (being just, impartial and fair)	Inclusiveness	Respect? Why?
Accountability	Top-down accountability	Citizens' accountability	Outsourced/private accountability
Transparency			
Pluralism	Rule of law	Participation	Small government
Mindfulness	Transparency	Co-creation	Efficiency
Reflexivity	Reliability	Resilience	Flexibility
Long-term orientation	Chain of command	Collaboration	
Cultural sensitivity	Clear division of tasks	Non-discrimination	Empowerment
Multi-sector	Independent oversight		
Multi-actor		Multi-actor	
Multi-level	Multi-level		
Multi-perspective			
Holistic			
Knowledge-based	Authoritative	Consensual	Cost-effective
Compatibility (in relation to values/ traditions)			
Redundancy			
Coherence	Coercion	Collaboration	Competition
Irony	Legitimacy	Empathy	Innovation
Additional principles for sustainable development governance			
Common but differentiated governance	Intergenerational equity	Leave no one behind	Subsidiarity

(Various sources)

Sørensen et al. (2015) distinguishes cooperation, coordination and collaboration as three strategies of network metagovernance, each with a hands-off and a hands-on version. These three strategies can be seen as optional tools to be used for the design, switching and maintaining strategies.

Strategy 1. Framing: making problem definitions match with the governance environment

When policy development or implementation is stalled, new perspectives are needed. To generate such perspectives, (re)framing is often used. Dominant frames in society have an immense influence on social decision making (Wesselink and Warner, 2010). Framing is an instrument that may play a role in all phases of policy. Voets et al. (2015) even note that the conclusion about success or failure of a policy is "shaped in framing contests between its advocates and critics". Such framing contests are also happening around success and failure of (meta)governance.

A metagovernance attitude suggests continuously scanning for 'places' or 'spaces' where new opportunities are possible or already emerging. A successful strategy can be to reframe a problem that seems unsolvable into one that is – wholly or partially – solvable. In highly urbanized areas, traffic congestion can become a 'wicked' problem for which there is no solution. In the Netherlands, this problem was reframed from 'insufficient road capacity' into 'reachability of inner cities'. This new perspective resulted in a multi-faceted policy approach which focused more on improving public transport than on building more roads. Another example from the Netherlands is the reframing of flood protection policy from building ever-higher dikes to providing "room for the river" with wide floodplains and retention basins. This example has been followed in many European countries (Wesselink and Warner, 2010).

A new frame can make all the difference in the world. Countries which are suffering from an increasing environmental and social pressure of (mass) tourism could take the strategic decision to switch from 'low value–high impact' to 'high value–low impact' tourism. Bhutan has implemented this frame this before tourism became a pressure. Numbers of tourists are being kept relatively low, but they have to pay a high 'entry fee' to visit the country. Even if a country does not want to go this far, such a reframing helps direct policies and practice towards a more sustainable path.

Moomaw (2013) suggested reframing of the main challenge of energy policy, from energy production to supplying energy services. He argues that the promoting the effectiveness of the energy services should be central, as they create income, wealth, and – I would add – well-being. Examples of these services are lighting, cooking, space comfort, safe water, electrical and mechanical work and mobility. Only at the second place comes the question how much and what kind of energy is needed to deliver these services in a sustainable way, and what this implies for energy production. The old frame is based on a misunderstanding (or a power constellation?): "The availability of energy is an important driver of economic growth and development, but the usual analysis fails to grasp why. It is not only because of the energy sector's contribution to the economy increases GDP". In a similar vein, Moomaw and Papa convincingly claim that the current way of addressing climate change "misdiagnoses the issue as a pollution problem by focusing on symptoms (emissions) and not on underlying causes (unsustainable development)" (Moomaw and Papa, 2012).

Another strategy is to start tackling a part of the problem that is 'solvable'. Solving such a part may have the result that the remaining (smaller) problem can be dealt with successfully – either with the original governance tools, or with different tools or by first reframing the problem. In any case, new perspectives will be visible. Taking the metaphor of travelling: when you need to travel from Munich to Moscow and you only have the resources to get to Warsaw, it might be worthwhile to start with the first leg of the trip. Upon arrival in Warsaw, new options may be available (switching from plane to train or bus, or finding additional resources, etc.). This kind of approach requires a certain level of willingness to take risks. Risk aversion and metagovernance do not go very well together. A less risky strategy which is often used in public policies is starting with a small pilot project. If successful, it may receive support for the more complex follow-up from the higher levels in hierarchy, and the necessary resources.

Strategy 2. Combining governance styles

The second metagovernance strategy is combining different governance approaches into arrangements of institutions, instruments, processes and actor constellations which are compatible enough with existing values and traditions to be accepted and at the same time different enough to push/pull/nudge towards change, and which are synergetic and not internally contradictory. This strategy is in the first place useful for the design phase, when a governance framework is being composed. When a framework is being implemented, re-combination can always be necessary.

Chapter 5 will show with 50 features of governance for which the three basic styles have distinct operational forms. This means that there are too many possible combinations of governance style features to give a full overview, so I will only give a few illustrative examples here. One style (hierarchy) can be used to solve conflicts and another to develop more solutions (network). Hierarchical governance can be used to stimulate the start and to mark the end of a network process. Market techniques like public-relations campaigns can be used to stimulate civil society involvement (network governance).

Spatial of land-use planning is, like sustainable development, a meta-policy area, where many different instruments and process tools are being used to create a future-oriented and widely accepted allocation of spatial functions. Metzger and Schmitt (2012) observe in the case of the EU Baltic Sea Strategy that from having been a primarily regulatory activity, spatial planning, which entails processes of policy integration, consultation and coordination, is increasingly becoming focused on metagovernance.

Strategy 3. Switching between governance styles

The third strategy is switching from one to another dominant governance style. Shifts in styles of governance happen all the time. They may emerge gradually, over a longer period, following broader societal trends, such as globalization has triggered more market-driven governance approaches. One of the first descriptions of such switches is the example of was public partnerships in the field of urban regeneration (Lowndes and Skelcher, 1998). Four phases were distinguished with different governance styles or combinations:

- pre-partnership collaboration (network governance);
- partnership creation and consolidation (hierarchy, e.g. to formalize authority in a partnership board and associated staff);

- partnership programme delivery (market governance with tendering and contractual agreements; hierarchical governance with regulation and supervision of contractors; network governance supporting production of bids and management of expenditure programmes);
- partnership termination and succession (networking between individuals and organizations as a means of maintaining commitment, involvement and staff employment).

These shifts of governance style can happen without a change of the underlying rationale of a policy. A research on governance style shifts in the Brazilian oil sector over a 60-year period showed that the changes did not affect the power of the state oil company Petrobas, which maintained its central role with regard to empowering and disempowering non-state actors (de Aguiar and Freire, 2017). Other changes are abrupt, sometimes unannounced, for example when a complex and contested topic for which a network approach was designed turns into a disaster and suddenly command and control (hierarchy) is needed. Shifts may also result from a new societal, economic, environmental and/or political reality (e.g. after an election, earthquake, flood, epidemic), of a result of decisions taken at a central/influential level of government.

Gradual, as well as abrupt shifts can occur independently of the mindful application of metagovernance. Experienced public managers tend to use the switching strategy in their daily work as a way to cope with changing challenges. Comparative research between cases of soil protection policy around 2000 in England, the Netherlands and Germany showed that, during the policy-making process, they 'switched away' from their original primary governance style – the one fitting best with their national culture (see Chapter 6) – to other styles which could help overcoming some of the problems they were facing (Meuleman, 2008a), as I have illustrated in the introduction to this chapter.

Switching from one to another style is rarely a complete shift. Lange et al. (2013) observed in two cases of environmental governance in the Netherlands, that switching was more like building on a style, than completely replacing the style with another.

Strategy 4. Maintenance of a governance framework

The metagovernance strategy is maintenance of a chosen governance framework or arrangement by, for example, protecting it against perverse/undermining influences in the governance environment. Maintenance complements the combining and switching strategies. This can be necessary at the micro level of internal metagovernance. For example, a maintenance strategy within a ministry might focus on keeping good working relations between different divisions or departments who have conflicting cultures and governance styles. In the example I mentioned in the introduction of this chapter, this was about three policy units working on the same theme (soil protection) who had developed their own governance style. In the same example, restoring trust with stakeholders was needed as maintenance intervention, after a conflict had emerged. Relation management is a key component of the maintenance strategy. Another example is maintenance of arrangements (formal or informal) to promote policy or institutional cohesion – having them in place does not automatically ensure that they work in an effective way.

Combining the four strategies

In the spirit of metagovernance, the four strategies can be also combined. In the field of inner safety, for example, the establishment of so-called 'Safety Regions' in the Netherlands is a good example. These arrangements between local authorities, police, fire brigades and first aid/health authorities have a built-in style shift, which needs some maintenance. When there is no crisis (e.g. natural disaster or a big accident), the style is network governance; as soon as there is a crisis, there is an immediate switch to 'command and control' (hierarchy), after which each of the organizations will try to 'clean up the mess' in an efficient way (market governance). That metagovernance strategies can be combined was also found in a research on metagovernance in Eurocities, a network of cities (Kroemer, 2010).

4.9 Multi-level and multi-scalar metagovernance

Multi-level governance is a good opportunity to apply metagovernance (Kull, 2016). In addition, it has been proposed to add multi-scalar governance to multi-level governance, to include the geographical component which does not always match administrative boundaries (Haughton and Allmendinger, 2008, Jessop, 2009). Jessop considers the 'Open Method of Coordination' of the EU as an example of multi-scalar (European) metagovernance: different institutions (the European Council, the European Commission, the Committee of the Regions and the member states) each play a specific role in the EU's metagovernance polity.

Sub-national governance for sustainable development is usually seen as a part of a wider multi-level governance process with nations and the international level 'above' and the local level 'below'. This assumes a hierarchy between the levels, which is only part of the picture in reality: sustainability multi-level governance requires combining top-down and bottom-up relations, and also horizontal relations. The specificities of each level – including cultural (values and traditions) preferences – necessitate a certain amount of flexibility, also in legal terms, on all levels, and even an 'exit clause' for the lower levels has been proposed in order to accommodate innovative initiatives. Moreover, the concept of governance 'levels' can be contested: global governance is rather a comprehensive arrangement of different governance forms on different decision-making levels. This may also apply to sub-national governance: the remits of sub-national governments vary widely across countries.

Each level of government should therefore have the possibility to develop sustainability governance arrangements which are tailor-made to both the area and to the type of challenges. Hierarchical governance is still very useful for framework-setting and ensuring level playing fields, but the complexity of sustainable development requires, at all levels and scales, a shift from a primarily hierarchical towards a more horizontal logic: more partnerships, new alliances, voluntary agreements, exchange of practices, capacity building and so on. The same applies to the 'management' of societal transformation. Sustainable transitions/transformations require dynamic mixtures of different governance logics, adapted to place-based values and traditions. Its leadership should be situational: sometimes steering, sometimes rowing and sometimes surfing the waves.

The above implies that sub-national governments are more than only instrumental to implementation of global or national policies, and should develop their own sustainability approach, based on a situational and dynamic mixture of different governance styles. This is a matter of mindful governance: not just copying successes from elsewhere but taking into account what works where and why (or why not). Sub-national governments can act as pilot recipients and implementers of programmes, and subsequently help scaling them up and define requirements for national or global policy frameworks, whereby sustainable solutions are piloted at a lower level for assessment of their replication and scale up potential (Christopoulos et al., 2012).

Metagovernance and long-term policy making

Considering the slow pace of societal transitions, SDG implementation should have a strong long-term view. Policies triggering transitions often have a long lead time: the results may be only visible after 5–10 years (Meuleman and In 't Veld, 2010). This requires excellent management of expectations. Frequent changes of political leadership may interrupt or even disrupt SDG implementation. In countries with a majoritarian system, such changes may happen every four of five years, when a new ruling party may change crucial staff in ministries and/or 'clean up' policies.

From a metagovernance perspective on sustainable development, it is logical to establish mechanisms to ensure/maintain the long-term dimension. Examples of successful strategies include:

- Involve the whole of government and the whole of parliament in key long-term policies;
- Show that even long-term policies will have also results in the short and medium terms;
- Build reflection mechanisms for keeping the long-term view, such as parliamentary SD commissions (Germany), or national SD councils (Finland, Belgium, Germany).

For maintaining a long-term view in a governance framework, a minimum of hierarchy is needed (institutional basis), broad involvement of stakeholders and layers of government (network governance), as well as effective marketing of policies (market governance).

4.10 Metagovernance, power and failure

Implementation of Sustainable Development Goals and targets is influenced by how (political/administrative) power is distributed in a society. This is a topic that is not regularly addressed in governance literature. One of the reasons may be that power relations and – networks are often hidden for the (public) eye. That said, at the same time not knowing them is a guarantee for governance failure.

Power – "the ability of groups and individuals to make others act in the interest of those groups and individuals and to bring about specific outcomes" (World Bank, 2017) – is inherent in governance (Kooiman and Jentoft, 2009), but it has different forms depending on how governance and consequently metagovernance are defined. Different forms of executing power are related to the different governance styles. Typical mechanisms are coercion (hierarchical governance), manipulation (network governance) and lobbying (market governance).

Power relations are often more informal than the formal institutional framework would suggest. This seems to imply that SDG governance and metagovernance practitioners should try to 'map' also informal power relations, not only formal ones.

Maybe the most problematic relation between governance and power is within the network governance paradigm. Network governance is, in a way, 'anti-power'. It is based on empathy and trust, and on partnerships on equal footing. Who defines governance as management of networks, and thus metagovernance as governance of network governance, will be tempted to consider hierarchical governance – which is then outside the scope of 'governance' as a power mechanism (maybe even a 'power network', to stay within the network paradigm), whereas hierarchy also has other characteristics such as structure, transparency, legitimacy and reliability. Thus, metagovernance by central government actors is seen as exercising power upon (subordinate) central organizations. This was the analytical frame through which Porras-Gómez (2013) analysed the increased power of the European Commission in the multi-level governance of the EU's structural funds.

Another case study is about water-sharing plans, where "water management committees were given the apparent responsibility, but not the power, to create water-sharing plans". The committees had only an advisory role, while the minister had the discretion to alter or change the plans (Walker, 2014).

Taking the broader lens of 'metagovernance of hierarchies, networks and markets' might have shown that power battles occur at all levels and between all actors operating in (or influenced by) a governance framework. An example is the metagovernance analysis of a cross-border cooperation programme between Estonia and Latvia, financed through EU structural funds, where (meta)governance failure occurred in the cooperation between the various public and non-public actors. This was not analysed as a power problem but as caused by weaknesses in the administrative structure of the programme, including a lack of (meta)governing skills such as cultural sensitivity (Lorvi et al., 2018).

A recent survey on how information and communication technology influences India's capacity for governance shows that electronic documentation and processing substantially reduce the possibility of manipulations as regards data management, which is one of the causes of and corruption (Roy, 2018). Application of ICT can therefore improve metagovernance as governance of governance by strengthening and better regulating the functioning of administrations and promotes network processes which can make governance more transparent, accountable, participatory and responsive.

This brings us to metagovernance failures. Metagovernance is not a purely technical matter that can be left to experts in organizational design or public administration (Jessop, 2009). However, from a network governance perspective, it has to be perceived as a metagovernance failure when the political dimension is too much visible. "The state may flank or support the emergence of specific governance structures" ... "However, these will always be monitored and influenced by partisan and political interests, introducing the risk of a governance structure being undermined by short-term political interests or conflicting priorities within the state".

Indeed, this is the risk of democracy. It reminds me of the rhetorical question the British permanent secretary Sir Humphry gave in the very first episode of the satirical BBC TV series *Yes, Minister*: "If the right people don't have power, do you know what happens? The wrong people get it. Politicians! Councilors! Ordinary voters!"

Metagovernance has been criticized as an approach which only helps governments to take back power they have lost during agentification, privatization, decentralization and devolvement of tasks based on the New Public Management ideology politicians. Indeed, as Boin et al. (2009) illustrate for the UK, there are examples of "the centre striking back" to regain some power it has lost during this process.

4.11 Conclusions

In this chapter metagovernance was introduced. It was concluded that:

- Metagovernance in the broadest sense, of hierarchical, network and market governance, offers the best model to analyse, design and manage governance style combinations for sustainability and other complex challenges.
- Metagovernance principles complement other governance principles.
- Four metagovernance strategies can be used: framing, combining, switching and maintenance.
- The complexity of multi-level and multi-scalar governance is a good opportunity to apply metagovernance.

Notes

1 The East African Community (EAC) is a regional intergovernmental organization of six partner states: the Republics of Burundi, Kenya, Rwanda, South Sudan, the United Republic of Tanzania and the Republic of Uganda.

2 Source: http://eur-lex.europa.eu/summary/glossary/open_method_coordination.html, retrieved on 28.01.2018
3 Around 26,000 staff, similar to European cities of around 1 million citizens such as Vienna.

References

Abdallah-Pretceille, M. 2006. Interculturalism as a paradigm for thinking about diversity. *Intercultural Education*, 17, 475–483.

Agu, S. U., Okeke, R. C. & Idike, A. N. 2014. Meta-governance and the metaphysics of political leadership in 21st Century Africa: A focus on election administration in Nigeria. *Mediterranean Journal of Social Sciences*, 5 (9), 177–182.

Amore, A. & Hall, C. M. 2017. Tourism recreation research from governance to meta-governance in tourism? Re-incorporating politics, interests and values in the analysis of tourism governance. *Tourism Recreation Research* 41 (2), 109–122.

Beisheim, M., Ellersiek, A., Goltermann, L., et al. 2017. Meta-governance of partnerships for sustainable development: Actors' perspectives from Kenya: Kenyan meta-governance for partnerships. *Public Administration and Development*. https://doi.org/10.1002/pad.1810

Bell, S. & Hindmoor, A. 2009. *Rethinking Governance: The Centrality of the State in Modern Society*. Cambridge: Cambridge University Press.

Bell, S. & Park, A. 2006. The problematic metagovernance of networks: Water reform in New South Wales. *Journal of Public Policy*, 26, 63–63

Boin, A., 't Hart, P. & McConnell, A. 2009. Crisis exploitation: political and policy impacts of framing contests. *Journal of European Public Policy*, 16, 81–106.

Bonivento, J. 2014. Del gobierno a la gobernanza local. Capacidades, instituciones y visiones de lo público en el proceso de descentralización en Colombia: estudio de casos (Doctoral dissertation, Universidad Complutense de Madrid).

Brandtner, C., Höllerer, M., A., Meyer, R., E. Meyer, et al. 2016. Enacting governance through strategy: A comparative study of governance configurations in Sydney and Vienna. *Urban Studies*, 54, 1075–1091.

Burau, V. 2007. The complexity of governance change: Reforming the governance of medical performance in Germany. *Health Economics, Policy and Law*, 2(4), 391–407.

Candel, J., Breeman, G. & Termeer, C. 2015. The European commission's ability to deal with wicked problems: An in-depth case study of the governance of food security. *Journal of European Public Policy*, 23(6), 789–813.

Capano, G., Howlett, M. & Ramesh, M. 2015. Bringing governments back in: Governance and governing in comparative policy analysis. *Journal of Comparative Policy Analysis: Research and Practice*, 17, 311–321.

Christopoulos, S., Horvath, B. & Kull, M. 2012. Advancing the governance of cross-sectoral policies for sustainable development: A metagovernance perspective. *Public Administration and Development*, 32, 305–323.

Cini, M. 2012. The Limits of Inter-Institutional Co-operation: Defining (Common) Rules of Conduct for EU Officials, Office-holders and Legislators. Paper prepared for the Conference of the Jean Monnet Multi-lateral Research Group DEUBAL, Leiden University, 3–4 November 2011.

Cini, M. 2015. Good governance and institutional change: Administrative ethics reform in the European Commission. *Journal of Contemporary European Research*, 12(1) (2016): Special Issue.

Daugbjerg, C. & Fawcett, P. 2015. Metagovernance, network structure, and legitimacy: Developing a heuristic for comparative governance analysis. *Administration & Society*, 49, 1223–1245.

De Aguiar, T. R. S. & Freire, F. D. S. 2017. Shifts in modes of governance and sustainable development in the Brazilian oil sector. *European Management Journal*, 1–10.

De Bruijn, H., Ten Heuvelhof, E. & In 'T Veld, R. 2010. *Process Management: Why Project Management Fails in Complex Decision Making Processes*. Berlin, Heidelberg: Springer.

Derkx, B. 2013. Meta-governance in the realm of voluntary sustainability standards: Early experiences and their implications. UNFSS discussion papers no. 1. New York: United Nations.

Doberstein, C. 2013. Metagovernance of urban governance networks in Canada: In pursuit of legitimacy and accountability. *Canadian Public Administration*, 56, 584–609.

Dommett, K. & Flinders, M. 2015. The centre strikes back: Meta-governance, delegation, and the core executive in the United Kingdom, 2010–14. *Public Administration*, 93(1), 1–16.

European Commission. 2017a. *Quality of Public Administration – A Toolbox for Practitioners*. 2017 edition, Abridged version. Brussels: European Commission.

European Commission. 2017b. *Quality of public administration. European Semester Thematic Factsheet.* [Online]. Available: https://ec.europa.eu/info/sites/info/files/file_import/european-semester_thematic-factsheet_quality-public-administration_en.pdf. Brussels: European Commission.

Fleming, J., Rhodes, J. & Fleming, J. 2005. Bureaucracy, contracts and networks: The unholy trinity and the police. *Australian and New Zealand Journal of Criminology*, 38, 192–205.

Fransen, L. 2018. Beyond regulatory governance? On the evolutionary trajectory of transnational private sustainability governance. *Ecological Economics*. 146, 772–777.

Glasbergen, P. 2011. Mechanisms of private meta-governance: An analysis of global private governance for sustainable development. *International Journal of Strategic Business Alliances*, 2, 189–189.

Grote, J. R. & Gbikpi, B. 2002. Participation and metagovernance: The white paper of the EU Commission. *In*: Grote, J. R. & Gbikpi, B. (eds.) *Participatory Governance: Political and Societal Implications*. Wiesbaden: VS Verlag für Sozialwissenschaften.

Gupta, J., Nilsson, M. & Young, O. R. 2017. Toward a multi-level action framework for sustainable development goals. *In: Governing Through Goals*. Cambridge, MA: MIT Press.

Haughton, G. & Allmendinger, P. 2008. The soft spaces of local economic development. *Local Economy*, 23, 138–148.

Haveri, A., Nyholm, I., Rølseland, A. R., et al. 2009. Governing collaboration: Practices of meta-governance in Finnish and Norwegian local governments. *Local Government Studies*, 35, 539–556.

Héritier, A. 2002. New modes of governance in Europe: Policy making without legislating? *IHS Political Science Series – No. 81. [Working Paper]*. Vienna: Institute for Advanced Studies.

Higdem, U. 2015. Assessing the impact of political partnerships on coordinated metagovernance of regional governance. *Scandinavian Journal of Public Administration*, 19(4), 89–109.

Hill, C. J. & Lynn, L. E. 2005. Is hierarchical governance in decline? Evidence from empirical research. *Journal of Public Administration Research and Theory*, 15(2), 173–195.

Hofstede, G. 1980. Motivation, leadership, and organization: Do American theories apply abroad?. *Organizational Dynamics*, 9(1), 42–63.

Jessop, B. 2002. Governance and meta-governance in the face of complexity: On the roles of requisite variety, reflexive observation, and romantic irony in participatory governance. In: Heinelt, H., Getimis, P., Kafkalas, G., Smith, R. & Swyngedouw, E. (eds.) *Participatory Governance in Multi-Level Context*. Wiesbaden: VS Verlag für Sozialwissenschaften.

Jessop, B. 2005. The political economy of scale and European governance. *Tijdschrift voor economische en sociale geografie*, 96, 225–230.

Jessop, B. 2008. *State Power: A Strategic-Relational Approach*. Cambridge: Polity Press.

Jessop, B. 2009. From governance to governance failure and from multi-level governance to multi-scalar meta-governance. In: Arts, B. & Al, E. (eds.) *The Disoriented State: Shifts in Governmentality, Territoriality and Governance*. Heidelberg: Springer.

Jessop, B. 2011. Metagovernance. In: Bevir, M. (ed.) *The Sage Handbook of Governance*. London: Sage,

Kirschbaum, C. 2015. Shifts in control disciplines and rescaling as a response to network governance failure: The BCJ case, Brazil. *Policy and Politics*, 43, 443–458.

Klijn, E. H. & Koppenjan, J. F. M. 2000. Public management and policy networks. *Public Management Review*, 2, 135–158.

Knill, C. & Lenschow, A. 2005. Compliance, competition and communication: Different approaches of European governance and their impact on national institutions. *Journal of Common Market Studies*, 43, 583–606.

Kooiman, J. 2003. *Governing as Governance*. London: Sage.

Kooiman, J. a. N. & Jentoft, S. 2009. Meta-governance: Values, norms and principles, and the making of hard choices. *Public Administration*, 87, 818–836.

Kroemer, J. 2010. *Meta-Governors and Their Influence on Network Functioning: A Study of Meta-Governance in the Case of the European City Network Eurocities*. Master International Public Management and Public Policy, Erasmus University Rotterdam.

Kull, M. 2016. *European Integration and Rural Development: Actors, Institutions and Power*. London: Routledge.

Kull, M., Pyysiäinen, J., Christo, G., & Christopoulos, S. 2017. Making sense of multilevel governance and governance coordination in Brazil: The case of the Bolsa Verde Programme. *Regional & Federal Studies*, 28(1), 47–78.

Lampropoulou, M. & Oikonomou, G. 2016. Theoretical models of public administration and patterns of state reform in Greece. *International Review of Administrative Sciences*, 84(1), 101–121.

Lange, P., Driessen, P. P. J., Sauer, A., et al. 2013. Governing towards sustainability-conceptualizing modes of governance. *Journal of Environmental Policy and Planning*, 15, 403–425.

Larsson, O. 2015. *The Governmentality of Meta-Governance. Identifying Theoretical and Empirical Challenges of Network Governance in the Political Field of Security and Beyond*. Skrifter utgivna av Statsvetenskapliga föreningen i Uppsala 193. 204 pp. Uppsala: Acta Universitatis Uppsaliensis. ISBN 978-91-554-9296-0.

Larsson, O. 2017. Meta-governance and collaborative crisis management – Competing rationalities in the management of the swedish security communications system. *Risk, Hazards & Crisis in Public Policy*, 8, 312–334.

Li, J. 2015. Governance and metagovernance in local government innovation. *Journal of Xiamen University (Arts & Social Sciences)*, No. 2015/3.

Lorvi, K., Kull, M. & Meuleman, L. 2018. Fostering cooperation along the borders in the EU: The Estonia-Latvia cross-border programme in metagovernance perspective. [*under review*]. *Journal of the Baltic Studies*.

Lowndes, V. & Skelcher, C. 1998. The dynamics of multi-organizational partnerships: an analysis of changing modes of governance. *Public Administration*, 76, 313–333.

Lujie, W. 2017. *The motivations and barriers for Chinese enterprises to participate in*. Doctoral dissertation, Xi'an Jiaotong-Liverpool University Suzhou, China 15.

Maas, W. & Caviedes, A. 2016. Sixty-five years of European governance. *Journal of Contemporary European Research*, 12(1): Special Issue.

Magalhaes, A., Veiga, A., Ribeiro, F., & Amaral, A. 2013. Governance and institutional autonomy: Governing and governance in portuguese higher education. *Higher Education Policy*, 26, 243–262.

Metzger, J. & Schmitt, P. 2012. When soft spaces harden: The EU strategy for the Baltic Sea Region. *Environment and Planning A*, 44, 263–280.

Meuleman, L. 2006. *Internal metagovernance as a new challenge for management development in public administration*. Paper presented at the 2006 EFMD Conference Post Bureaucratic Management: A New Age for Public Services? Aix-en-Provence, 14–16 June 2006.

Meuleman, L. 2008a. *Public Management and the Metagoverance of Hierarchies, Networks, and Markets*. Heidelberg: Springer.

Meuleman, L. 2008b. Reflections on metagovernance and community policing: The Utrecht case in the Netherlands and questions about the cultural transferability of governance approaches and metagovernance. In: Bissessar, A.M. (ed.). *Governance and Institutional Re-engineering*. Cambridge: Cambridge Scholars Publishing in Association with GSE Research.

Meuleman, L. 2011. Metagoverning governance styles – broadening the public manager's action perspective. In: Torfing, J. & Triantafillou, P. (eds.). *Interactive Policy Making, Metagovernance and Democracy*. Colchester; ECPR Press.

Meuleman, L. 2013. Cultural diversity and sustainability metagovernance. In: Meuleman, L. (ed.). *Transgovernance – Advancing Sustainability Governance*. Berlin/Heidelberg: Springer Verlag.

Meuleman, L. & In 't Veld, R. J. 2010. *Sustainable Development and the Governance of Long-Term Decisions*. In: In 't Veld, R.J. (ed.). *Knowledge Democracy*. Berlin/Heidelberg: Springer.

Meuleman, L. & Niestroy, I. 2015. Common but differentiated governance: A metagovernance approach to make the SDGs work. *Sustainability*, 12295–12321.

Moomaw, W. 2013. Can the international treaty system address climate change? *The Fletcher Forum of World Affairs*, 37(1), 105–118.

Moomaw, W. & Papa, M. 2012. Creating a mutual gains climate regime through universal clean energy services. *Climate Policy*, 12, 505–520.

Mundle, L., Beisheim, M. & Berger, L. 2017. How private meta-governance helps standard-setting partnerships deliver. *Sustainability Accounting, Management and Policy Journal*, 8, 525–546.

Murakami, Y., Yokoyama, Y. & Hirashi, A. 2010. *A Case Study on FCV Policy in Japan: A Meta-Governor's Role and Its Limits*. Paper presented at the World Conference on Transport Research in Lisbon, 13 July 2010.

Niestroy, I. 2000. *Die strategische UVP als Instrument zur Integration von Umweltbelangen in andere Politikbereiche. Fallstudien im Bereich Wasserstrassenplanung an Elbe und San Francisco Bay*. Berlin: VWF Verlag.

Niestroy, I. 2014. Governance for sustainable development: How to support the implementation of SDGs? In: (ASEF), A.-E. F. (ed.) *ASEF Outlook Report 2014/2015 – Facts & Perspectives. Volume II: Perspectives on Sustainable Development*. Singapore: Asia-Europe Foundation (ASEF).

Niestroy, I. 2016. How are we getting ready? The 2030 agenda for sustainable development in the EU and its member states: Analysis and action so far. DIE discussion paper 9/2016. Bonn.

Nooteboom, S. & Teisman, G. 2003. Sustainable development: Impact assessment in the age of networking. *Journal of Environmental Policy & Planning*, 5, 285–308.

Porras-Gómez, A.-M. 2013. Metagovernance and control of multi-level governance frameworks: The case of the EU structural funds financial execution. *Regional & Federal Studies*, 24, 173–188.

Rachel, P. & Damian, H. 2013. Enterprise policy and the metagovernance of Firm capabilities. *Administration & Society*, 47, 656–679.

Radaelli, C. 2003. The open method of coordination: A new governance architecture for the European Union?: Stockholm: Swedish Institute for European Policy Studies.

Rayner, J. 2015. The past and future of governance studies: From governance to metagovernance? In: Capano, G., Howlett, M. & Ramesh, M. (eds.). *Varieties of Governance*. Basingstoke: Palgrave Macmillan.

Renaud, E. N. C., Bueren, E. T. L. V. & Jiggins, J. 2016. The meta-governance of organic seed regulation in the USA, European Union and Mexico. *International Journal of Agricultural Resources, Governance and Ecology*, 12, 262–291.

Roy, M. N. 2018. Electronic technology and the changing dynamism of governance. In: Sekher, M., Parasuraman, S. & Kattumuri, R. (eds.) *Governance and Governed: Multi-Country Perspectives on State, Society and Development*. Singapore: Springer.

Schuppert, G. F. 2007. Was ist und wozu Governance? *Extended Version of a Paper Presented on 20 March 2007 at the Wissenschaftszentrum Berlin.*

SDG Fund. 2017. *Business and Peace: How the Private Sector Can Contribute to SDG 16.* New York: SDG Fund.

Sowman, M. 2015. Governability challenges facing small-scale fishers living adjacent to marine protected areas in South Africa. In: Jentoft, S. & Chuenpagdee, R. (eds.), *Interactive Governance for Small-Scale Fisheries: Global Reflections*. Cham: Springer International Publishing.

Sørensen, E. 2006. Metagovernance: The changing role of politicians in processes of democratic governance. *The American Review of Public Administration*, 36, 98–114.

Sørensen, E., Torfing, J. & Rhodes, R. A. W. 2005. The democratic anchorage of governance networks the new governance: Governing without government. *Scandinavian Political Studies*, 28, 195–218.

Sørensen, E., Triantafillou, P. & Damgaard, B. 2015. Governing EU employment policy: Does collaborative governance scale up? *Policy and Politics*, 43, 331–347.

Stark, A. 2015. More micro than meta? Competing concepts of metagovernance in the European Union. *Public Policy and Administration*, 30, 73–91.

Steurer, R. 2012. *The Governance of Sustainable Development: Putting the Pieces of Regulation Together.* Vienna: University of Natural Resources and Life Sciences.

Stevens, V. & Verhoest, K. 2015. Theorizing on the metagovernance of collaborative policy innovations. Paper presented at the 2nd ICPP Conference, 1–4 July, 2015, Milan, Italy.

Struyven, L. & Van Parys, L. 2016. *How to act? Implementation and evolution of the PES conductor role: The Belgian PES in Flanders as a case study.* Luxembourg: Publications Office of the European Union.

Thang, H. V. 2018. *Rethinking Fisheries Governance: The Role of States and Meta-Governance.* Basingstoke: Palgrave Macmillan.

Tömmel, I. 2016. EU governance of governance: Political steering in a non-hierarchical multilevel system. *Journal of Contemporary European Research*, 12(1): Special Issue.

Torfing, J. & Triantafillou, P. 2011. Introduction to interactive policy making, metagovernance and democracy. In: Torfing, J. & Triantafillou, P. (eds.). *Interactive Policy Making, Metagovernance And Democracy*. Colchester: ECPR Press.

Tsheola, J. P. 2013. State capitalism, meta-governance, state-capital nexus and state activism paradoxes in the Gauteng freeways. *Journal of Public Administration*, 48, 35–50.

United Nations. 2015. *Transforming Our World: The 2030 Agenda for Sustainable Development*. New York: United Nations.

United Nations. 2016. *Synthesis of Voluntary National Reviews 2016*. New York: United Nations.

United Nations. 2017. *Towards a Set of Internationally Recognized Principles of Responsible and Effective Governance*. Committee of Experts on Public Administration. New York: United Nations.

United Nations. 2018. *Elaborating Principles of Effective Governance for Sustainable Development*. Committee of Experts on Public Administration. New York: United Nations.

UNOSD. 2016. Incheon Communiqué: Summary of the 2016 sustainable development transition forum, 27 October 2016, Incheon, Republic of Korea.

UNPAN. 2017. Symposium on 'Implementing the 2030 Agenda for Sustainable Development and the SAMOA Pathway in Small Island Developing States – SIDS: Equipping Public Institutions and Mobilizing Partnerships' 21–23 February 2017, Nassau, Commonwealth of the Bahamas. Informal communique.

Venghaus, S. & Hake, J. F. 2018. Nexus thinking in current EU policies – The interdependencies among food, energy and water resources. *Environmental Science & Policy*. https://doi.org/10.1016/j.envsci.2017.12.014

Voets, J., Verhoest, K. & Molenveld, A. 2015. Coordinating for integrated youth care: The need for smart metagovernance. *Public Management Review*, 17, 981–1001.

Walker, G. 2014. Water scarcity in England and Wales as a failure of (meta)governance. *Water Alternatives*, 7, 388–413.

Wan, Y. K. P. & Bramwell, B. 2015. Political economy and the emergence of a hybrid mode of governance of tourism planning. *Tourism Management*, 50, 316–327.

Wesselink, A. & Warner, J. 2010. Reframing floods: Proposals and politics. *Nature and Culture*, 5, 1–14.

Whitehead, M. 2003. 'In the shadow of hierarchy': Meta-governance, policy reform and urban regeneration in the West Midlands. *Area*, 35, 6–14.

World Bank. 2017. *World Development Report 2017: Governance and the Law*. Washington, DC: World Bank.

Yanwei, L., Homburg, V., De Jong, M., et al. 2016. Government responses to environmental conflicts in urban China: The case of the Panyu waste incineration power plant in Guangzhou. *Journal of Cleaner Production*, 134, 354–361.

Yoo, J. W. & Kim, S. E. 2012. Understanding the mixture of governance modes in korean local governments: An empirical analysis. *Public Administration*, 90, 816–828.

Yu, J. & He, Z. 2012. The Tension Between Governance and State-Building.In: Yu J., Guo S. (eds.). *Civil Society and Governance in China*. New York: Palgrave Macmillan.

Part 2
Features of metagovernance

Chapter 5. Fifty shades of governance: a toolbox
Chapter 6. How values, traditions and geography shape the feasibility of governance approaches
Chapter 7. Mind-sets and mental silos: rise and fall of simple switches

5 Fifty shades of governance
A toolbox

> Even if enforcement becomes redundant, its existence continues in the shade of hierarchy.
> – Höpner, 2008

> Some governmental action or a 'shadow of hierarchy' in the form of initiating steps to legislate or take executive action is a precondition for self-regulation to develop.
> – Héritier and Eckert, 2008

> The evidence gathered here suggests . . . that it is time to take persuasion out of 'the shadow of networks', and to study it as a steering mechanism that plays obviously an important role in various governance settings.
> – Steurer, 2011

> Perhaps here we might talk about hierarchies and networks operating in the 'shadow of markets'.
> – Atkinson and Klausen, 2011

Hierarchical, network and market governance operate in mixed forms, but even when one style appears to be absent, there is often still a shade, or a shadow with a potentially great impact. This impact can be positive or negative.

Chapter 4 introduced metagovernance as an approach to prevent or tackle problems with and between governance styles. This chapter presents a 'toolbox' for those who apply a metagovernance approach when analyzing, designing or managing a governance framework in its context. The toolbox consists of short descriptions of 50 'shades of governance' with each three subtypes or 'operational forms': one for each basic governance style.

5.1 Differences between governance styles

The three ideal-typical governance styles discussed in Chapter 2 which are the building blocks of metagovernance are expressions of different logics. Grouping typical features per style helps to illustrate this (Table 5.1).

It is easy to see how the three governance styles can undermine each other. For example, it is impossible to maximize stability (hierarchy) and flexibility

110 Features of metagovernance

Table 5.1 Hierarchical, network and market governance: three distinct logics

Governance styles	Examples of typical features of the styles
Hierarchical governance	Rational, reliable, stability, legitimacy, justice, accountability, risk averse, government-centred, centralized, planning and design, authoritative, instructions, one-way communication, dependency, subordinates, obedience, rules-based, command and control
Network governance	Partnerships, collaborative learning, co-creation for innovation, informal arrangements, trust-based, harmony, communication as dialogue, process management, diplomacy, mutual dependence, mutual gains approach, consensus, voluntary agreements, covenants
Market governance	Rational, cost-driven, flexible, competition as driver for innovation, price, marketing, decentralized, bottom-up, individualist, autonomy, self-determination, empowering, services, contracts, incentives, awards and other market-based instruments

(market) at the same time, or diplomacy and competition. Different styles can also support each other. Rules (hierarchy) can make informal network arrangements more democratic and accountable. In order to generate a deeper insight in the three logics, we need a more structured overview of all these governance features. This is the purpose of this chapter.

There is no objective way to determine on how many different features (or characteristics, or dimensions) of governance the three basic governance styles have distinct operational forms. Earlier, I distinguished 35 features (Meuleman, 2008), but meanwhile I have collected 50 of such 'shades of governance'. Some of them characterize internal (meta)governance, others external (meta)governance; both types are necessary as they can influence each other. It would take a separate book to discuss all features in-depth. However, because these features present so concretely the richness of the metagovernance toolbox, I will present them all, although in a concise way. Some features are closely related and I could have merged them. On the other hand, it makes sense to have several 'hammers' and a range of 'screwdrivers' in the governance toolbox. Several features will be discussed more in-depth elsewhere in this book. The 50 governance features with operational forms per governance style stand for 150 characteristics of governance.

In 2008, I clustered the features in five groups: vision and strategy, orientation, structure, processes, people and results. For the meanwhile grown collection, I have merged the first two clusters (vision, strategy, orientation) (Section 5.2). The other clusters cover institutions and instruments (Section 5.3), processes and roles of actors (Section 5.4), and different types of problems and outcomes (Section 5.5).

5.2 How people deal with context: vision, strategy and orientation

The first cluster covers 16 features expressing basic assumptions in public administration organizations about reality, virtues and motives, key values, role of government, views of what strategy means and orientation with regard to 'the others' (Table 5.2).

These 16 features will be introduced in a very condensed way on the following pages, with literature sources for further reading and with concrete links to metagovernance and to sustainable development.

Feature	Hierarchical governance	Network governance	Market governance
1. Cultures/'Ways of life'	Hierarchism	Egalitarianism	Individualism

Maybe the best 'proof' that sustainability (meta)governance is normative and that we need three different governance styles as building blocks of governance frameworks is that the three basic styles are rooted in human cultures and are in many ways reflections of them. They are congruent with the three most 'socially active' *ways of life* distinguished in cultural theory: hierarchism, egalitarism and individualism (Thompson et al., 1990). The other human cultures, fatalism and autonomism, are socially less active and therefore less relevant in governance frameworks. Fatalism is a 'no trust' culture existing usually in the shadow of extreme hierarchy, whereas in autonomism, people put themselves outside the direct influence of societal relations altogether.

The three ways of life compete with each other, often in a hostile way, but on the other hand require one another and therefore continue to co-exist (Thompson, 2003b, p. 50, 257). They also reflect different political cultures (Thompson, 1999)(see Section 6.5). Chapter 6 will show that the values and traditions of a country are important dimensions to be taken into account for the design and management of governance frameworks, which implies that copying a governance approach from one to another country (of even within a country) is risking to fail unless adaptation has taken place. This deep cultural dimension explains why metagovernance as combining and switching between different governance styles can meet strong resistance: people may identify themselves with a specific governance style. It also underpins the claim that governance for the SDGs should be "common but differentiated" (Meuleman and Niestroy, 2015).

Feature	Hierarchical governance	Network governance	Market governance
2. Relational values	Hegemonism	Tolerance, pluralism	Indifference

Relational values are about how people relate to each other in a public administration organization or in a society (In 't Veld, 2010, In 't Veld et al., 2011).

112 Features of metagovernance

Table 5.2 Governance style differences (1): vision, strategy and orientation

Features	Hierarchical governance	Network governance	Market governance
1. Cultures/ 'Ways of life'	Hierarchism	Egalitarianism	Individualism
2. Relational values	Hegemonism	Tolerance, pluralism	Indifference
3. Theoretical background	Rational, positivist	Socio-constructivist, social configuration theory	Rational choice, public choice, principal-agent theory
4. Key concepts	Public goods	Public value	Public choice
5. Mode of calculation	*Homo hierarchicus*	*Homo politicus*	*Homo economicus*
6. Primary virtues	Reliable	High level of discretion, flexible	Cost-driven
7. Common motive	Minimizing risk and predictability	Satisfying identity	Maximizing competitive advantage
8. Motive of subordinate actors	Fear of punishment	Belonging to group	Material benefit
9. Roles of government	Government rules society	Government is partner in a network society	Government delivers societal services
10. Metaphors	Machine, stick, iron fist	Brain, sermon, word, dialogue	Flux, carrot, invisible hand
11. Strategy styles	Planning and design style; power style; compliance to rules and control procedures; state and producer-centred	Learning style; chaos style: coping with unpredictability; deliberation; shaped by civil society	Power style; getting competitive advantage; market and customer-centred
12. Response to resistance	Use of power	Persuasion of rebels	Negotiated deals, using incentives
13. Orientation of organizations	Top-down, formal, internal	Horizontal, informal, open-minded, external and internal	Bottom-up, suspicious, external and internal
14. How actors are perceived	Subjects, voters	Partners	Clients, customers
15. Selection of actors	Controlled by written rules (instructions)	Free, ruled by trust and reciprocity	Free, ruled by price and negotiation
16. Aim of stock-taking of actors	Anticipating protest/ obstruction/ opponents	Involving stakeholders to get better results and acceptance	Finding reliable contract partners, know competitors

They are a sub-set of values in a culture. Relational values are highly relevant for governance because governance as it is defined in this book is a relational concept. In Section 6.2 the linkages between the three governance styles and specific relational values will be discussed.

This feature is important for sustainability governance because it can explain why conflicts may arise between on the one hand the overall, uniform narrative of participation, inclusion, partnerships across the SDGs and in particular in SDG 17, and on the other hand the relational values dominant in a (national or governmental) culture. Hegemonic thinking does not have much affinity with partnerships, for example (see also Section 6.2).

Feature	Hierarchical governance	Network governance	Market governance
3. Theoretical background	Rational, positivist	Socio-constructivist, social configuration theory	Rational choice, public choice, principal-agent theory

The three governance styles have a different theoretical background (Bevir and Rhodes, 2001, Bevir, 2011). The even have "incompatible contentions about what is knowable in the social world and what does or can exist – the nature of being – in the social world" (Dixon and Dogan, 2002, p. 191). In neo-institutional theory, rational choice institutionalism is seen as an important source of New Public Management (a combination of market governance and hierarchical accountability and control mechanisms), while sociological institutionalism has been a source for actor-oriented institutional approaches such as network governance (Schedler, 2006). New Public Management changed the control of administrative organizations from input control (budget) to output control (measuring performance and results). In the actor-centred view, control is achieved in a much more informal way, through policy networks (Scharpf, 1997).

For practitioners of governance, this feature may seem too abstract. However, its relevance becomes clearer when we translate it into the different academic disciplines policy makers have had their education in. Policy makers with a natural science or macro-economics degree may find the 'constructivist' attitude of colleagues with a social science background difficult: too irrational, illogical and even manipulative. Constructivists can talk forever about the best road to reach an objective; rationalists believe they know what the best road is: a straight line ahead from the start to the finish with 'objective' indicators and targets. The education of the majority of a policy making project team can determine which kinds of solutions are selected and which are left aside. From a metagovernance perspective a certain balance would help generating a broader set of policy options. Constructivists have, for example, more affinity with what happens in the governance environment outside of the team. Rationalists may consider this as being unfocused. The latter group may be the best to keep the project on track; the first group tends to involve everything and everybody. Therefore, understanding and taking into account the theoretical background of those who are or will be responsible for policies and (meta)governance to implement the SDGs, is important to observe governance failures in an early stage and to propose solutions.

114 Features of metagovernance

Feature	Hierarchical governance	Network governance	Market governance
4. Key concepts	Public goods	Public value	Public choice

The feature 'key concepts' is about how the output and outcome of the activities of governments should generally be measured. Public goods, value and choice are parameters stemming from different theories (Hartley, 2005). Benington (2011) observed that the two competing social-economic theories of "public goods" (Keynesian post–Second World War reforms; focus on hierarchical governance) and of "public choice" (New Public Management; related to rational choice theory; focus on market governance) have guided policy development, institutional and cultural reform, and organizational and cultural change for a long time. Hierarchy is often legitimized with the argument that it is a necessary approach for the protection and delivery of public goods. Market thinking bases relations on individualism, autonomy and maximization of self-interest as the main determinant of both economic and social behaviour, which is underpinned by the economic theory of public choice (in which 'public' points at the type of choices) (Dunleavy, 2014). With the emergence of network governance as a third way, specific criteria were needed to judge the value of the outcomes. Network governance is more driven by relational values than the other styles: the outcome of a networking process should reflect the shared values of the community that forms a network. For this, the term "public value" is appropriate, which was framed by Moore (1995) and concerns "what the public values, and what adds value to the public sphere" (Benington, 2011).

This feature is relevant for the discussion on indicators, monitoring and reporting on implementation of the SDGs. There is clearly not one right way to approach this, and some sort of combined approach will probably be needed. Some of the indicator sets published since the adoption of the SDGs are largely based on a rational approach. Examples are the indicator set of SDSN/Bertelsmann Foundation (Kroll, 2015), and the SDG indicators prepared by Eurostat (Eurostat, 2017). Indicators have been selected as proxies for an SDG or a part of it, based more on the availability of hard/numerical/monetizable data, than on the availability of qualitative assessment criteria, which could be adapted to different circumstances. Achieving a better balance will be a great challenge, especially because the selection of indicators and how to use them usually has a normative dimension. For more on this topic, see Features 27 and 29 on knowledge and on impact assessment, respectively.

Feature	Hierarchical governance	Network governance	Market governance
5. Mode of calculation	*Homo hierarchicus*	*Homo politicus*	*Homo economicus*

Three stylized 'modes of calculation' about how people judge public sector work can be distinguished (Jessop, 2003). The *homo hierarchicus* judges based on effective goal attainment and legitimacy. The *homo politicus*, who has a network orientation, uses reflexivity and dialogue in order to achieve an estimation of what would be a wise decision or action. The *homo economicus* calculates primarily with criteria such as efficiency of resource allocation.

Just like feature four, the fifth feature, 'mode of calculation', is important for indicators, monitoring and reporting of the SDGs. The modes of calculation are also formative for SDG (meta)governance as a whole: it shows three different methods to reach conclusions about which policies to pursue and which governance frameworks should ensure implementation.

Feature	Hierarchical governance	Network governance	Market governance
6. Primary virtues	Reliable	High level of discretion, flexible	Cost-driven

All governance styles have their 'virtues' (Considine and Lewis, 2003): characteristics which are usually appreciated across proponents of all governance styles. Hierarchies are reliable and markets are cost-driven. Networks have the benefits of greater discretion and flexibility than hierarchies, but may be less reliable. Market governance has more flexibility than hierarchies, but market flexibility is a means to become more cost-efficient, it should not be an end in itself. The three different operational forms of the feature 'primary virtues' show a great conflict potential. Maximizing reliability and flexibility at the same time is impossible and trade-offs should be made on this issue. Reliability may come with high costs, which conflicts with being cost-driven. The great discretion of 'street-level bureaucrats' like police officers, environmental inspectors or teachers can make them seen unreliable and not cost-efficient.

For SDG (meta)governance this means that trade-offs which are inevitable because several options are at the same time requisite and mutually undermining should not be delayed, hidden, or ignored. Such trade-offs may be necessary between any of the following objectives (which are just examples): reliability, discretionary space, flexibility, cost-effectiveness and inclusiveness.

Feature	Hierarchical governance	Network governance	Market governance
7. Common motive	Minimizing risk and predictability	Satisfying identity	Maximizing competitive advantage

Which motives do people in public administrations responsible for (sustainability) governance share? In a hierarchical organization or culture this might be minimizing risk, which is typically ensured by having in place clear lines of command and clear division of tasks into secure 'silos'. A network governance approach, based on empathy and trust, is motivated by the achievement of a satisfying joint identity of the group/community. Proponents of market governance have in common that they are motivated by maximizing (competitive) advantage (Streeck and Schmitter, 1985, p. 122).

For the implementation of the SDGs, this means that an approach to motivate governmental and other organizations as well as citizens needs to be 'common but differentiated'. It should be based on the same, universal key messages, but where necessary translated into different narratives, which are appealing to different publics. This is not rocket science: it is common practices among advertising companies to tell different tales under different circumstances to different customer groups, about essentially the same products.

Feature	Hierarchical governance	Network governance	Market governance
8. **Motive of subordinate actors**	Fear of punishment	Belonging to group	Material benefit

Although the difference between leaders and followers is much clearer in a hierarchical than in a network context, under every governance framework there are people with more and with less power and influence. What can be said about the motives of 'subordinate' people? A hierarchical culture produces not only law and order but may also result in fear. According to (Streeck and Schmitter, 1985, p. 122), hierarchical leaders motivate their subordinates through fear of punishment; stepping out of the lines is risky. In a military organization, this is partly compensated by a strong sense of belonging to a group. Generally, hierarchical governments may try to 'buy' acceptance through material benefits. Participants in network governance are motivated by the reward of belonging to a specific group (social inclusion) (Van Twist and Termeer, 1991). Customers of market governance select their providers based on material/financial benefits.

Motivation is an extremely important driver for sustainability. This feature helps deciding what ways could work best in a given situation, to motivate actors to become the transition processes towards sustainable development.

Feature	Hierarchical governance	Network governance	Market governance
9. **Roles of government**	Government rules society	Government is partner in a network society	Government delivers societal services

The implicit assumptions about what is the main role of government in society can lead to misunderstanding in international collaboration. For a Dane, government is a (special) partner and should behave as such; government is not above society but an equal partner with other key actors. For a German, it is natural that governments are on top of what is happening, and steer a country from a cockpit. A Brit believes that governments should deliver services and products. Of course, also a Danish public sector organization delivers products, such as passports, and the German constitution forces the federal government to find solutions in partnership with the *Laender*. The British government does not work only with market-based instruments but also relies on laws.

The differences in views on the role of government have a great impact on the kind of policies and of governance frameworks, which are preferred, and are related to 'national' cultures (see also Section 6.1). The impact of views about the role of government is also visible at the organizational level. Units at the bottom of the organizational pyramid have less influence in a hierarchical setting than in an administration also with network characteristics (e.g. cross-departmental project teams). In such a network approach, power mechanisms are fuzzier; this is the key argument for hierarchical leaders to oppose to the introduction of matrix organizations, which are a mixture of vertical and horizontal organization principles.

Network governance takes the position that all internal parties are, in principle and ultimately, equal. This is quite impossible to achieve in a political organization like a national ministry – but elements of it could be tried out. Early experiments in the 1990s in the Netherlands showed that a joint project between the Ministries of Agriculture and Environment not only created better policy solutions but also better mutual understanding. A market governance approach creates competition between administrative units, uses financial and other incentives and disincentives, and defines relations in terms of producers and buyers of internal products and services.

This feature is important because it gives countries different starting points for implementation of the SDGs. In a country with a strong hierarchical orientation in the administration (which is the case in most UN member countries), metagovernance would lead to paying special attention to integrate characteristics of network and market governance, to cater for the needs of societal stakeholders.

Feature	*Hierarchical governance*	*Network governance*	*Market governance*
10. Metaphors	Machine, stick, iron fist	Brain, sermon, word, dialogue	Flux, carrot, invisible hand

Metaphors contribute highly to understanding complex organizations, and Morgan's (Morgan, 1986) image of the hierarchical organization as a well-oiled machine is famous. Morgan has compared network organizations with brains, because of the reflexive variety, learning and feedback mechanisms. Market governance relates more or less to his metaphor of organizations as flux and transformation. Governance literature uses various metaphors for the three basic governance styles. Jessop (2003) named the ways hierarchy, networks and markets exercise influence respectively the stick, the sermon and the carrot. The 'stick' and the 'carrot' are clear, but for network governance the 'sermon' might be too top-down. Therefore, I tend to use 'dialogue' instead. In the example of the EU's Environmental Implementation Review (see Section 9.1, Box 9.1), I have noticed that everybody immediately understood what was meant when I said that we needed a dialogue process to add to the stick of legal enforcement and the carrot of funding, when the stick and carrot together did not deliver sufficiently. Elsewhere, Jessop characterized the governance styles with the metaphors of, respectively, the 'iron fist' (perhaps in a 'velvet glove'), 'dialogue' and 'anarchy', showing the normative significance of this trichotomy (Jessop, 2002).

Metaphors can play an important role for implementing the SDGs, because they speak to people's imagination. However, there is not per definition congruency between the metaphor that is preached and the key governance style that is practiced. Political leaders may speak about partnerships, about doing things together, and at the same time focus on hierarchical or market-based instruments. Other leaders may talk hierarchically about the strength needed to face hostile opponents (domestic or external), but at the same time chose non-hierarchical tools to achieve their policy objectives. When New Public Management was very popular among politicians from various parties in the Netherlands in the 1990s, I could not help noticing that many politicians thought hierarchy, talked network and ultimately chose market instruments. Maybe they were metagovernors *avant la lettre*, but political opportunism may have played a role too.

Feature	Hierarchical governance	Network governance	Market governance
11. Strategy styles	Planning and design style; power style; compliance to rules and control procedures; state- and producer-centred	Learning style; chaos style: coping with unpredictability; deliberation; shaped by civil society	Power style; getting competitive advantage; market and customer-centred

Strategy can be defined as a course of action leading to the allocation of an organization's finite resources to reach identified goals. Mintzberg et al. (1998) distinguished 10 strategy 'schools'. An 11th relevant approach, the 'chaos school', argues that management has to address complexity and unpredictability (Stacey, 1992). Hierarchical governance has most affinity with the design, the planning, the power, and the positioning schools. These strategy types are useful for structured, manageable, clear problems with a low external impact. Targets can be reached by planning and design, and by focusing on how to develop an optimal (hierarchical) position. These types of strategy are often found with line and project managers who consider their environment as relatively simple and stable. Financial and legal experts may prefer this approach. Their internal and external organizational environment is relatively rational and stable.

The chaos school of strategy helps dealing with the unpredictability of complex, target-led policy processes. The internal and external problems are so complex that the whole system cannot be 'steered'. The solution to this challenge is a strategy that aims at determining (or 'discovering') the rules under which the system can operate successfully. Besides the chaos school of strategy, the learning school is also useful for network governance: the idea that actors are mutually dependent makes them open to collective learning. Proponents of network governance have defined strategy as a discourse, and strategy texts as "powerful discursive governance devices deployed by organizations to plan, plot and project their futures" (Brandtner et al., 2016). From a market governance perspective, the best school of strategy might be the power school, as this enables creating competitive advantage. The chaos school may also provide useful strategic insight in self-regulation processes. Furthermore, the learning school is important: the main aim of public market governance is not to influence but to stimulate and enable societal actors to collectively learn from experiences and apply these experiences autonomously.

Strategy styles vary depending on which actors they focus. Hierarchical strategies may be state- and producer-centred, network governance strategies shaped by civil society, and market governance strategies market- and customer-centred (Benington and Hartley, 2001). A metagovernance approach to strategy development requires sufficient awareness that different strategy approaches exist for different purposes (including governance styles) may support better understanding within and across public sector organizations, as well as in their relations with stakeholders. Different views on strategy do not exclude each other: a balanced synthesis may create the best results (De Wit and Meyer, 1999).

National sustainable development strategies are already often of a mixed type. It is clear that a holistic set of policies cannot be realized with only legislation, only market-based instruments, or only discursive, bottom-up network initiatives. Such SD strategies may have a legal basis – in a number of countries, sustainable development is even anchored in the constitution – but are usually mainly a combination of agenda-setting for the medium to long term and a short-term action plan. In 2002, the UN issued guidance suggesting that the national SD strategies should be in the first place a "coordinated, participatory and iterative process of thoughts and actions to achieve economic, environmental and social objectives in a balanced and integrated manner" . . . "It is a cyclical and interactive process" . . . "rather than producing a 'plan' as an end product" (United Nations Commission on Sustainable Development (UNCSD), 2002). This shows a clear preference for a network form of strategy and not a blueprint planning type, which is understandable considering the complexity of the challenge. In 2005, a comparative study on nine European examples concluded that coordination should be at the highest level (prime minister) but the strategy at the same time should foster ownership, actions and commitment in all parts of society (Niestroy, 2005). Around the same time, Meadowcroft concluded in an evaluation of such strategies rather pessimistically that "most strategies gravitated towards the 'cosmetic' rather than the 'ideal' pole" and that a long list of shortcomings was identified in the literature (Meadowcroft, 2007). Meanwhile, in some countries, SD strategies are already in their third or fourth round, and other countries have started adopting such strategies, along with their Voluntary National Reviews for the HLPF.

Feature	Hierarchical governance	Network governance	Market governance
12. **Response to resistance**	Use of power	Persuasion of rebels	Negotiated deals, using incentives

The term 'resistance' has a negative connotation. I have always tried to see its potential, though: resistance is 'frozen energy'. It may be better to deal with resistance, which might be turned around into support, than with indifference. Dixon and Dogan (2002, pp. 184–186) showed that the commitment of 'governors' (those who are responsible for governing) to one of the governance styles leads to different strategies to deal with 'rebels'. A hierarchical governor's first response to resistance may be to try to use legitimate power to force rebels to compliance. If that does not work, a fall-back strategy is to threaten with sanctions or try other ways to make non-compliance impossible. When a network governor cannot persuade rebels to engage in a network, their last resort is to expel them from the network. A market governor might first try to bargain an agreement with 'rebels', using rewards and other incentives. If that does not succeed, the hard-liner counter response he may try is to use economic power to make it difficult for rebels to not comply.

Sustainability transitions are change processes in society as well as in public sector organizations; resistance will always be there, and it is useful to realize that there are different approaches to deal with resistance – and that the first reflex is not always the approach with long-term effects.

Feature	Hierarchical governance	Network governance	Market governance
13. **Orientation of organizations**	Top-down, formal, internal	Horizontal, informal, open-minded, external and internal	Bottom-up, suspicious, external and internal

The orientation of public sector organizations towards their external environment differs with their dominant governance style (Streeck and Schmitter, 1985, Jessop, 2003). When hierarchical governance dominates, the general orientation will be top-down, formal and internal. Network governance is informal, empathic, open-minded and based on mutual respect. A network organization is both externally and internally oriented; formal organizational boundaries are not that important and proponents of network governance know that often at these fuzzy boundaries, new ideas are generated. The latter also applies to a market governance style.

In addition, the market orientation of autonomy brings about a bottom-up orientation, self-interest as a main value, and the predominance of competition makes the organization to a certain extent suspicious.

Feature	Hierarchical governance	Network governance	Market governance
14. **How actors are perceived**	Subjects, voters	Partners	Clients, customers

How societal actors are perceived uncovers the preferred governance style of governmental actors. Before the 1960s, at least in the Western world, citizens were never called 'partners' or 'clients'. They were subjects, ruled and protected by a political class, and voters (in democracies, that is). From the 1980s onwards, with the emergence of New Public Management and starting in Anglo-Saxon countries, citizens became known as 'clients' of the service-providing public sector. After a while, they started to live up to their new profile: unsatisfied citizens were complaining about lack of service, rather than the about lack of democratic legitimacy, for example. The label 'partners' only came up with the emergence of network governance.

For sustainability metagovernance it is important to realize that the perception of actors by government is only one side of the coin: how are stakeholders perceiving themselves? Their self-perception may be conditioned by government practice in the past. Where business or civil society were never involved as partners, it may take considerable efforts to develop the trust needed for engaging in multi-stakeholder partnerships as promoted by SDG 17.

From a metagovernance perspective, effective governance can be based on all governance styles. Although stakeholder involvement is popular in contemporary governance studies, not all citizens or stakeholders want to spend their time helping the state and its institutions to perform public tasks. Depending on the problem type, any mix from hierarchical, network and market governance can work, as long as it is adapted to the specific situation.

From a sustainability perspective, a long-term, generational view is important. This may imply a trade-off between close involvement of stakeholders, which can result in government policies serving short-term interests and top-down, hierarchical policy decisions with a long-term goal. In the Netherlands, the already existing network ('polder') culture has evolved in a general preference for network governance, which has led commentators to suggest that huge long-term investments such as the Delta Works to protect the country against the sea (finished after almost 50 years in 1997) are not possible anymore.

Feature	Hierarchical governance	Network governance	Market governance
15. Selection of actors	Controlled by written rules (instructions)	Free, ruled by trust and reciprocity	Free, ruled by price and negotiation

How free or constrained are the choices a governmental organization can make to select societal actors to be involved in policy making or implementation (Assens and Baroncelli, 2004)? The logic of hierarchy suggests choosing actors according to written rules. Preferably, there is a law that prescribes that certain NGOs or businesses, or governmental levels shall be involved. In a network logic, the rules for choosing partners are looser. There are strict criteria but of a different type: trustworthiness and reciprocity, the latter meaning that partners must be willing to share ideas, knowledge and information. A market governance logic is different. Partners are chosen which can – in the end – contribute to one's competitive advantage, and price is an important variable.

Feature	Hierarchical governance	Network governance	Market governance
16. **Aim of stock-taking of actors**	Anticipating protest/ obstruction/ opponents	Involving stakeholders to get better results and acceptance	Finding reliable contract partners, know competitors

Mapping or stock-taking of the relevant stakeholders in a policy process looks like a typical network governance approach. This is indeed indispensable for interactive or participative processes, not only in order to know who should be involved but also to understand about other (hidden or not) coalitions.

Having made a detailed stakeholder relation map with my team in the Netherlands Planning Ministry helped me understand the potential when the director of an environmental NGO called me to tell that he had succeeded in forging a secret alliance with farmers and a recreation lobby organization, united against urban developers wanting to build a new city in a quiet rural area. We updated such stakeholder maps on a monthly basis during this one-year project, because we knew that things were very dynamic.

Stock-taking of stakeholders is also useful in a hierarchical or market governance context; it is therefore also a requisite of metagovernance as method (see Chapter 9), and in particular for making partnerships (SDG 17) successful. For hierarchical governance to be successful it is relevant to know what 'opponents' are doing and who their 'friends' are. Market-type governors will want to know with whom they could close contracts, who their competitors are, and what their plans are.

5.3 Institutions, instruments and tools

The second group of 15 governance features is related to institutions, instruments and tools (Table 5.3). Institutions are usually seen as organizational arrangements in or with public administration organizations. Examples are a department, a policy unit, a commission, an agency, a local authority. Most formal institutions – which is globally the default – have a legal basis, which determines their remit, budget and other resources. In academic circles, institutions are normally defined as having a much broader scope: "Institutions are stable, valued, recurring patterns of behaviour" (Huntington, 1965). Huntington adds that organizations and procedures "vary in their degree of institutionalization" (p. 394). Institutional change or adaptation is a key theme for implementation of any societal transition, and in particular for implementation of the SDGs. SDG 16 focuses on institutions, and since the adoption of the SDGs in 2015, numerous conferences have been organized to discuss institutional reform. In this section the challenge of institutional reform will be approached through the lens of governance styles and metagovernance.

In daily public administration language, laws and procedures are seen as 'instruments' or 'tools', not as institutions like political scientists tend to do. In

Table 5.3 Governance style differences (2): institutions, instruments and tools

Features	Hierarchical governance	Network governance	Market governance
17. Institutional logic	Line organization, centralized control systems, project teams, stable/fixed	Soft structure, with a minimum level of rules and regulations	Decentralized, semi-autonomous units/agencies/teams; contracts
18. Addressing organizational silos	Keep silos for structure	Teach silos to dance	Break down the silos
19. Typical policy instruments	Law-making, control mechanisms, penalties, fees	Networks, stakeholder involvement	Incentives, awards
20. Unit of decision making	Public authority (person, institution)	Group	Individual
21. Main control mechanism	Authority	Trust	Price
22. Coordination mechanism	Imperatives; ex-ante coordination	Diplomacy; self-organized coordination	Competition; ex-post coordination
23. Transaction types	Unilateral	Multilateral	Bi- and multilateral
24. Degree of flexibility	Low to medium	Medium	High
25. Commitment among parties/partnerships	Medium to high/public-private partnerships	Medium to high/multi-stakeholder partnerships	Low/public-private partnerships
26. Communication styles	Communication about policy: giving information	Communication for policy: organizing effective dialogue, connecting	Communication as policy: influencing, incentives, PR campaigns
27. Roles of knowledge	Expertise for effectiveness of ruling; authoritative knowledge	Knowledge as a shared good; agreed knowledge	Knowledge for competitive advantage; cost-effective knowledge
28. Type of science-policy interface	Chief scientist; embedded science model	Partnership model: dialogue	'Speaking truth to power' model
29. Approaches to impact assessment	Evidence-based policy making	Inclusive assessment of policy options	Cost-benefit analysis
30. Access to information	Partial: segregated information	Partial: fragmented information	Total, determined by price
31. Accountability style/tools	Order and observance	Interactive persuasion, participation and co-working	Market competition

this section a variety of policy tools will be discussed which have different uses depending on the governance styles they are part of. The appropriateness of tools and instruments depends strongly on the policy objectives and the governance environment. For example, peer support and peer review are tools which are natural in a consensus culture, but might be looked at suspiciously in more hierarchical cultures. Impact assessment is a tool to assess ex-ante the possible impacts of a policy option. How impact assessment functions and how inclusive it is, for example, differs according to different preferential governance style mixtures in a given situation (Meuleman, 2015).

Policy and governance tools can be used in many ways, just as a hammer can be used to nail, to hit, or as a dead weight to prevent paper from blowing away. New tools create their own history and their use may change over time. When the first fax machine was installed at my then-workplace at a provincial administration around 1988, it was kept in the main library and a form had to be filled in before the 'guardian' of the machine was able to use it. New inventions can be used for their principal purpose but also for other uses. The first solar panels introduced in developing countries were also used as building material to create shelter against rains – which was logical because shelter was a basic need.

Feature	Hierarchical governance	Network governance	Market governance
17. Institutional logic	Line organization, centralized control systems, project teams, stable/fixed	Soft structure, with a minimum level of rules and regulations	Decentralized, semi-autonomous units/agencies/teams; contracts

Politics, policies and polity are all normative. For politics and policies this is obvious, but not all political practitioners are aware that political and administrative institutions – the 'hardware' or polity – are crystallizations of a normative vision on societal problems and their solutions which was powerful at a certain time and place. As their establishment and abolition can take many years, institutions are almost per definition not optimally adapted to their remit. There is a certain analogy with the famous 'Peter Principle' (Peter and Hull, 1969) that predicts that in hierarchies people are always promoted to the level of their incompetence: institution-building takes so much time that institutions are often already incompetent when finally in place. Politicians also usually have an embedded preference for a specific governance approach, which can lead to conflicts with situation-driven leadership. Their preferred governance approach fits with their world views and, according to Rayner (2012), knowledge which is in tension or outright contradiction with those versions will be expunged. Such 'uncomfortable knowledge' is kept at bay through denial, dismissal, diversion and displacement.

Three different institutional logics can be distinguished: the logic of hierarchies, the logic of networks and the logic of markets (Meuleman, 2008, 2013). The logic of hierarchies produces centralized institutions, which work on the basis of authority, with rules and regulations and imperatives. Institutions are, for example, legally based agreements. This aligns with classical representative forms of democracy, but also with authoritarian types of ruling. Decisions are made top-down. Government is central. Blueprinting is an engineering term that aligns well with the hierarchical logic. Hierarchical institutions are often best suited for dealing with emergencies and disasters, and for control tasks.

The logic of networks tends to produce more informal institutions in which trust and empathy are key values. Examples are covenants and Internet communities. This shares a logic with deliberative forms of democracy. Decisions are made together. Government is a partner in society. Network institutions have proven to be able to lead to ways out of 'wicked', complex and disputed problems. The logic of markets aims at small, decentralized government, and at using market types of institutions such as contracts, incentives and public-private partnerships as well as other hybrid organizations. Decisions are made bottom-up, through mechanisms like the invisible hand of the market. Government is a societal service provider. Market institutions with their focus on autonomy and efficiency are best for routine problems.

The logic of metagovernance suggests that situationally effective combinations of the three prototypical logics should be made. It is therefore impossible to describe an optional institutional framework for sustainable development.

Feature	Hierarchical governance	Network governance	Market governance
18. Addressing organizational silos	Keep silos for structure	Teach silos to dance	Break down the silos

The existence of 'silos' in public administration organizations is a logical consequence of a Weberian organizational principle that tasks should be divided into small, clearly distinguishable sub-tasks. In social sciences and management advice, the word 'silo' has a negative connotation. It emphasizes that departments have closed borders, which hamper cross-border collaboration and therefore policy coherence. Silos are characteristic for fragmented organizations, and such organizations tend to produce fragmented policies. However, silos also have a positive side. They support structure, reliability, accountability and transparency on who is responsible for what. Fragmentation has also a positive neighbour, namely specialization. The general call to 'break down silos' is NPM heritage and may destroy structures in public sector organizations that are necessary to be effective; it is often better to break down mental silos and 'teach silos to dance' (Niestroy and Meuleman, 2016).

Because coherence is important to achieve sustainability and for this aim the SDGs are designed to be 'indivisible', the question of how to deal with existing silos is a key issue for sustainability metagovernance. The hierarchical approach is too strict, while the market governance approach is too drastic. The network approach of 'teaching silos to dance' might be a good connector between the other styles. Organizing informal contacts between people who work in different silos could even be a way to find out what the right balance should be in a given situation.

Feature	Hierarchical governance	Network governance	Market governance
19. **Typical policy instruments**	Law-making, control mechanisms, penalties, fees	Networks, stakeholder involvement	Incentives, awards

The instrumental toolbox of each of the three styles is distinct. Hierarchical governance has a strong affinity with legal and other regulatory instruments: law-making, control mechanisms, penalties, fees. The tools of network governance include networks, communities and stakeholder involvement. Market governance typically works with incentives such as awards, which stimulate competition (see Hartley, 2005).

Feature	Hierarchical governance	Network governance	Market governance
20. **Unit of decision making**	Public authority (person, institution)	Group	Individual

Different cultures may clash on the question of who takes decisions. The main unit of decision making according to market governance is the individual: individual players compete against each other. Network governance presumes collective decision making among a group of actors. Hierarchical governance requires a (public, if the issue is a public issue) authority to take decisions (Arentsen, 2001). In some aspects, this seems to contradict with what cultural theory would predict. In collectivist cultures, such as in Southern and Eastern Europe, it is not the group but an authority who tends to be the decision maker. In individualist cultures such as the Netherlands, the preferred decision making unit is a group. Only in countries with a general market governance culture – which are across the board also individualist cultures, people find it normal that much decision power rests with the individual.

For implementing sustainability programmes such as the transition to the circular economy, the preferred unit of decision making is important because numerous decisions have to be made, on investments, incentives, regulatory tools such as taxes and funds for applied research. France and the Netherlands are both relative forerunners on the circular economy, but in France decisions are much more centralized while in the Netherlands the relevant decisions are based on consensus, for example in the more than 150 'green deals' between government, private sector and civil society. This should not discourage learning from each other: in 2018, both countries engaged in a bilateral exchange of experiences with drafting circular economy programmes.

Feature	Hierarchical governance	Network governance	Market governance
21. Main control mechanism	Authority	Trust	Price

Authority, trust and price are the main control mechanisms of the three governance styles (Davis and Rhodes, 2000, p. 18). Authority is the main control mechanism of hierarchies. It concerns the formal roles individuals have been given in their official capacities. Authority is something else than power. Power is more informal and about individuals pursuing values, interests and goals of their own choosing (Hutchcroft, 2001). In the real world power and authority interact constantly, and informal networks of power mix with the formal structure of authority. The main control mechanism of network governance is trust. Market governance realizes its objectives primarily through price.

Hierarchy controls through oversight (inspections, directives, legal powers of intervention), network governance controls through mutuality (cooperative interaction, informal consultations, negotiating), and market governance relies on control through rivalry (competition, benchmarking) (Lodge and Wegrich, 2005). Acceptance by citizens and stakeholders of oversight depends on the perceived authority, power and legitimacy of it – and on the extent to which a culture accepts a power distance between people in a society (see Chapter 6).

Feature	Hierarchical governance	Network governance	Market governance
22. Coordination mechanism	Imperatives; ex-ante coordination	Diplomacy; self-organized coordination	Competition; ex-post coordination

Coordination is one of the key mechanisms to promote coherence of policies and institutions (see Section 10.2). The three governance styles use different coordination mechanisms: hierarchies coordinate activities via rules, networks via diplomacy, and market governance uses competition (Kaufman, 1986, Thompson, 1991, Davis and Rhodes, 2000, Jessop, 2002, Thompson, 2003a). Coordination within government is typically top-down steering, but there are also examples of informal coordination and coordination by competition.

Hierarchical coordination is ex-ante coordination through imperatives, using rules and regulations. In market governance, coordination can be achieved 'by the invisible hand of the self-interest of participants'; this works well when 'buyers' and 'sellers' can be differentiated. Market coordination – ex-post coordination through exchange, with a key role for competition – is usually not accepted in countries with strongly legalistic administrative cultures (Peters, 1998). Network governance coordinates through reflexive self-organization. Because coercion is weak or absent, diplomacy is an important prerequisite of network coordination. The composition of networks influences how effectively they can coordinate:

networks with a pluriform composition are more likely to coordinate effectively than other networks.

For sustainable development policies, central coordination is often seen as a good practice. This makes sense because government administrations in most countries are predominantly hierarchical – but it is also logical because coordination should ensure collaboration among departments or sectors representing different interests. On the other hand, softer coordination mechanisms have also proven to be beneficial. The national sustainable development councils or commissions that exist in many countries constitute a good example.

Feature	Hierarchical governance	Network governance	Market governance
23. Transaction types	Unilateral	Multilateral	Bi- and multilateral

Hierarchy is characterized by unilateral transactions, while network governance prefers multilateral transactions; the market governance logic may result in bilateral or multilateral transactions, typically in the form of contracts. A hierarchical manager gives instructions to his or her subordinates, who may interpret them in a narrow way to prevent risks, when failure is not an option. This kind of transaction may result in a risk-averse culture of fear. A well-known and effective type of negotiation to reach multilateral agreement is the mutual gains approach (MGA) to reach consensus, based on creating value according to the interests of the participants (Susskind and Field, 1996). Essential are the differentiation between interests (what each participant in a group process seeks to achieve) and positions or demands (what people say they must have). Consensus sounds good from an equality perspective, but it is important to note that networks are not per definition democratic (Sørensen, 2006).

From a sustainability metagovernance perspective, public sector organizations should be able to engage effectively in both types of transactions.

Feature	Hierarchical governance	Network governance	Market governance
24. Degree of flexibility	Low to medium	Medium	High

The culture and structure of a public sector organization can be very flexible (market governance), medium flexible (network governance) or inflexible or medium flexible (hierarchical governance) (Powell, 1990). An organization characterized by its flexibility will typically produce flexible governance frameworks. However, flexibility is a feature that is difficult to combine with certain other features. For example, the virtues of discipline and stability of a hierarchical organization are generally accompanied by a low level of flexibility. Public sector organizations tend to be less flexible and less efficient than non-public organizations, because they are dependent on politics, independent from market performance, and have specific restrictions in employee policy, the public budget system and problematic performance measurement (Mayntz, 1985). In a network context, flexibility may be medium to high, depending on the strictness of the agreed network rules. Market governance can be associated with high levels of flexibility: an entrepreneurial attitude implies constant looking for new opportunities. However, contracts may lessen this flexibility.

There are different types of flexibility. Hierarchy may be inflexible in its mindset, but can be very flexible as regards implementation, as this can be legally obliged and enforced. Network governance may be flexible in its mind-set, taking many views into account, but inflexible as regards decision making and implementation, when members of a network need to achieve consensus about it. Market governance is flexible in terms of the choice of instruments, but is notoriously inflexible with regard to the (economic) assumptions behind cost-benefit models, such as whether long-term sustainability investments should be discounted at the same level as business investments.

The New Public Management mantra of flexibility and the related preference of discretionary power at lower levels ('empowerment') is one of the clearest causes of conflicts with the rigidities created by complicated civil service laws and regulations in hierarchies (Ingraham, 1996). Exceptions seem to apply: the European Commission, once characterized as being "half way between a French Ministry and the German Economics Ministry" (Dimitrakopoulos and Page, 2003), has under its hierarchical framework a professional network culture and is known for its practice of metagovernance, as we saw in Section 4.6.

Achieving the SDGs cannot be realized without innovation at many levels and sectors and by many actors. Einstein's famous aphorism applies, that we cannot solve our problems with the same thinking we used when we created them. Therefore, sustainability metagovernance would try to maximize the flexibility needed to stimulate innovations, including through flexible financial and legal instruments, while ensuring sufficiently stable ('inflexible') framework conditions prevent innovative ideas from undermining existing sustainable practices. Too much flexibility is one of the reasons why the Netherlands is a latecomer as regards renewable energy: funding and other incentives changed so often that they became disruptive: "frequent changes occurred in regulations and subsidies . . . from 1995 onwards. Shifts in policy, almost every few years, created uncertainties and hampered private investments" (Verbong et al., 2008; see also Niestroy, 2005). Not the Dutch, known for their ages-old windmills, but the Danes became leaders in wind turbine technology and industry.

Feature	Hierarchical governance	Network governance	Market governance
25. Commitment among parties/partnerships	Moderate to high/public-private partnerships	Moderate to high/multi-stakeholder partnerships	Low/public-private partnerships

Actors involved in hierarchical or in network governance may have a medium to high commitment in partnerships, while market governance generally requires a lower level of commitment (Powell, 1990). In hierarchies, commitment of people may be moderate when their creativity and knowledge is not fully utilized, and high when they have accepted the authority of superiors. Network partners may also have a moderate to high level of mutual commitment: the level of commitment depends on the interests at stake, their 'BATNA' ('Best Alternative to a Negotiated Agreement') (Fisher et al., 2011), and the level of trust in the network. Market governance actors may have a low level of commitment. Their commitment may depend on the competitive advantage that the cooperation with (contract) partners produces.

In Agenda 2030, partnerships are considered as an important instrument and their promotion is part of SDG 17. While public-private partnerships have a long history under hierarchical and market governance – with critical failures mentioned in Section 7.6 – multi-stakeholder partnerships will probably benefit more from network governance conditions, in particular when administrations, business and civil society are meant to work together on equal footing (see 'ABC' partnerships, Section 11.5).

Feature	Hierarchical governance	Network governance	Market governance
26. Communication styles	Communication about policy: giving information	Communication for policy: organizing effective dialogue, connecting	Communication as policy: influencing, incentives, PR campaigns

Three communication styles can be distinguished, each following the logic of one of the three basic governance styles (Rijnja and Meuleman, 2004, Meuleman, 2008). Under hierarchical governance, stakeholders are kept outside the decision-making process. Communication is defined as giving information *about* policy. In a market governance approach, communication may be used *as* a policy instrument, for example a public relations campaign in order to 'sell' a policy. In a network approach governments try to involve societal stakeholders in the making and execution of policies. This involvement can range from influencing the decision to real co-decision. In this case, communication is a means to improve the quality of the participation process: communication *for* policy.

Social media are no exception to the general practice that governments tend to see communication as presenting information. As a tool for public policy communication their first use was – and still predominantly is – to send out information. This is a strong habit: I also use webpages, Twitter and Facebook to increase the outreach for my messages. Two-way communication may meet hostility because it risks disturbing the 'order' in an administrative organization. There are good examples though, where governments have leapfrogged from this 'sending messages' use to engage in two-way communication.

Because the three communication styles are so different, conflicts may arise and management of expectations is therefore essential. When stakeholders expect to be listened to, the two rational approaches (hierarchy and market governance) may cause disappointment and ultimately disengagement. Designing the right mix of communication styles is a typical metagovernance challenge. Between 2004 and 2009 I have conducted more than 30 workshops on this topic for communication and policy officers of most of the ministries, as trainer at the Dutch Academy of Public Communication. An early adopter of this approach was the project manager of the German Soil Protection Law project around 2000. Knowing that soil protection is a politically highly disputed topic, he created expert platforms (e.g. through establishment of expert magazines) and motivated discussions at schools through a dedicated children's book. The campaign part (market governance) of his communication strategy included a reach-out to the general public through a series of postal stamps (Meuleman, 2008).

Feature	Hierarchical governance	Network governance	Market governance
27. Roles of knowledge	Expertise for effectiveness of ruling; authoritative knowledge	Knowledge as a shared good; agreed knowledge	Knowledge for competitive advantage; cost-effective knowledge

The three basic governance styles have different views about what is 'usable' knowledge for policy making or policy implementation (Adler, 2001, Meuleman and Tromp, 2010). Authoritative and undisputed knowledge ('expertise') is highly valued in the context of hierarchical governance. In such a context, politicians and civil servants believe that knowledge produced by authoritative experts or institutions is the most usable. Critics will be told that the experts involved have a long-standing reputation in the scientific world, have produced many articles in peer reviewed academic journals, and have been proven "right" in many occasions. From a hierarchical point of view, it makes sense that a knowledge provider has authority. Hierarchy builds on clear divisions of tasks, and experts are responsible for producing evidence. When a network approach dominates, trust, empathy and consensus are important criteria to determine the usefulness of knowledge. Knowledge commonly agreed upon, produced by trusted experts and lay people in a transdisciplinary approach, will be valued most. "Fact finding" should be a joint process of governmental actors and non-governmental stakeholders.

Market governance proponents consider costs crucial. This may lead to pressure to producing studies in a short period and relying on relatively simple models. In the competitive culture of market governance, being fast is better than being rigorous. The logic of network governance, which appreciates dialogue and evaluating different views, achieving consensus on knowledge is likely to be valued higher than scientific authority (Meuleman, 2013). In a policy process in which market governance is the dominant approach, efficiency, price and

132 *Features of metagovernance*

competition are highly valued. When the usability of knowledge is discussed, the cost-benefit ratio will play an important role. Knowledge (producers) may be authoritative or broadly accepted, but if the price is too high, it is not considered to be very usable. This attitude can lead to late and expensive lessons.

Knowledge is not always welcome, and the three governance styles have different approaches and arguments to undermine unwelcome knowledge. Because hierarchy leans on authoritative expertise, it may try to undermine the authority of the experts. The satirical BBC series *Yes, Minister* illustrated this with a four-step method to discredit any study or research without even having to read the reports.[1] As network governance is based on trust, undermining a consensus about a research could be a feasible way to disqualify research. From a market governance perspective, criticizing the price-quality ratio could have undermine a study and use (or commission) another one.

SDG Target 17.6 calls for sharing knowledge between North and South, South and South, and for 'triangular regional and international cooperation'. Target 17.16 states that multi-stakeholder partnerships should mobilize and share knowledge, expertise and technology to achieve the SDGs. Furthermore, "Processes to develop and facilitate the availability of appropriate knowledge and technologies globally, as well as capacity building, are also critical" (United Nations, 2015). For the implementation of the SDGs, a Technology Facilitation Mechanism was launched in 2015, among others to develop 'knowledge societies'. Implementation of these ambitions, which seem to be inspired by a network governance logic, will need to recognize the different views on knowledge in different cultures and traditions, and consequently under different governance styles. Knowledge sharing is logical when knowledge is considered a common good, but many influential actors will consider it rather as a strategic asset (under hierarchical governance) or private property (under market governance). In addition, the 'knowledge society' is not an isolated construct. It operates in what has been called the 'knowledge democracy' (In 't Veld, 2010, 2013), which is characterized by coexistence of and interrelations between old and new forms of democracy (representative and deliberative), science (disciplinary and transdisciplinary) and media (traditional and social media).

Feature	Hierarchical governance	Network governance	Market governance
28. **Type of science-policy interface**	Chief scientist; embedded science model	Partnership model: dialogue	'Speaking truth to power' model

Science is not the only but an essential supplier to public policy making. Policy making is the process of solving challenges or creating opportunities in the public domain. Public governance is about how such processes are organized, in terms of their design and management. The delivery to and the demand for science in public policy is not without problems, and the often-used metaphor 'bridging the gap' is appropriate for the challenge of improving science-policy relations.

Science-policy relations differ not only for the type of science production (mono-, multi-, inter- or transdisciplinary) or for the type of discipline(s) at stake (natural science, social science). A metagovernance perspective on science-policy relations takes value-laden assumptions on both sides of the equation into account, such as on the validity and usability of data, the credibility and accountability of science providers as well as of science users.

Science-policy relations are consumed through language, and language contains many hidden normative bullets. In the paragraph above, for example, I used economic jargon often used in market governance: I wrote that scientist are 'suppliers', 'providers' and 'producers' and 'deliver' data on 'demand'. This suggests that science is a market and that science-policy relations are ruled by market principles. This is a popular approach, but not the only one and in fact a relatively recent invention. It emerged with the popularity of market principles and instruments in public governance – including the governance of knowledge organizations. A more traditional, hierarchical approach would be to speak about the 'authoritativeness' of scientific evidence 'underpinning' policy decisions, etcetera. Moreover and more recently, some would prefer to frame science-policy relations in terms of 'collaboration' between 'partners' and value 'joint fact finding' processes and 'transdisciplinarity' highly.

In line with the three 'linguistic' models above, three different science-policy arrangements can be distinguished. The first is the model where a government appoints a 'chief scientist' at a level close to the centre of power; this is a hierarchical approach. Having a chief scientist embedded close to the political centre means that, if deemed appropriate, control can be executed over it. This model emerged in the slipstream of New Public Management in Anglo-Saxon countries: NPM combines hierarchical and market governance concepts. It has been popular in the last two decades but seems to have lost its attraction already. The European Commission uses this approach by having an in-house think tank, the European Political Strategy Centre (EPSC), which reports directly to the President. Tasked with "a mission to innovate and disrupt, the EPSC provides the President and the College of Commissioners with strategic, evidence-based analysis and forward-looking policy advice".[2]

A second model is a partnership model in which dialogue is important. In the Netherlands, for several decades special advisory bodies to the national government have existed which focused on science-policy relations. Contrary to policy advisory councils whose main intervention was giving policy advice ('giving answers'), these so-called "sector councils" produced knowledge agendas; their main intervention type was 'asking questions'. Before issuing advice, they brought together various academic disciplines and societal stakeholders, as well as policy makers, to discuss what would be the main knowledge gaps. This model was abolished in the mid-2000s, with the last surviving being abolished end of 2009 – a science-policy council on nature, environment and spatial planning (RMNO), of which I then was the director.

A more antagonist model which seems to fit better in market governance thinking is the 'speaking truth to power' model. It considers that "politicians and citizens are at a loss to judge (the) nature and quality (of the) market-place of political ideas and arguments" (Hoppe, 1999); "For scientific policy analysis to have any political impact under such conditions, it should be able somehow to continue 'speaking truth' to political élites". The speaking truth to power model is a linear model that assumes one-way traffic of truth from science to policy and separate domains of production and use of knowledge (In 't Veld, 2010): "Under this linear logic, deviations are often considered as resulting from a lack of information and communication".

Feature	Hierarchical governance	Network governance	Market governance
29. Approaches to impact assessment	Evidence-based policy making	Inclusive assessment of policy options	Cost-benefit analysis

When we buy a product, we expect that it was tested thoroughly on possible health impacts or risks before it arrived in the shop. Similarly, we probably also expect that government decisions are tested beforehand, as they can result in unexpected damage during their implementation. However, fact is that governance decisions are often not put to the test in some sort of a 'policy wind tunnel'. In many, but not a majority of the nations who endorsed Agenda 2030, legally binding procedures exist for such ex-ante assessments. Most of them focus on the environmental dimension, like the Strategic Environmental Assessment (SEA) and Environmental Impact Assessment (EIA) Directives, which apply in all EU countries and are promoted on a voluntary basis in other nations. More integrated Sustainability Impact Assessment (SIA) mechanisms are also available. The experience with all these assessments is that they lead to much better decisions that take into account environmental and other sustainability impacts, to broader and earlier participation, to less litigation and other delays after the decision, and therefore to saving money and time. Such assessments do not guarantee the most sustainable decisions, but at least that the decisions are more transparent and more evidence-based. Moreover, they trigger decision makers to think more long-term and consider indirect and cumulative impacts. This of course applies to all governmental decisions with possible sustainability impacts, not only Rio+20 implementation actions.

Impact assessment (IA) is a forward-looking instrument that seeks to advise decision makers proactively on the potential advantages and disadvantages of a proposed action (Partidário et al., 2012). The three governance styles emphasize different aspects of IA (Meuleman, 2015). From a hierarchical governance perspective, IA will be focusing on hard 'evidence'. Because hierarchical governance is a low-trust style, when this style dominates an IA process, decision makers may, democratically legitimized, act in ways which increase distrust of stakeholders and the general public of the decision process and the decision makers. Examples are IA procedures with inadequate publicity, restricted access to documents, short consultation deadlines, hearings organized as only information meetings, or which discredit evidence from stakeholders as 'not authoritative'.

Under network governance, IA processes will be relatively inclusive, in the phase of gathering information as well as in the phase of generating and comparing options. Market governance may tend to focus on the methods of assessment, with a preference on the rational-sounding method of cost-benefit analysis, rational because it is often presented as an objective method while the assumptions (such as discount rates which have a huge impact on the results) are sometimes hidden in an annex.

Impact assessments are used to bring together relevant knowledge, in order to assess the benefits and costs of various options. A comparison of five case studies of complex environmental and infrastructural problems in the Netherlands (In 't Veld, 2000) concluded that in such cases impact assessment methods should be inclusive and have the character of joint fact-finding processes. Cost-benefit analysis (CBA) is an often-used method in impact assessment. A review of costs-benefit analysis for environmental policy concluded that in order to prevent disputes about the results, it would be recommendable to integrate some stakeholder participation during the process (De Zeeuw et al., 2008). Knowledge disputes may emerge during impact assessment processes because of the different paradigms of impact assessment scholars and practitioners coming from different academic disciplines, such as law, economics, geography, engineering and natural and social sciences (Meuleman, 2015). Some may have, by training, a stronger preference for using statistics and will utilize a different definition of what is 'usable knowledge' for political decisions than others.

Addressing governance in IA such as strategic environmental assessments can play a pivotal role in defining goals, setting priorities and making choices. However, a recent case study in Portugal shows that although national guidance recommends linking IA and governance, this is not yet fully being explored (Monteiro and Partidário, 2017). Only one of 60 cases applied a full 'governance-inclusive approach', focusing on elements from network governance mainly. For example, the IA process was organized to be a discussion arena, managing different expectations, and as an empowerment tool, with participation of a wide range of stakeholders in dialogues, which created a sense of ownership.

A metagovernance approach to IA could be to include tools from other styles if one governance style dominates in the design of the IA process, as this can bring about a more rounded approach. For example, dialogues (network governance) need some structure (hierarchical governance); efficiency (market governance) is useless without effectiveness (hierarchy); and authority (hierarchy) erodes without trust (network) (Meuleman, 2015). This is important, because comparative research on SEA has shown that the capacity of this tool to deliver depends on its level of adaptation to its governance environment: cultural and institutional values influence how SEA is interpreted and carried out (Monteiro et al., 2018).

Table 5.4 illustrates how different typical challenges of strategic environmental assessment can be dealt with under different governance styles.

Table 5.4 Challenges of strategic environmental assessment and typical governance style reactions

SEA problem	Governance topics	Hierarchical governance reflex	Network governance reflex	Market governance reflex
Insufficient scoping	Dealing with uncertainty	Scoping is the responsibility of government	Include stakeholders in scoping phase	Scoping is the responsibility of the planning agency
No cumulative and synergistic effects	Complexity, dealing with uncertainty	Keep it manageable and facts-based	Create consent on difficult topics like cumulation and synergy	Keep it simple, focus on direct impacts
Insufficient consideration of alternatives and scenarios	Reflexivity, compliance, flexibility	Rule of law approach: focus on existing legal provisions	Maximalist approach: open to alternatives, including those proposed by stakeholders	Flexible attitude but ruled by cost of assessing alternatives
Insufficient quality of the IA report	Reporting, review	Focus on statistics	Focus on consensus on data	Focus on monetization
Insufficient explanation of uncertainties and other difficulties	Complexity, dealing with uncertainty	Rule out uncertainty by defining a narrow scope of the plan	Joint fact finding and participative scenario process	Select no regret options
Unclear impact of public participation	Participation	Public will be informed in due time	Public should be informed during the whole process	Public should be informed on a need to know basis
Insufficient consideration given to monitoring	Monitoring	Monitor impacts of measures	Monitor change	Monitor costs and benefits
Insufficient capacity of authorities	Capacity building	Establish a quality control authority	Establish a learning community	Establish a competitive training centre

(After Meuleman, 2015)

Impact assessment can play an important role in achieving effective implementation of the SDGs. Especially in developing countries there is scope to do more with environmental impact assessments (McCullough, 2017), social impact assessments (Vanclay, 2003) and the broader sustainability assessments (Niestroy, 2008). The European Commission utilizes for all its policy and legal proposals a combination of regulatory and sustainability impact assessment (European Commission, 2017), which is considered a good practice (Ferretti et al., 2012).

Feature	Hierarchical governance	Network governance	Market governance
30. Access to information	Partial: segregated information	Partial: fragmented information	Total, determined by price

In a hierarchical approach context, 'subjects' have limited access to information. The power of the hierarchical governor is partly based on exclusive access to certain information. In a network governance setting, information is in principle shared among the partners; however, it is often fragmented: there is no procedure or mechanism that guarantees that *all* relevant information is shared (Assens and Baroncelli, 2004). Market governance does not exclude power games with information, but the main difference with the two other styles is that information has a price, and if one is prepared to pay that price, the 'buyer' may have unlimited access to information.

Access to information is mentioned in SDG 9.c with regard to access to information and communication technology, including the Internet, and in SDG 16.10 in a more general meaning. The Aarhus Convention on Access to Information, Public Participation and Access to Justice in Environmental Matters (United Nations Economic Commission for Europe, 1998) is an important tool but it has its weaknesses. For example, Convention Parties have ample discretion in interpreting Aarhus rights; private entities are excluded from mandatory information disclosure duties (Mason, 2010).

Feature	Hierarchical governance	Network governance	Market governance
31. Accountability style/tools	Order and observance	Interactive persuasion, participation and co-working	Market competition

The term 'accountability' refers to the ability to explain or give an account of one's actions to another interested party. Public accountability is inherently a relational concept, and because different governance styles assume relation types between actors, the notion of public accountability needs to be redefined in each new governance setting (Kang and Groetelaers, 2017). How public accountability is embedded in governmental processes depends on the governance approach chosen, which may be hierarchical (top-down), horizontal (network governance), market-oriented or various combinations of the above; these approaches are characterized by 'order and observance', 'interactive persuasion, participation and co-working', or 'market competition' (Kang and Groetelaers, 2017), see Table 5.5.

Also accountability systems in education governance can be linked to governance styles. The responsibility for the quality monitoring and improvement of schools can be organized in a hierarchical public or state control model (employers or political power holders are in charge of monitoring), a network governance-related partnership model (based on a partnership between the

138 Features of metagovernance

Table 5.5 Three governance strategies and accountability relationships

Governance styles Difference in three dimensions	Hierarchical governance	Network governance	Market governance
Approaches to assure governmentality in a specific policy arena	Through imposing rules based on a formal authority	Through co-shaping and co-producing by participating in a network of actors	Through defining property rights, fostering competition and enforcing contract law
Focus of accountability criteria	Conformance with pre-defined rules and standards	Mutually constructed understanding, common notions and agreements	Profits or losses of service providers, consumer preferences or compliance with voluntary contracts
Presumed relationship between acountee and accountor	Independent and unilateral	Dependent, reciprocal and coordinated	More autonomous, discrete, individual and spontaneous

(After Kang and Groetelaers, 2017)

parents and/or students and teachers), or a free market model (break away from public control and replace it with the control of the individual consumer). A fourth model merges elements of market governance (empowerment) and network governance (networks): the professional accountability model (the professional community, e.g. the teaching force, is in charge of monitoring) (De Grauwe, 2007, pp. 14–16).

Realizing accountability in network governance implies adding some elements of hierarchical governance. In a case study on urban metagovernance (of networks) in Canada, Doberstein observed accountability in the form of networks reporting to (public sector) metagovernors, with metagovernors directly supervising the performance of networks and issuing sanctions or corrective action when required. In order to retain an oversight role on networks, metagovernors may join them as members – a common practice in economic development, but less in the social sector (Doberstein, 2013).

5.4 Processes, people and partnerships

This section discusses 14 features about how to find the appropriate and dynamic balance between process management (focus on open boundaries), project management (focus on control) and line management (focus on hierarchical steering), among others. What role are formal procedures playing and when can informal procedures be applied successfully?

Fifty shades of governance 139

Table 5.6 Governance style differences (3): processes, people and partnerships

Features	Hierarchical governance	Network governance	Market governance
32. Context	Stable	Continuously changing	Competitive
33. Process and project management	Project management (control)	Process management (contextuality)	Project management (flexibility)
34. Public sector reform approach	Top-down	Inclusive	Outsourced expertise
35. Innovation	Large-scale, national and universal innovation	Innovation at both central and local levels	Innovations in organizational form more than content
36. Relation types	Dependent	Interdependent	Independent
37. Societal interactions	Interventions	Interplays	Interferences
38. Roles of public managers	Clerks and martyrs	Explorers producing public value	Efficiency and market maximizers
39. Leadership styles	Command and control	Coaching and supporting	Delegating, enabling
40. Degree of empowerment inside organizations	Low	Empowered lower officials	Empowered senior managers
41. Values of civil servants	Law of jungle	Community	Self-determination
42. Key competences of civil servants	Legal, financial, project management, information management	Network moderation, process management, communication	Economy, marketing, public relations
43. Objectives of management development	Training is an alternative form of control over subordinates	Training helps 'muddling through'	Training helps making more efficient decisions
44. Dealing with power	Coercion	Manipulation	Competition, lobbying
45. Conflict resolution types	Classical negotiation, power-based (win-lose)	Mutual gains approach to negotiation (win-win); diplomacy	Classical negotiation, competition-based (win-lose)

The order of the features is from a broad perspective to relations between policy makers and with stakeholders, including what type of conflict management tools are typical for each governance style.

Feature	Hierarchical governance	Network governance	Market governance
32. Context	Stable	Continuously changing	Competitive

The dynamics of the context in which governance can be successful – the governance environment – is an important factor to grasp. Stability is essential for hierarchical governance. A stable context (or rigid, or reliable, depending on the perception) is achieved by issuing clear and detailed instructions, rules and procedures. For a network governance approach it is not per definition beneficial when the context continuously changes, but this is considered as advantage compared to rigidity, as it offers more new and unexpected opportunities. The preferred context of a market governor is flexible and dynamic; however, it is 'ruled' by the mechanism of competition. A market approach might decline the option to invest time and resources in consensus building (Benington and Hartley, 2001).

Feature	Hierarchical governance	Network governance	Market governance
33. Process and project management	Project management (control)	Process management (contextuality)	Project management (flexibility)

Not knowing the difference between project management and process management can result in massive governance failures. In a hierarchical setting, the organizational form of change processes seldom has the form of an "adhocracy" (Mintzberg and McHugh, 1985): change is achieved by making use of the classical line organization, and management of innovative policy or organization processes is primarily line management. A project organization, a temporary team consisting of representatives of units of the line organization, is more flexible. Project management is a threat to the line organization, for example, because it complicates decision making as it leads to confusion and battles between line and project managers. However, line and project management share the rational logic of hierarchy – project management just adds the features of flexibility and striving for efficiency from market governance. Project management is in the first place about planning and control. It aims to oversee and manage resources and typically works in meticulously planned phases towards a result that was defined in the beginning. The logic of market governance does not prescribe a form of organization, although often a project organization will be chosen because its flexibility.

The logic of network governance requires a network form of organization with an emphasis on process management (De Bruijn et al., 2010). This is a flexible form of management, which is designed to deal with complexity, unpredictability and disputes, and supports the idea that a solution for a certain problem can only be achieved when relevant actors are involved in the process in all phases, from the definition of the problem to the choice of a solution. Much of research and guidance on process management stems from consensus-oriented cultures.

None of these types of management (line, project or process management) is better than the other: it depends on the circumstances, such as the type of problem, the relative influence of internal and external actors, and the organizational and wider societal culture. Table 5.7 highlights some of the main differences between project and process management.

Table 5.7 Differences between project and process management

Project management	Process management
Focus on content	Focus on stakeholders
Clear objectives; good plan	Good process; objectives and plans result from this process
Push for action: quick and clear decision making creates better results	Keep options open: stakeholders must continue to find the initiative attractive
Communication with stakeholders is mainly explaining and convincing of the quality of the plan, and follows after decision making	Communication is a process of discussion and negotiating; decision making is the result
Focus on execution of the decision; dynamics make the execution difficult	Focus on generating a win-win situation, resulting in dealing with dynamics

(After De Bruijn et al., 1999)

Metagovernance implies mixing the three forms of management in a situationally optimal way. It seems that a successful metagovernor can be a line manager (who has the advantage of clear defined resources), a project manager (who has the advantage of flexibility in the shadow of a robust line organization) or a process manager (who has the advantage of being allowed to bring together all actors that have vested interests in an issue). Experience with process management is requisite for sustainability governance, because it is usually multi-actor, multi-sector and multi-level.

Feature	Hierarchical governance	Network governance	Market governance
34. Public sector reform approach	Top-down	Inclusive	Outsourced expertise

For a traditional, old-fashioned bureaucracy, public sector reform is a contradiction in terms. Not changing is one of the strengths of a bureaucracy. However, the New Public Management movement, nourished by budgetary constraints, has resulted in massive change operations, first in Western countries and then in other countries through e.g. conditions linked to development cooperation funds and World Bank loans. Different types of reform have taken place, which may be grouped in several types: Anglo-Saxon countries combine the 'marketiser' (introduction of market mechanisms) and 'minimiser' (small state) type. Continental European countries have chosen a combination of the 'preserver' (marginal changes) and the 'modernizer' (integrated changes) types. Comparative research has linked the reform pathways to the different history, traditions and culture in different countries (see e.g. Stevens and Verhoest, 2015, Pollitt and Bouckaert, 2017, Thijs et al., 2018).

Public sector reform has become so popular that it has almost become a virtue in itself. It seems that public organizations feel obliged to provide evidence and arguments that they are 'modernizing' and 'improving' – independent of an actual need (Hartley, 2005). A more elaborate discussion will be given in Chapter 10. Here it suffices to say that reform is normally organized in a way that matches with the dominant governance style. In a hierarchical culture, it is normal that reform is organized top-down; employees and other stakeholders are informed about the results afterwards. In a network culture, reform processes will be inclusive, incorporating ideas from those who volunteer to be involved. I experienced the difference when I switched job from a Dutch Ministry to the European Commission, from a network to a hierarchical culture. In a market governance perspective, it is not a question of top-down or inclusiveness: whatever works best and is the most cost-efficient will be done. This typically includes outsourcing of the management of the reform process, as well as outsourcing of tasks as one of the preferred solutions to organizational challenges.

For the implementation of the SDGs it seems recommendable to strike a balance between disruptive reform for leap-frogging to new systems (top-down, hierarchical), and more gradual reform processes which are inclusive (network governance), depending on the situation. Market governance ideas can help keeping the costs at a reasonable level, but this has its limits, especially when it becomes dogmatic.

Feature	Hierarchical governance	Network governance	Market governance
35. Innovation	Large-scale, national and universal innovation	Innovation at both central and local levels	Innovations in organizational form more than content

What was argued above (Feature 34) about public sector reform also applies to innovation. "The world is littered with examples of innovations that led either to few, if any, improvements, or which had unintended consequences (for example high-rise housing and out-of-town supermarkets)" (Hartley, 2005).

Because each governance style has its particular set of assumptions, innovation will take different forms (Hartley, 2005):

- Hierarchical governance assumes a legislative, bureaucratic and rule-based approach to public service provision. The population is assumed to be fairly homogeneous, and the definition of needs and problems is undertaken by professionals, who provide standardized services for the population.
- Market governance assumptions are based on neo-liberal economics and a particular form of management theory. Innovations focus particularly on organizational forms and processes such as executive agencies in central government, the purchaser/provider splits seen in health, education and local government, and a 'customer' focus.
- In a network governance setting, the role of the state is to steer action within complex social systems rather than control solely through hierarchy or market mechanisms. It is about supporting innovation through enabling legislation or providing resources for experiments and collaboration and orchestrating the interests of different stakeholders.

Fifty shades of governance 143

Feature	Hierarchical governance	Network governance	Market governance
36. Relation types	Dependent	Interdependent	Independent

The three governance styles represent different relation types as regards the dependency of actors, the type of societal interactions and the type of coordination mechanism (Kickert, 2003). Hierarchical governance puts public administration in a central role, and other actors are seen as *dependent*. Market governance is the opposite: societal actors are in principle *independent* and autonomous and government should deliver to them appropriate services. In network governance, actors are *interdependent*; this includes also public sector organizations. In the governance feature of relation types, hierarchy and market are two extremes, with the network mode more or less in between.

Feature	Hierarchical governance	Network governance	Market governance
37. Societal interactions	Interventions	Interplays	Interferences

Three different social interaction types match with the relation types mentioned in Feature 36: interventions (hierarchy), interplays (co-governance, network governance) and interferences (self-governance, market governance) (Kooiman, 2003).

Feature	Hierarchical governance	Network governance	Market governance
38. Roles of public managers	Clerks and martyrs	Explorers producing public value	Efficiency and market maximizers

Public manager is a term that only emerged with New Public Management: before that, bureaucrats were not defined by the fact that they manage, but by their responsibilities, which required a high level of expertise. NPM has in most countries resulted in the appointment of generalists as managers at all levels but in particular the higher levels of administration. Thirty years ago, when a director-general in one of the ministries gave a speech, this was considered an important event, also among experts in the field: the DG was more or less the most knowledgeable person in the area of his or her responsibility in the country. Nowadays, such senior officials are experts in management and communication but not per definition in the policy sector they manage.

Hierarchical public managers are classical bureaucrats: they are servants, responsible for legitimacy and for administrating ('clerks'), and are 'martyrs' because, for the 'public cause', they have to fulfil their job in a command and control culture which may only to a certain extent require their creativity and entrepreneurship (Hartley, 2005). In a network context, they are explorers, producing public value and stimulating new approaches. Under market governance, public managers focus on efficiency and maximization of their organization's 'market'.

For sustainability metagovernance, this poses a dilemma: the coordinating responsibility of sustainable development should be at the highest level, because of the inherent interlinkages between the different transition pathways and between the SDGs; however, the highest level may not be the most knowledgeable level. Being conscious of this dilemma is the first step to solve it, as there are many ways to strike a balance between the height of the level in the public sector and the level of expertise needed.

Feature	Hierarchical governance	Network governance	Market governance
39. Leadership styles	Command and control	Coaching and supporting	Delegating, enabling

Most public sector managers and politicians have a preferred style of leadership. Hersey and Blanchard (1988) differentiated four styles of leadership for different situations: directing (S1 type), coaching (S2), supporting (S3) and delegating (or enabling; S4) that should be applied situationally. These styles can be related to the three governance styles. A command and control leadership style relates well to a hierarchical governance style. A network style of governance relates to coaching and supporting styles of leadership. A market governance style typically relates to an empowering and delegating leadership style. This means, for example, that in a hierarchical organization or governance framework, more command and control leadership can be expected than in a network or market governance culture. The concept of situational leadership styles is now 30 years old but still as important as then. A recent research on leadership style of government ministers showed that they use (and mix) approaches, including 'the master' (hierarchy) and 'the consultor' (network governance) (Lees-Marshment and Smolović Jones, 2018).

Hersey and Blanchard's 'situational leadership' is a very early version of metagovernance on the dimension of leadership. Their approach suggests that leaders should be able to use, and switch between, all leadership styles, depending on the situation. Effective leadership is, like authority, something that depends on the acceptance by those who are being led. Sustainability depends a lot on leadership in all its forms, and its leadership is no exception to the general rule that it should be carefully adapted to the circumstances, and that it should be dynamic and flexible leadership.

Feature	Hierarchical governance	Network governance	Market governance
40. Degree of empowerment inside organizations	Low	Empowered lower officials	Empowered senior managers

Different leadership styles (Feature 39) result in different degrees of empowerment (Peters, 2005). A hierarchical organization can delegate tasks, but only to a certain degree and with sufficient control. Using network governance requires that officers involved in network processes are empowered to take decisions within a relatively broad range: they must possess a high level of discretion. In an organization with primarily market governance, also the senior managers must have the freedom to take far-reaching decisions that make the organization more efficient.

The SDGs assume that public sector organizations work across sectors and in partnerships and with stakeholders. This requires a quite high level of empowerment of civil servants, which is difficult to realize in hierarchical organizations – and we know that most government authorities are primarily the hierarchical type. Therefore, it will be a challenge for political and administrative leaders to find an appropriate balance between hierarchical control and accountability, and empowerment. Nothing is more frustrating for a public officer engaging in partnerships with business and civil society than having no discretionary space whatsoever.

Feature	Hierarchical governance	Network governance	Market governance
41. Values of civil servants	Law of jungle	Community	Self-determination

Public sector organizations form their employees to adapt to the internal governance style. Newcomers will often be selected because of their match with the existing culture. Key values of civil servants may vary from law of the jungle (hierarchy), community (network governance) and self-determination (market governance) (Laske, 2006). These values are also congruent with different stages of personal development as distinguished by Laske.

Another system of values (at the personal as well as the organizational level) that corresponds to governance styles is the one developed by Graves (1965). Hierarchical people are authoritarian, obedient, disciplined and use a pyramidal organization structure ('blue' level). People with a network attitude are communitarian and highly value equality, learning from others, openness and trust ('green' level). Market type leaders are driven by other values: being entrepreneurial and rational, striving for personal success, emphasizing money rather than loyalty, valuing competition and autonomy ('orange' level). Both Laske and Graves distinguish a level where people are generally willing and able to take multiple perspectives. This is a quality needed for metagovernance, as it allows taking perspectives beyond one's own personal preferences or values.

Feature	Hierarchical governance	Network governance	Market governance
42. Key competences of civil servants	Legal, financial, project management, information management	Network moderation, process management, communication	Economy, marketing, public relations

Competence management is a dimension of human resource management. Hierarchical governance tends to prefer staff with a legal or financial background and line managers or project managers who focus on risk management and clear lines of command and control. Network abilities have meanwhile been added to the list of competencies public managers have to master. This includes not only being informed about the findings on the functioning of policy networks in public administration literature, but also about general characteristics of complex networks, such as the meaning of 'strong' and 'weak' ties in networks (Granovetter, 1973). It is important to have 'weak ties' (acquaintances) besides 'strong ties' (friends) because the former are able to provide information from distant parts of the social systems. In addition, 'hubs' – people or places with many network links – are important structuring (hierarchical) elements in networks (Barabási, 2003). As market governance requires that people focus on efficiency, a business administration or marketing background is useful.

Feature	Hierarchical governance	Network governance	Market governance
43. Objectives of management development	Training is an alternative form of control over subordinates	Training helps 'muddling through'	Training helps making more efficient decisions

Management development is, like the term management in the public sector, a result of New Public Management ideas. Training existed already before that, and already in the first editions of Simon's seminal book *Administrative Behaviour* (1947) he claimed that from the viewpoint of hierarchical governance, training procedures are alternatives to the exercise of authority or advice as means of control over the subordinate's decisions (Simon, 2013). Market governance needs management development programmes that provide management tools to make efficient decisions. For network governance, training should help public managers to learn how to cope with an irrational and permanently changing multi-actor organizational environment.

The increase of empowerment in public sector organizations accompanying New Public Management has made public employees 'customers' of the training departments; the latter need to 'sell' training packages. This increased freedom of choice may have one disadvantage, namely that employees tend to select training that strengthens the competences they already possess and that matches with their personal character and with the main culture of the organization. Less popular competences will not be chosen. I have conducted several trainings on the use of the various governance styles in the daily work of colleagues at the European Commission, and was a bit frustrated that those who subscribed to the training were the ones who needed it least. Those who were the most comfortable with the dominating hierarchical culture of the organization generally did not want to shop from the shelves of multi-perspective thinking. The then–training coordinator was also somewhat nervous when she saw my training materials on metagovernance. She asked me to ensure that participants would not have fundamentally changed their mind about the importance of hierarchy after having followed the course. At the same time, the European Commission is one of the best examples of organizations applying metagovernance in practice, as illustrated in Section 4.6.

Feature	Hierarchical governance	Network governance	Market governance
44. Dealing with power	Coercion	Manipulation	Competition, lobbying

Power can be defined as "the ability of groups and individuals to make others act in the interest of those groups and individuals and to bring about specific outcomes" (World Bank, 2017). Different forms of executing power are related to the different governance styles. Typical mechanisms are coercion (hierarchical governance), manipulation (network governance) and lobbying (market governance). For more on power and metagovernance, see Section 4.10.

Feature	Hierarchical governance	Network governance	Market governance
45. Conflict resolution types	Classical negotiation, power-based (win-lose)	Mutual gains approach to negotiation (win-win); diplomacy	Classical negotiation, competition-based (win-lose)

Conflicts can be solved in many ways, and it is no surprise that the three basic governance styles have different preferences as regards conflict resolution. The two rational styles will try preventing the fuzziness of personal relations, empathy and trust that is linked to diplomacy, and in particular to the Mutual Gains Approach, which is also called the 'win-win' approach. A hierarchical approach could be to first try to win the conflict by applying coercion of rules, or other power mechanisms. Alternatively the approach could be to at least not lose the 'battle' by blocking unfavourable results. Market governance would try the same, but use competitive advantage – if available – and financial arguments.

Most theories of conflicts and conflict resolution focus on struggle and power distribution, meeting human needs, or even optimization exercise (Mostert, 1998). Such literature advocates specific methods and procedures for conflict resolution without setting them in a proper framework and explicating their limitations in specific contexts. Indeed, not even conflict resolution escapes from the need to take into account the context, including the cultural, social and economic dimensions. Conflict is not just about who is right and who is wrong.

When neighbouring countries are in conflict with each other or have at least no diplomatic relations, is this then a governance failure? This is difficult to say: there are many reasons for war and peace. Serious governance failure could cause an armed conflict – but probably a serious policy failure is a more frequent cause of conflict. Geography and natural resources are examples of other causes. Transboundary relations are more often the object of governance, through legal agreements (hierarchical tool), trade agreements or contracts (partially hierarchy, but mainly market governance), or informal agreements (network governance – in about all policy fields). And where there is governance, there is governance failure.

The field of environmental protection is a good example, where many international agreements exist with legal requirements on transboundary relations. The environment does not recognize political boundaries. Hence, there is a need for shared management of natural resources between countries, including countries that do not have strong political alliances or even worse: diplomatic tensions can arise between neighbouring countries that rely on shared ecosystems. The case of conflict associated to environmental issues in transboundary contexts is a good example of how situations of conflict in the absence of diplomatic relations can be successfully dealt with through governance interventions. Still, conflicts can endure between two neighbouring countries, leading to, for example, non-implementation of the requirements of the UN Convention on transboundary environmental impact assessment (Espoo Convention) (Kolar-Planinsic et al., 2013).

Disasters, conflict and other crisis events have short- and long-term impacts. Opinions differ on whether crises cause, or are simply a catalyst, for these impacts. But it is clear that impact assessments that do not consider possible disasters and crises may produce inaccurate results. The consequences of inaccurate assessments range from inconvenience in project implementation to, in the extreme, mortality rising above pre-disaster levels. While lack of diplomatic relationships between states weakens international efforts to tackle environmental challenges, examples of some neighbouring countries demonstrate that environmental challenges may trigger collaboration and develop peace-building processes: the environment provides opportunities for initiating dialogue and with that may be the catalyst for cooperation also in policy fields without diplomatic relationships.

All over the world, examples can be found of more or less fruitful transboundary collaboration under politically tense circumstances. In Europe, such examples exist in the Danube River Basin, in Africa in the Nile Basin area and the Benguela Current Large Marine Ecosystem with Namibia and Angola, in Asia situations are known regarding the Yellow Sea and the South China Sea, and the Humboldt Current (Kolar-Planinsic et al., 2013). From these examples three lessons can be learned. Firstly, that environmental peace-making processes may start from initiatives of NGOs, scientists, business sector, media, religious and other societal groups, also when governments are in conflict with each other. Secondly, environmental conflict resolution may not necessarily require that countries formalize diplomatic relationships. Thirdly, the political will of decision makers to establish mutually beneficial cooperation over environmental issues is key to enhancing the cooperation.

What does this mean in terms of governance styles? When two or more neighbouring states have no direct diplomatic, trade and/or cultural relations, we could call this a *hierarchical lock-in* situation. Making use of the other governance styles can create ways around the lock-in. Such initiatives do not have to be 'metagoverned' by government bodies but this could be done through partnerships between or among civil society and private organizations. It would be interesting to know what would happen with the informal governance framework of non-state actors when a hierarchical lock-in between states is solved – especially if the state 'governance religion' is hierarchy. The example of the falling of the Berlin Wall and the reunification of Germany is not a good example, because the informal exchanges between East and West Germany had been kept at a very minimum.

Figure 5.1 distinguishes four different situations:

1 Diplomatic relationships and no conflicts ('Peaceful');
2 Diplomatic relationships and conflicts ('Negotiable');
3 No diplomatic relationships and conflicts ('Frozen');
4 No diplomatic relationships and no conflicts ('Collaborative').

The challenge is how to move from situation 3 to situation 4, assuming that it may very well be possible to establish environmental collaboration and communication without (re)establishing diplomatic relationships between countries.

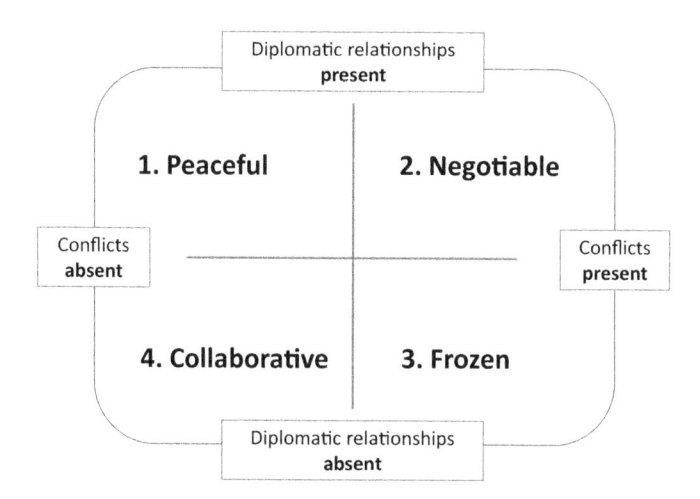

Figure 5.1 Breaking the ice? How to move towards collaboration
(After Kolar-Planinsic et al., 2013)

5.5 Problems, solutions and their linkages

The last five features (Table 5.8) are about how problem types and governance style relate, how problems are framed, and what the typical instruments and tools

150 *Features of metagovernance*

Table 5.8 Governance style differences (4): problems, solutions and their linkages

Features	Hierarchical governance	Network governance	Market governance
46. Suitability for problem types	Crises, disasters, problems that can be solved by executing force	Complex, unstructured, multi-actor issues	Routine issues, non-sensitive issues
47. Framing of problems	A relevant problem is framed as disorder	A relevant problem is framed as lack of consent	A relevant problem is framed as inefficiency
48. Typical governance failures	Ineffectiveness; red tape	Never-ending talks, no decision, undemocratic	Economic inefficiency, market failures, efficiency killing effectiveness
49. Role of public procurement	To establish stable relations with suppliers; use as leverage for government policy	To stimulate innovation partnerships	To stimulate competition among suppliers; create new markets
50. Typical output and outcome	Laws, regulations, control, procedures, accountancy reports, decisions, compliance	Expert networks, consensus, voluntary agreements, covenants	Services, products, contracts, outsourcing, voluntary agreements

and expected results of each governance style are. First, I would like to make a short excursion to the relations between problems and solutions.

Decisions are not made in linear or cyclical process but by "problems, solutions and participants moving from one choice opportunity to another" Cohen et al. (1972). Organizations can therefore be conceptualized as "crossroads of time-dependent flows of four distinct classes of objects: 'participants', 'opportunities', 'solutions' and 'problems'. Collisions among the different objects generate events called 'decisions'" (Fioretti and Lomi, 2008). This 'Garbage Can Theory' (GCT) was an innovation as it acknowledged the role of coincidence and introduced the decisive role of 'participants' in decision making. We would call them meanwhile 'stakeholders' or, more neutrally, 'actors'.

The GCT was an attempt to integrate uncertainty and irrationality in the modelling of decision-making processes in organizations, but its impact reached far beyond modelling. Where the GCT pays tribute to the role of coincidence in decision making, governance is basically intentional: it is about making choices about institutions, instruments, processes and roles of actors in the implementation of political goals. When the GCT was formulated in 1972, most national governments were still solid hierarchies, with differences, of course, across countries; this implies that 'authoritative' knowledge is valued most. Market governance and network governance were still absent or at least not dominating public governance. From the 1980s onwards, market governance and network governance started to challenge hierarchical governance.

Public decision makers found the GCT way of framing the process of preparing decisions attractive, because it was closer to their daily reality than the linear and even the circular models. However, the model became less popular under New Public Management, which is a rational style. Meanwhile, there is again more (indirect) support for the concept because there is more knowledge about the non-linearity of decision making. The main lesson from the Garbage Can Theory important for contemporary governance is that decision making is to a large extent based on

> a temporal understanding of events, in contrast with an intentional or consequential one. The framing of decisions may be to a considerable extent determined by temporally unfolding processes of participation and able extent determined by temporally unfolding processes of participation and attention.
>
> (Cohen et al., 2012)

This implies for (sustainability) metagovernance: decision making only based on hierarchical and/or market governance – which both assume that decision making is (or should be) mostly rational – may lead to missing important opportunities, that emerge and may disappear again in an unpredictable way.

Feature	Hierarchical governance	Network governance	Market governance
46. Suitability for problem types	Crises, disasters, problems that can be solved by executing force	Complex, unstructured, multi-actor issues	Routine issues, non-sensitive issues

Each of the governance styles is more usable for dealing with certain types of problems than with the other types. Hierarchical governance, with its carefully defined division of tasks, is successful in dealing with problems that can also be divided into clear sub-problems. Because hierarchy "can only deal comfortably with problems whose value dimension is not openly contested, and because they are used to imposing their values on others, they are ill disposed to treat moderately structured problems/means" (Hoppe, 2002, p. 313). In other words: don't ask a hierarchical manager to solve a 'wicked problem'. Government agencies who are responsible for safety, for instance, usually work with detailed emergency plans and a clear command structure.

Complex, unstructured problems are better dealt with through some form of network governance. Dealing with uncertainty requires trust, empathy and dialogue between partners. Routine issues are the best example of problems that can be tackled successfully through market governance. Efficiency is then a prominent objective.

152 Features of metagovernance

A metagovernance approach would be applying a governance style that matches the problem type, or reframe the problem to make it manageable under a certain given governance style. Implementation of the SDGs presents challenges of all types. Combating climate change may require legal emission standards (hierarchy), adopting flood protection measures (hierarchical planning style and market-style efficient implementation), and behavioural change (network and market governance elements).

Feature	Hierarchical governance	Network governance	Market governance
47. Framing of problems	A relevant problem is framed as disorder	A relevant problem is framed as lack of consent	A relevant problem is framed as inefficiency

Framing problems in a way that they can be tackled with the feasible governance approach is one of the four metagovernance strategies (see Section 4.8). As regards the three styles, the stereotypical framing needs are as follows: Hierarchical governance needs 'disorder' before something can be called a problem. Problems will be redefined until they look like a lack of rules, structure or order. Network governance has such a drive to create consensus that a problem is best defined in terms of dissent. Market governance needs to identify inefficiency before it can become a priority. In reality this is usually more subtle. But how problems are framed is extremely important, because this determines whether they can be solved within the governance environment as given, or the limits to where and how it may be stretched.

Feature	Hierarchical governance	Network governance	Market governance
48. Typical governance failures	Ineffectiveness; red tape	Never-ending talks, no decision, undemocratic	Economic inefficiency, market failures, efficiency killing effectiveness

Governance failures are extensively treated in Chapter 3, but at this place I would like to mention the failures that are the most characteristic for each of the governance styles (Jessop, 2003): ineffectiveness because of bureaucratic behaviour and 'red tape' (hierarchy); endless discussions without clear conclusions, accompanied by undemocratic procedures (network); and economic inefficiency, market-type failures and ineffectiveness due to over-emphasizing efficiency measures (market).

Feature	Hierarchical governance	Network governance	Market governance
49. Role of public procurement	To establish stable relations with suppliers; use as leverage for government policy	To stimulate innovation partnerships	To stimulate competition among suppliers; create new markets

Public sector organizations can make an important contribution to sustainable consumption and production, using their purchasing power to choose goods and services with lower impacts on the environment. In the EU, public authorities are major consumers, spending approximately 2 trillion Euros annually, equivalent to approximately 19% of the EU's gross domestic product (European Commission, 2012). Across the OECD, this percentage varies between 8% and 25% of the gross domestic product (GDP) (OECD, 2009). At the UN level, the Ten-Year Framework Programme on Sustainable Public Procurement (SPP) is a global platform that supports the implementation of SPP around the world. It brings together a variety of stakeholders, building synergies and leveraging resources to achieve its objectives. In 2015, Sustainable Public Procurement Principles were published.

A review of SPP in over 280 public sector organizations from across the public sector globally concluded that leadership is a significant factor in SPP being implemented by public sector organizations (Brammer and Walker, 2011). Financial concerns still remain the biggest barrier to SPP, with public sector procurers resistant to paying more to buy sustainably. The review observed that SPP concentrates on the environmental aspects in Europe, while in other countries local social topics, such as the need to empower minority groups within society, are an important part of the SPP.

Sustainable or green public procurement is a so-called market-based instrument such as taxation. In this sense, it is a market governance instrument. Globally, it seems that it is only lightly regulated. In the European Union, typical political discussions range between countries that would prefer a more regulated approach (hierarchy) and those who prefer the voluntary approach, with the Netherlands as a consensus culture with sympathy for market governance preferring 'green deals': voluntary agreements between government, private sector, sometimes with civil society involvement. The regulatory dimension is usually about setting standards for product groups, in order to determine which products are sustainable and which not. Such standards can also be negotiated with private sector organizations in the lead (see: private metagovernance, Section 4.4).

Promoting SPP may have different priorities depending on the nationally dominating governance style. Under hierarchical governance, SPP can be used as leverage for government policy but only under the condition that stable relations with suppliers are maintained or developed; this can lead to tensions in a product area where innovation is fast and competition among suppliers is high. Under network governance, establishing partnerships based on mutual interests may be more important than full implementation of an SPP policy. Market governance typically sympathizes with SPP because it is a tool that uses market mechanisms.

Feature	Hierarchical governance	Network governance	Market governance
50. Typical output and outcome	Laws, regulations, control, procedures, accountancy reports, decisions, compliance	Expert networks, consensus, voluntary agreements, covenants	Services, products, contracts, outsourcing, voluntary agreements

There is an overlap between instruments (Feature 19) and output: a law can be a policy instrument and it can also be the result of a decision-making process. In addition, it is useful to distinguish between output, results and outcome.

Typical products of a hierarchy are laws, regulations, control mechanisms, procedures, accountancy reports, decisions and compliance. These products are 'output' that can be measured, but are not necessarily strongly related to the real 'outcome'. Network governance does not aim at producing 'products' but at creating change in a consensual way and is generally more outcome-oriented. Market governance typically produces services and products. Sometimes the result of a market governance approach is the outsourcing of some of these services and products. Other results of market governance include 'market-based instruments' such as subsidies, taxes and charges, and public procurement policies and standards. Voluntary agreements of governments and market parties are a mixture of network and market governance.

5.6 Conclusions

This chapter showed that

- Hierarchical, network and market governance represent different operational characteristics on around 50 features of governance. These characteristics represent different ways of dealing with policy making, law-making, policy and law implementation, organizational and relational challenges.
- Within each governance style, the set of operational characteristics shows a great deal of internal consistency: they represent three different logics.
- Taking all governance features together means that there are in principle three well-filled governance toolboxes available for (meta)governance. Together they constitute the 'metagovernance toolbox', because it enables metagovernance to use, reject, combine and replace or switch parts of specific governance frameworks, taking into account conflicts and indicators of governance failure.

In the next chapters, context is discussed which influences the feasibility of the operational forms of the 50 features of governance in practice, because of culture and traditions (Chapter 6), or because of the dominant 'mantras' from New Public Management (Chapter 7).

Notes

1 *Yes, Minister* episode 'The Greasy Pole', see on Wikipedia https://en.wikipedia.org/wiki/The_Greasy_Pole
2 Website EPSC: http://ec.europa.eu/epsc/about_en

References

Adler, P. S. 2001. Market, hierarchy, and trust: The knowledge economy and the future of capitalism. *Organization Science*, 12, 215–234.

Arentsen, M. J. 2001. Negotiated environmental governance in The Netherlands: Logic and illustration. *Policy Studies Journal*, 29, 499–513.
Assens, C. & Baroncelli, A. 2004. Marché – Réseau – Hiérarchie: A la recherche de l'organisation idéale. *La Revue des Sciences de Gestion : Direction et Gestion*, 39, 43–55.
Atkinson, R. & Klausen, J. E. 2011. Understanding sustainability policy: Governance, knowledge and the search for integration. *Journal of Environmental Policy & Planning*, 13, 231–251.
Barabási, A.-L. 2003. *Linked: The New Science of Networks*. New York: Plume.
Benington, J. 2011. From private choice to public value. In: Benington, J. & Moore, M. (eds.) *Public Value – Theory & Practice*. Basingstoke: Palgrave Macmillan.
Benington, J. & Hartley, J. 2001. Pilots, paradigms and paradoxes: Changes in public sector governance and management in the UK. Barcelona, 25–37.
Bevir, M. 2011. Governance as theory, practice, and dilemma. In: Bevir, M. (ed.) *The SAGE Handbook of Governance*. London: Sage.
Bevir, M. & Rhodes, R. A. W. 2001. A decentred theory of governance: Rational choice, institutionalism, and interpretation. Available: https://escholarship.org/uc/item/0bw2p1gp. UC Berkeley: Institute of Governmental Studies.
Brammer, S. & Walker, H. 2011. Sustainable procurement in the public sector: An international comparative study. *International Journal of Operations & Production Management*, 31, 452–476.
Brandtner, C., Höllerer, M., A., Meyer, R., E. Meyer, et al. 2016. Enacting governance through strategy: A comparative study of governance configurations in Sydney and Vienna. *Urban Studies*, 54, 1075–1091.
Cohen, M. D., March, J. G. & Olsen, J. P. 1972. A garbage can model of organizational choice. *Administrative Science Quarterly*, 17, 1–25.
Cohen, M. D., March, J. G. & Olsen, J. P. 2012. 'A Garbage Can Model' at forty: A solution that still attracts problems. In: Lomi, J. & Harrison, R. (eds.). *The Garbage Can Model of Organizational Choice: Looking Forward at Forty*. Boston: Emerald Group Publishing Limited.
Considine, M. & Lewis, J. 2003. Bureaucracy, network, or enterprise? Comparing models of governance in Australia, Britain, the Netherlands, and New Zealand. *Public Administration Review*, 63 (2), 132–140.
Davis, G. & Rhodes, R. A. W. 2000. From hierarchy to contracts and back again: Reforming the Australian public service. *Institutions on the Edge*, 74–98.
De Bruijn, H., Ten Heuvelhof, E. & In 't Veld, R. 2010. *Process Management: Why Project Management Fails in Complex Decision Making Processes*. Berlin, Heidelberg: Springer: Imprint Springer.
De Bruijn, H., Ten Heuvelhof, E. & Kuit, M. 1999. *Sport 7. De opkomst en ondergang van een Nederlandse sportzender*. Rotterdam: Aristos.
De Grauwe, A. 2007. Alternative models in reforming school supervision. *Reforming school supervision for quality improvement, Module 7* [Online].
De Wit, B. & Meyer, R. 1999. *Strategy Synthesis: Resolving Strategy Paradoxes to Create Competitive Advantage*. London: Thomson Learning.
De Zeeuw, A., In 't Veld, R. J., Van Soest, D., Meuleman, L. & Hoogewoning, P. 2008. *Social cost benefit analysis for environmental policy-making*. The Hague: RMNO.
Dimitrakopoulos, D. & Page, E. 2003. Paradoxes in EU administration. In: Hesse, J. J., Hood, C. & Peters, B. G. (eds.) *Paradoxes in Public Sector Reform: An International Comparison*. Berlin, Germany: Duncker and Humblot.
Dixon, J. & Dogan, R. 2002. Hierarchies, networks and markets: Responses to societal governance failure. *Administrative Theory & Praxis*, 24, 175–196.

Doberstein, C. 2013. Metagovernance of urban governance networks in Canada: In pursuit of legitimacy and accountability. *Canadian Public Administration*, 56, 584–609.
Dunleavy, P. 2014. *Democracy, Bureaucracy and Public Choice: Economic Approaches in Political Science*. London: Routledge.
European Commission. 2012. *Green public procurement: A collection of good practices*. Luxembourg: Publications Office of the European Union.
European Commission. 2017. *Better Regulation Guidelines*. European Commission Staff Working Document SWD (2017) 350. Brussels: European Commission.
Eurostat. 2017. *Sustainable Development in the European Union. Monitoring Report on Progress Towards the SDGs in an EU Context*. 2017 Edition. Brussels: Eurostat.
Ferretti, J., Podhora, A., Weingarten, E., et al. 2012. Moeglichkeiten und Hemmnisse zur Beruecksichtigung von Umweltbelangen im Folgenabschaetzungssystem der Europaeischen Kommission. *UVP-report*, 26.
Fioretti, G. & Lomi, A. 2008. An agent-based representation of the garbage can model of organizational choice. *Journal of Artificial Societies and Social Simulation*, 11(1), 1.
Fisher, R., Ury, W. L. & Patton, B. 2011. *Getting to Yes: Negotiating Agreement Without Giving In*. New York: Penguin.
Granovetter, M. S. 1973. The strength of weak ties. *American Journal of Sociology*, 78, 1360–1380.
Graves, C. W. 1965. Value systems and their relation to managerial controls and organizational viability. In: *(2001) From the Historical Collection of the Work of Dr Clare W. Graves*. Great Falls, VA: Clare Graves Society, 1–12.
Hartley, J. 2005. Innovation in governance and public services: Past and present. *Public Money & Management*, 25, 27–34.
Héritier, A. & Eckert, S. 2008. New modes of governance in the shadow of hierarchy: Self-regulation by Industry in Europe. *Journal of Public Policy*, 28, 113–138.
Hersey, P. & Blanchard, K. H. 1988. *Management of Organizational Behavior: Utilizing Human Resources*. New York: Prentice Hall.
Höpner, M. 2008. *Coordination and Organization: The Two Dimensions of Nonliberal Capitalism*. Cologne: Max Planck Institute for the Study of Societies.
Hoppe, R. 1999. Policy analysis, science and politics: From 'speaking truth to power' to 'making sense together'. *Science and Public Policy*, 26, 201–210.
Hoppe, R. 2002. Cultures of public policy problems. *Journal of Comparative Policy Analysis: Research and Practice*, 4, 305–326.
Huntington, S. P. 1965. Political development and political decay. *World Politics*, 17, 386–430.
Hutchcroft, P. D. 2001. Centralization and decentralization in administration and politics: Assessing territorial dimensions of authority and power. *Governance*, 14, 23–53.
Ingraham, P. W. 1996. The reform agenda for national civil service systems: External stress and internal strains. In: *Civil Service Systems in Comparative Perspective*. Bloomington: Indiana University Press, 247–267.
In 't Veld, R. 2010. *Knowledge Democracy: Consequences for Science, Politics, and Media*. Heidelberg: Springer.
In 't Veld, R. J. 2000. *Willingly and Knowingly: The Roles of Knowledge About Nature and the Environment in Policy Processes*. Utrecht: Lemma.
In 't Veld, R. J. 2013. Sustainable development within knowledge democracies: An emerging governance problem. In: *Transgovernance*. Heidelberg: Springer.
In 't Veld, R. J., Töpfer, K., Meuleman, L., et al. 2011. *Transgovernance: The Quest for Governance of Sustainable Development*. Potsdam: Institute for Advanced Sustainability Studies (IASS).

Jessop, B. 2002. Governance and meta-governance in the face of complexity: On the roles of requisite variety, reflexive observation, and romantic irony in participatory governance. In: Heinelt, H., Getimis, P., Kafkalas, G., Smith, R. & Swyngedouw, E. (eds.) *Participatory Governance in Multi-Level Context*. Wiesbaden: VS Verlag für Sozialwissenschaften.

Jessop, B. 2003. Governance and metagovernance: On reflexivity, requisite variety, and requisite irony. In: Bang, H. (ed.). *Governance as Social and Political Communication*. Manchester: Manchester University Press.

Kang, V. & Groetelaers, D. A. 2017. Regional governance and public accountability in planning for new housing: A new approach in South Holland, the Netherlands. *Environment and Planning C: Politics and Space*, 2399654417733748.

Kaufman, F. -X. 1986. The relationship between guidance, control and evaluation. In: Kaufmann, F. -X., Majone, G., Ostrom, V. & Wirth, W. (eds.) *Guidance, Control and Evaluation in the Public Sector*. Berlin: Walter de Gruyter.

Kickert, W. J. M. 2003. Beneath consensual corporatism: Traditions of governance in the Netherlands. *Public Administration*, 81, 119–140.

Kolar-Planinsic, V., Partidario, M. & Meuleman, L. 2013. Background paper on good practice on communication, cooperation and conflict resolution. *Note prepared for the Working Group on Environmental Impact Assessment and Strategic Environmental Assessment of the UNECE Espoo Convention at its second meeting (ECE/MP.EIA/WG.2/2013/2)*, with input from Margarida Barata Monteiro and Yuliya Rashchupkina. Geneva: UNECE.

Kooiman, J. 2003. *Governing as Governance*. London: Sage.

Kroll, C. 2015. *Sustainable Development Goals: Are the Rich Countries Ready?* Gütersloh Germany: Bertelsmann Stiftung.

Laske, O. 2006. *Measuring Hidden Dimensions. The Art and Science of Fully Engaging Adults*. Medford, MA: Interdevelopment Institute Press.

Lees-Marshment, J. & Smolović Jones, O. 2018. Being more with less: Exploring the flexible political leadership identities of government ministers. *Leadership*, 1742715016687815.

Lodge, M. & Wegrich, K. 2005. Control over government: Institutional isomorphism and governance dynamics in German public administration. *Policy Studies Journal*, 33, 213–233.

Mason, M. 2010. Information disclosure and environmental rights: The Aarhus Convention. *Global Environmental Politics*, 10, 10–31.

Mayntz, R. 1985. *Soziologie der öffentlichen Verwaltung*. Heidelberg: Müller.

McCullough, A. 2017. Environmental Impact Assessments in developing countries: We need to talk about politics. *The Extractive Industries and Society*, 4, 448–452.

Meadowcroft, J. 2007. National sustainable development strategies: Features, challenges and reflexivity. *European Environment*, 17, 152–163.

Meuleman, L. 2008. *Public Management and the Metagoverance of Hierarchies, Networks, and Markets*. Heidelberg: Springer.

Meuleman, L. 2013. Cultural diversity and sustainability metagovernance. In: Meuleman, L. (ed.). *Transgovernance – Advancing Sustainability Governance*. Berlin/Heidelberg: Springer Verlag.

Meuleman, L. 2015. Owl meets beehive: How impact assessment and governance relate. *Impact Assessment and Project Appraisal*, 33, 4–15.

Meuleman, L. & Niestroy, I. 2015. Common but differentiated governance: A metagovernance approach to make the SDGs work. *Sustainability*, 12295–12321.

Meuleman, L. & Tromp, H. 2010. The governance of usable and welcome knowledge, two perspectives. In: In 't Veld, R. J. (ed.). *Knowledge Democracy*. Berlin/Heidelberg: Springer.

Mintzberg, H., Alhstrand, B. & Lampel, J. 1998. *Strategy Safari. A Guided Tour Through the Wilderness of Strategic Management.* New York: Simon & Schuster.
Mintzberg, H. & McHugh, A. 1985. Strategy formation in an adhocracy. *Administrative Science Quarterly,* 160–197.
Monteiro, M. B. & Partidário, M. R. 2017. Governance in strategic environmental assessment: Lessons from the Portuguese practice. *Environmental Impact Assessment Review,* 65, 125–138.
Monteiro, M. B., Partidario, M. R. & Meuleman, L. 2018. A comparative analysis on how different governance contexts may influence Strategic Environmental Assessment. *Environmental Impact Assessment Review,* 72, 79–87.
Moore, M. H. 1995. *Creating Public Value: Strategic Management in Government.* Cambridge, MA: Harvard University Press.
Morgan, G. 1986. *Images of Organization.* Thousand Oaks: Sage.
Mostert, E. 1998. A framework for conflict resolution. *Water International,* 23, 206–215.
Niestroy, I. 2005. *Sustaining Sustainability: A Benchmark Study on National Strategies.* Den Haag, Utrecht: RMNO, Lemma.
Niestroy, I. 2008. Sustainability impact assessment and regulatory impact assessment. In: OECD *Sustainable Development Studies Conducting Sustainability Assessments,* 41. Paris: OECD.
Niestroy, I. & Meuleman, L. 2016. Teaching silos to dance: A condition to implement the SDGs. *IISD SD Policy & Practice. Guest Article, posted on 21 July 2016* [Online]. Available: http://sd.iisd.org/guest-articles/teaching-silos-to-dance-a-condition-to-implement-the-sdgs/.
OECD. 2009. *OECD in Figures.* Paris: OECD.
Partidário, M., Den Broeder, L., Croal, P., et al. 2012. *Impact assessment.* [Online]. Available: http://www.iaia.org/uploads/pdf/Fastips_1%20Impact%20Assessment.pdf. Fargo: International Association for Impact Assessment.o.
Peter, L. J. & Hull, R. 1969. *The Peter Principle.* New York: William Morrow & Company.
Peters, B. G. 1998. Managing horizontal government: The politics of co-ordination. *Public Administration,* 76, 295–311.
Peters, B. G. 2005. The search for coordination and coherence in public policy. Return to the Center? Unpublished paper. Department of Political Science, University of Pittsburgh.
Pollitt, C. & Bouckaert, G. 2017. *Public Management Reform: A Comparative Analysis-Into the Age of Austerity.* Oxford: Oxford University Press.
Powell, W. 1990. Neither market nor hierarchy: Network forms of organization. *Research in Organizational Behavior,* 12, 295–336.
Rayner, S. 2012. Uncomfortable knowledge: The social construction of ignorance in science and environmental policy discourses. *Economy and Society,* 41(1), 107–125.
Rijnja, G. & Meuleman, L. 2004. Maken we beleid begrijpelijk of maken we begrijpelijk beleid? In: *Communicatie in het hart van beleid.* The Hague: Rijksvoorlichtingsdienst.
Scharpf, F. W. 1997. *Games Real Actors Play: Actor-Centered Institutionalism in Policy Research.* London: Routledge.
Schedler, K. 2006. Networked policing: Towards a public marketing approach to urban safety. *German Policy Studies/Politikfeldanalyse,* 3, 112–136.
Simon, H. A. 2013. *Administrative Behavior.* New York: Simon & Schuster.
Sørensen, E. 2006. Metagovernance: The changing role of politicians in processes of democratic governance. *The American Review of Public Administration,* 36, 98–114.
Stacey, R. 1992. *Managing Chaos: Dynamic Business Strategies in an Unpredictable World.* London: Kogan Page.

Steurer, R. 2011. Soft instruments, few networks: How 'New Governance' materializes in public policies on corporate social responsibility across Europe. *Environmental Policy and Governance*, 21, 270–290.

Stevens, V. & Verhoest, K. 2015. Theorizing on the metagovernance of collaborative policy innovations. Paper presented at the 2nd ICPP Conference, 1–4 July, 2015, Milan, Italy.

Streeck, W. & Schmitter, P. C. 1985. Community, market, the prospective contribution of interest governance to social order. *European Sociological Review*, 1, 119–138.

Susskind, L. & Field, P. 1996. *Dealing with an Angry Public: The Mutual Gains Approach to Resolving Disputes*. New York: Free Press.

Thijs, N., Hammerschmid, G. & Palaric, E. 2018. *A Comparative Overview of Public Administration Characteristics and Performance in EU28*. Brussels: European Commission.

Thompson, G. 1991. *Markets, Hierarchies and Networks: The Coordination of Social Life*. London: Sage.

Thompson, M. 1999. *Cultural Theory as Political Science (European Political Science)*. London: Routledge.

Thompson, G. 2003a. *Between Hierarchies and Markets: The Logic and Limits of Network Forms of Organization*. Oxford: Oxford University Press.

Thompson, M. 2003b. Cultural theory, climate change and clumsiness. *Economic & Political Weekly*, 38, 5107–5112.

Thompson, M., Ellis, R. & Wildavsky, A. 1990. *Cultural Theory*. Boulder, CO: Westview Press.

United Nations. 2015. Transforming our world: The 2030 agenda for sustainable development. New York: United Nations.

United Nations Commission on Sustainable Development (UNCSD). 2002. *Guidance in Preparing a National Sustainable Development Strategy: Managing Sustainable Development in the New Millennium New York*. New York: United Nations.

United Nations Economic Commission for Europe. 1998. *Convention on Access to Information, Public Participation in Decision-Making and Access to Justice in Environmental Matters (Aarhus Convention)*. Geneva: UNECE.

Van Twist, M. & Termeer, C. 1991. Introduction to configuration approach: A process theory for societal steering. *In*: 't Veld, R. J., Schaap, L., Termeer, C. J. A. M. & Van Twist, M. J. W. (eds.). *Autopoiesis and Configuration Theory: New Approaches to Societal Steering*. Dordrecht: Springer.

Vanclay, F. 2003. International principles for social impact assessment. *Impact Assessment and Project Appraisal*, 21, 5–12.

Verbong, G., Geels, F. W. & Raven, R. 2008. Multi-niche analysis of dynamics and policies in Dutch renewable energy innovation journeys (1970–2006): hype-cycles, closed networks and technology-focused learning. *Technology Analysis & Strategic Management*, 20, 555–573.

World Bank. 2017. *World Development Report 2017: Governance and the Law*. Washington, DC: World Bank.

6 How values, traditions and geography shape the feasibility of governance approaches

> Every society needs to be bound together by common values, so that its members know what to expect of each other, and have some shared principles by which to manage their differences without resorting to violence. . . . So, at the international level, we need mechanisms of cooperation strong enough to insist on universal values, but flexible enough to help people realize those values in ways that they can actually apply in their specific circumstances.
> – Former UN Secretary-General Kofi Annan on 12 December 2003 in Tubingen, Germany

Chapter 2 formulated three problems resulting from the interactions between different governance styles. Firstly, that governance styles have typical, built-in failures, weaknesses, or even perversions; secondly, that they sometimes undermine each other. Thirdly, that the internal logic of each governance style is so attractive that some public managers and politicians may adopt one of the styles as a truth and solution to every problem that should be accepted without proof. This chapter explores the third pitfall by linking it to the cultural dimension of governance. It analyses how external factors such as values, traditions and geography influence which governance approach is feasible in a certain situation. These factors are also internal; values and traditions are part of everybody's views, emotions and behaviour. Even geography influences our behaviour.

In political science it has been disputed whether governance has a cultural dimension; it was even a taboo among scholars and students. I saw conference presentations of cross-country comparative political science research that completely ignored the cultural factor. This was rooted, inter alia, in the tendency to consider one's own culture as 'best' (and as norm) and to promote the own approaches as panacea. Meanwhile there is a broad agreement among scholars and practitioners that the specific social, cultural and geographical context is a key factor for the feasibility of a specific governance approach: there is not one size fits-all. This is also recognized in the Agenda 2030. But it not an easy task to find a balance between universality (e.g. human rights and the Sustainable Development Goals) and the need to take into account different contexts, as I have argued more elaborately earlier (Meuleman, 2010, 2013).

Another dispute is about whether culture should be seen as driver of human behaviour – as expressed in governance styles – or as phenomena or objects (e.g. art) which should be protected and celebrated as objective of sustainable development – or even as the fourth dimension of sustainability. A recent article takes the latter approach and argues that preservation of cultures is missing as a key objective of the SDGs (Throsby, 2017). I consider these approaches as being complementary: values and traditions are essential as basis of effective governance, but can also be worthwhile to preserve and defend – as long as they do not conflict with generally accepted values such as rule of law.

6.1 Values and traditions: the cultural dimension of SDG governance

Values and traditions, together making out what we call cultures, have a huge impact on the effectiveness of governance within, between and across countries. Cultures determine what kind of behaviour is seen as appropriate, how people communicate effectively, who are considered as 'peers' and also what kind of solutions for certain problems are seen as 'right'. Hall (1977) defined culture as "the values, attitudes, beliefs, orientations and underlying assumptions prevalent among people in a society". Cultures include identity, language, history and, generally, 'ways of doing things'. Dealing with cultural differences is indispensable for human relations and also for governance, because governance is a normative and relational concept:

> Culture economizes action by relieving people of the impossible task of interpreting afresh every situation before acting. Culture also renders interactions between people, sometimes complete strangers, predictable through conventions, habits, rules, routines, and institutions.
> (Hoppe, 2002, p. 306)

Understanding cultures helps applying metagovernance in an appropriate way, namely by designing and applying governance frameworks that are contextualized. Existing preferences for hierarchical, network and market governance are based on different sets of values and 'produce' different behaviour. The three styles express different 'ways of life' as distinguished in cultural theory (Thompson et al., 1990), with hierarchy related to 'hierarchism', network governance to 'egalitarism' and market governance to 'individualism' (see also feature 1 in Section 5.2). Moreover, the 'ways of life' identified in cultural theory can be linked to different strategies to define or solve problems (Hoppe, 2002): 'Hierarchists' will impose a clear structure on any problem, no matter what the cost. 'Egalitarians' will define any policy problem as an issue of fairness and distributive justice. 'Individualists' will exploit any bit of usable knowledge to improve a problematic situation.

The cultural dimension of governance is relevant at personal level, organizational level and administration level, where a preference for one governance style, or for a specific mixture, exists. Social pressure at all levels ensures that behaviour is more or less adjusted to what the dominant culture expects.

A good example of the fact that governance recipes cannot 'travel' without adjustment to a national culture is that China and Vietnam consistently score low on the Good Governance indicators of the World Bank. This does not mean that the governance in these countries is per definition 'bad' – it might just be the result of implementing governance models and prescriptions without taking into account the context, according to Painter (2014).

Implementing the 2030 Agenda requires addressing informal institutions such as culture, norms and values (United Nations, 2017a, p. 18). In addition, the SDGs should be implemented in an inclusive and participatory way. This makes recognition of cultural diversity an imperative. At the same time the SDGs, including the principle of rule of law in SDG 16, are universal. They assume that all countries and people should be working (jointly) on the common goals – but of course people are different and will want different things and achieve them in different ways. They have different starting points and different pathways to achieving the SDGs.

Unity and diversity are difficult to combine (Kao, 2011), but it is a necessity to find ways to reconcile them. Not for nothing, "unity in diversity" is the motto of both the European Union and the USA. As a more concrete principle for the implementation of the SDGs, this challenge was formulated as creating 'common but differentiated governance' (Meuleman and Niestroy, 2015) (see also Section 11.2).

In various areas the cultural dimension is beginning to be taken seriously as precondition of successful governance. An example is the paradigm shift' in water management from a technical, rational approach focusing on dikes, dams and reservoirs, to integrating belief systems, human attitudes and collective behaviour into integrated water resources management (Pahl-Wostl et al., 2008).

The values and traditions behind governance styles mean that each country has a preference for one of the styles. I called such national preferences the 'default style' in my comparative research of similar policy projects in different countries: the first style to be tried – and to be rejected when it is not feasible. Of course such national characteristics are generalized and may sound like stereotypes. Still, they contain useful hints, as Section 6.3 will show.

6.2 Relational values

In Chapter 2 I defined governance as an intentional, normative and relational concept. As a relational concept, hierarchical governance needs dependent subjects, network governance requires interdependency between partners, and market governance needs independent relationships (Kickert, 2003, p. 27). Hence, it is plausible to assume that different governance styles also express in different ways how people consider other people's values. In 't Veld (2011) distinguishes five relational values, which I have linked to different relation types:

Values, traditions, geography and governance 163

- Hegemony: "My values are superior to those of other people".
- Separatism: "I don't want to be confronted with the implications of other people's values".
- Pluralism: "Other people's values may be valuable, and I am co-responsible for protecting them".
- Tolerance: "I find my values superior to other people's values, but I abstain from interventions because of sympathy".
- Indifference: "I find my values superior to other people's values, but I abstain from interventions because I am not interested".

Hegemony and separatism are related to the top-down and authoritarian thinking of hierarchical governance. If hierarchical governance is chosen as the main style, its congruency with hegemony and separatism should be taken into account: it can destroy trust and innovation power. When the complexity of a specific sustainability challenge leads to choosing network governance, pluralism or at least tolerance are relational values to be expected. When a market-based approach is chosen, the indifference towards values and traditions related to market governance can become a bottleneck for implementation.

We have already seen that the 'wickedness' of many sustainability problems necessitates a strong network governance touch in the sustainability governance mixture. This suggests that for sustainable development at least tolerance, but even more so pluralism are relevant values and should be expected to support its governance. In any case, "the design of governance should always take into account the relevant relational values in the specific situation" (In´t Veld et al., 2011).

Tolerance is a multi-faceted concept. Kickert (2003) concluded that this well-known 'image of the Dutch' has indifference as its mirror-side: "simply not caring at all about some-body else". He also mentioned another aspect of tolerance, which is clearly related to network governance, namely the typical Dutch concept of '*gedogen*' (permissiveness); that is, "the state does have regulations and rules, but officially permits citizens to deviate from these rules".

6.3 Typologies of national cultures

Understanding what characterizes national traditions is essential in order to be able to design tailor-made (meta)governance arrangements adapted to what works well in each country:

> Depending on the governance traditions of the State and the national priorities with respect to the Sustainable Development Goals, it may be helpful to consider greater centralization or greater decentralization; to adopt policies that seek to adjust the public, private or civil society responsibilities; to enhance public accountability mechanisms; or to re-allocate resources in order to create an enabling environment that reinforces efforts to implement all of the Sustainable Development Goals in a country.
>
> (United Nations, 2018)

164 *Features of metagovernance*

Typologies of cultures will always be disputed and have resulted in heated scholarly and political debates. This is not surprising, because it is about what people believe in and what drives them, and this is very personal. This personal dimension also comes into play when metagovernance is discussed. I have frequently encountered emotional, even angry reactions when I presented the concept of metagovernance, because it takes all perspectives seriously and does not chose *a priori* what is the most practicable one for a certain governance challenge. This may be seen as an offense by those who believe that their preferred governance style is a *panacea*. The mere point that metagovernance is a concept 'above' governance styles, with a bird's-eye perspective, has rendered me the accusation "Do you think you are God?". Another problem is the suspicion that cultural typologies must lead to stereotyping and having prejudices about certain cultural groups.

Two often-used typologies of national cultures are developed by Geert Hofstede (sociologist) and Erin Meyer (business administration). Their approaches are already disputed because of the suggestion that national cultures do exist to some extent. Both models have their merits and risks. They are generalized proxies and not meant to predict individual behaviour or that of a concrete group of people. Nevertheless, they can provide useful information to support decision making during design and management of a governance framework, and to support the analysis of governance failure.

Hofstede's cultural dimensions

The sociologist Hofstede (2003) distinguished five dimensions of national cultures: power distance, the degree of individualism or collectivism, gender differences, uncertainty avoidance and long- versus short-term orientation. In Table 6.1

Table 6.1 Governance styles and Hofstede's cultural dimensions

Cultural dimensions	Hierarchical governance	Network governance	Market governance
Accepted power distance	High	Low	Medium
Individualism vs. collectivism	Both	Collectivist	Individualist
Uncertainty avoidance	High (many rules, low trust)	Low (high trust)	Rather low
Gender differences	High or low	Low ('feminine', consensus-oriented)	High ('masculine', performance, competition)
Long vs. short-term orientation	Could be both; often more long-term	Long (personal networks; empathy for next generations)	Short (achievement, freedom) or long (return on investments with long lead time)
Examples	Germany, France	Netherlands, Scandinavian countries	Australia, Canada, UK, USA, New Zealand

(adapted from Meuleman, 2010) the five Hofstede dimensions are linked with hierarchical, network and market governance. The Dutch (low accepted power distance: consensus oriented) are different from the Germans (relatively high power distance: hierarchical), and this is reflected in the stronger affinity of the Dutch with network governance. A culturally mixed country like Trinidad and Tobago has a mixed culture on the individualism-collectivism dimension, but at the same time a national culture of low power distance and low uncertainty avoidance (Tidwell Jr, 2001).

Cases of environmental policy making in the UK, the Netherlands and Germany show that the governance style closest to the national culture was the first to be tried as the dominant style. Only when this did not work, other styles were considered and a situational mixture emerged, which helped making the projects a success (Meuleman, 2008). The cultural dimension of governance is not only relevant on the side of the 'governors', but has also implications for assessing which approach works where and why (or why not). Table 6.2 gives examples from more countries in different global regions.

Hofstede's cultural dimensions were used in a recent comparative study on the quality of public administration in 28 EU member states (Thijs et al., 2018). The country analysis shows the relevance of taking into account the "often strongly varying national cultures to obtain a better understanding of the societal embeddedness of public administration and the dynamics of administrative reform".

A study exploring the relationship between international cultures and income inequality, using data from 75 countries, found that two of Hofstede's cultural dimensions, namely low level of individualism and low long-term orientation, are negatively correlated with the most commonly used measure of inequality,

Table 6.2 Four cultural dimensions: examples across countries

Country	Acceptance of power distance (PDI)	Individualist (high scores)/collectivist (IDV)	Masculine (high scores)/feminine (MAS)	Uncertainty avoidance (UAI)
Sweden/Denmark	31 / 18	71 / 74	5 / 16	29 / 23
Finland	33	63	26	59
Netherlands	38	80	14	53
UK/USA	35 / 40	89 / 91	66 / 62	35 / 46
Germany	35	67	66	65
Belgium/France	65 / 68	75 / 71	54 / 43	94 / 86
Colombia	67	13	64	80
Costa Rica	35	15	21	86
Rep. of Korea	60	18	39	85
Ghana	80	15	40	65
Mozambique	85	15	38	44
Vietnam	70	20	40	30
Indonesia	78	14	46	48
Bangladesh	80	20	55	60

(https://geert-hofstede.com/national-culture.html)

166 Features of metagovernance

the Gini coefficient of a country (Malinoski, 2012). According to the study cultures that exhibit a collectivist or a short-term orientation – or both – may suffer from higher levels of income inequality, due to the inefficiencies of a collectivist society, and the potentially low savings rates of short term–oriented countries. Although these findings are correlations rather than causal relations, they can be relevant for policy makers.

Meyer's eight scales

In *The Culture Map*, Erin Meyer (2015) distinguishes eight scales representing areas where national cultures differ distinctly. Some of these overlap with Hofstede's dimensions, but others are based on additional insights. The first scale is about communication. Everyone who has worked in intercultural meetings, working groups or project teams has experienced that we 'tick' differently. Miscommunication often occurs between people from a low-context culture (very direct, straight) and a high-context culture (indirect; reading between the lines is needed). These differences are even noticeable in international organizations where a joint professional culture is laid out as a 'new skin' covering personal cultures: the skin is thin and any crisis can bring back the original values and behaviour. Table 6.3 relates the eight scales tentatively to the three governance styles.

Table 6.3 Tentative relations: governance styles and Meyer's cultural scales

Meyer's culture map	Scales	Hierarchical governance	Network governance	Market governance
Communicating	High – Low context	High-context e.g. China, Japan, Russia,	Low-context e.g. Netherlands, Denmark	Low-context e.g. United Kingdom, USA
Evaluating (negative feedback)	Direct – Indirect	Direct	Indirect	Direct
Persuading: first on	Principles – Applications	Principles	Applications	Applications
Leading	Egalitarian – Hierarchical	Hierarchical	Egalitarian	(Individualist)
Deciding	Consensual – Top-down	Top-down	Consensual	(Top-down)
Trusting based on:	Tasks – Relationships	Tasks	Relationship	(Competition)
Disagreeing	Confrontational – Avoiding confrontation	Confrontational	Avoiding confrontation	Confrontational
Scheduling	Linear time – Flexible time	Linear time	Flexible time	Linear or flexible time

Relevance of culture typologies for sustainability metagovernance

Databases such as those by Hofstede and Meyer can help a better understanding of the specific features and challenges when it comes to implementing goals and policies across countries, with in a country or at sub-national level.

It is surprising that the insights presented and underpinned by Hofstede, Meyer and others are more widely recognized as important in the private sector than in the public sector. Some public officials might assume that they are part of an international professional working culture, which is superimposed upon their national culture, but this is only a reality for a small group. In any case, interdisciplinary research on the cultural dimension of sustainable development is needed, bringing together among others cultural anthropology, sociology, organization psychology with political science. As long as the current silo approach in social sciences prevails, political decisions regarding sustainability will remain ill-informed with knowledge about values, traditions and practices, and therefore also ill-informed about the possibility of implementation (Meuleman, 2013).

6.4 Cultural relativism or pluralism?

Cultural diversity is inherent to metagovernance, because combining the three basic governance styles automatically brings different values and traditions closely together. Cultural pluralism is therefore a principle of metagovernance: there is a need to deal with these differences.

However, accepting or even stimulating cultural diversity doesn't mean that anything goes. We have universal rights as common basis and "most important cross-cultural moral concept" (Kao, 2011). Humankind can be happy about the adoption of the human rights, even if not all countries adhere to them. While their universality has been subject to discussion, especially with respect to culturally sanctioned practice, some are recognized by international human rights law as "absolute in that they can neither be traded away nor overridden even in times of emergency" (Kao, 2011). It is an achievement that Agenda 2030 links to some of them in the SDGs, for example and importantly in SDG 5 on gender equality.

Cultural diversity or pluralism is not the same as cultural relativism. The latter is the idea that beliefs, values and practices should only be judged against their own criteria. Donnelly (2007), argues for the "relative universality" of internationally recognized human rights: "universal human rights, properly understood, leave considerable space for national, regional, cultural particularity and other forms of diversity and relativity", which has sparked some discussion among scholars. However, he also assessed that "protecting internationally recognized human rights is increasingly seen as a precondition of full political legitimacy". This will hopefully be fostered again by the implementation of the SDGs.

With respect to dealing with cultural differences within one country, the attitude often is the 'assimilation' of cultural and ethnical minorities (Verweel and De Ruijter, 2003). The predominantly used term 'integration' might be meant in this way, namely that the minority needs to adapt. Such a policy of assimilation has created social tensions between different cultural groups in many countries. An alternative approach is to not focus on achieving communality – commonly shared values – but on compatibility (De Ruijter, 1995). The compatibility principle recognizes that cultural differences may cause tensions and incompatibilities, but suggests that these differences should be regulated in one way or another, including to agree on a common basis (such as the values enshrined in the constitution of the respective country). Developing a positive attitude towards cultural diversity in sustainability governance and investing in compatibility of values and practices rather than in assimilation may lead to a rich variety of solutions to similar problems, instead of current governance practice in which centrally proposed solutions are accepted in some cultures and rejected in others (Meuleman, 2013).

6.5 Political cultures and flip-flop governance

Political cultures are "assumptions about the political world" which "focus attention on certain features of events, institutions and behaviour, define the realm of the possible, identify the problems deemed pertinent, and set the range of alternatives among which members of the population make decisions" (Elkins, 1979). These cultural assumptions can have the form of collective properties of groups such as nations or classes, without individuals of these groups realizing themselves that they have a joint culture: this "may be clear only in contrast to those of another culture" (Elkins, 1979). The concept 'political culture' points out that political behaviour has to be seen in a cultural context; political behaviour is directed by interpretations and preferences, not by bare facts and interests (Mamadouh, 1997).

Already in 1971, Jessop argued that political orientations and political behaviour are more influenced by political cultures than by institutions (Jessop, 1971). This is an important insight as regards the question which governance approaches are effective under certain circumstances. Political cultures, like national, organizational or individual cultures, influence what is acceptable. However, I would argue that political institutions are founded in a political culture – often from the past – and such institutional settings do have an impact on political behaviour, too. Therefore, history co-determines which state systems exist in countries and how this influences the feasibility of governance approaches.

A good example is the impact of having a majoritarian type of democracy. Majoritarian democracies – which make out most of the world's democracies, although there seems to be a trend towards political fragmentation and coalition governments – have inherent problems with long-term decision making. Because they tend to have only two large political parties who, after they win an election, may undo policies and institutional arrangements designed by the former ruling party and introduce different institutional arrangements. This 'flip-flop' effect may hamper the long-term orientation needed for the implementation of the SDGs.

The governance question is which institutional mechanisms could be introduced that are 'flip-flop resilient'. One good practice from Jamaica mentioned at a UN symposium on implementation of the SDGs in Small Island Developing

States (SIDS) was that the government responsible for the SDG implementation should involve the whole of parliament in planning and executing SDG actions (UN, 2017). This helps ensuring that the opposition also has ownership. Other mechanisms one could think of – and are good practice in some countries – include having external arrangements such as (stakeholder) sustainable development commissions or councils, creating a continuous stream of awareness-building and capacity development, and keeping the SDGs on the agenda.

6.6 The impact of geography on challenges and feasibility of governance

A country's social and physical geography may strongly influence its past, present and future political, economic, social and environmental challenges and constraints. Everybody understands how important geopolitics is – but not all realize that the 'geo' in this contraction is an abbreviation of 'geography'. Physical geography may block or enhance policy implementation; it is a fundamental part of the why (i.e. objectives) as well as of the what (i.e. political action) (Marshall, 2015). Mountain ranges between China and India have ensured non-intervention, and Europe's economy could kick-start thanks to its flat land and navigable rivers. Drawing lines on maps while disregarding topography and geographical cultures in an area is a good predictor of trouble (Marshall, 2015). A high proportion of the world's weakest economies is landlocked (Kaplan, 2013). Kickert (2003) as well as Hendriks (2017) have argued that the consensus culture of the Netherlands – commonly referred to as the 'polder model' – can be linked to the centuries of joint battle against the water to make this low-laying country survive.

Agenda 2030 recognizes the geographical dimension of sustainability. For example, recital 64 of the Agenda states that the UN supports the implementation of relevant strategies and programmes of action, including among others the SIDS Accelerated Modalities of Action (SAMOA) Pathway and the Vienna Programme of Action for Landlocked Developing Countries for the Decade 2014–2024. The 57 members of the SIDS group within the United Nations share special geographical conditions that influence institutional arrangements, e.g. small, fragmented, isolated and threatened by climate change. This has advantages and disadvantages:

- People know each other: There are short lines between decision makers – but also: it is difficult to enforce laws upon your friends/family.
- Because of the small scale, upscaling expertise is often impossible (e.g. waste recycling and incineration; specialist knowledge on chemicals, etc.).
- The existing heavy administrative burden due to the geography of fragmented island states is further aggravated by donor fragmentation.

Geographical imbalances play an important role for many countries when ensuring that no one is left behind. In the Voluntary National Reviews (VNRs) about SDG implementation, Indonesia referred to the challenge of this principle in the largest archipelagic country characterized by cultural, ethnic and religious diversity. Afghanistan, Bangladesh, Belgium, Botswana, India and Tajikistan highlighted the need to consider rural communities in this context. For some island states, geography poses similar challenges. Maldives noted that due to its population dispersion, additional efforts must be made to reach everyone (United Nations, 2017b).

Geography also plays an increasing role at the global level with respect to planetary boundaries and resource constraints:

> Ignoring physical constraints imposed by planetary limits is anti-poor because with fewer resources to go around, the lowest income people will lack the financial means to shield themselves from resource constraints, whether it is food price shocks, weather calamities, or energy and water shortages.
>
> (Wackernagel et al., 2017)

6.7 Conclusions

This chapter showed that

- Values and traditions, political cultures and geography help explain and in some ways predict which governance approach works well in a certain situation and why.
- Using this 'circumstantial evidence' is hugely important to prevent and remediate governance failure.
- For the implementation of the SDGs this means that for the implementation of the universal Goals and their targets, culturally and geographically sensitive governance frameworks are necessary. General recipes need to be adapted, translated or reframed in order to make them work.
- Cultural diversity across and within countries is an opportunity and not a burden for implementation of the SDGs, because (1) it connects the SDGs to people's beliefs/values, and (2) it creates a broader set of possible solutions. Therefore, striving for compatibility may be a better way to deal with cultural diversity than integration or even assimilation.

The next chapter introduces another set of soft rules that co-determine which governance emerges in concrete situations. This is about how certain standard governance solutions have settled in the minds of many people across the world. Most of them became popular during the heydays of the New Public Management movement, in the 1990s and 2000s.

References

De Ruijter, A. 1995. Cultural pluralism and citizenship. *Cult Dynamics*, 7(2), 215–231.
Donnelly, J. 2007. The relative universality of human rights. *Human Rights Quarterly*, 29, 281–306.

Elkins, D. J. 1979. A cause in search of its effect, or what does political culture explain? *Comparative Politics*, 11, 127–145.
Hall, E. T. 1977. *Beyond Culture*. New York: Anchor Books.
Hendriks, F. 2017. *Polder Politics: The Re-Invention of Consensus Democracy in the Netherlands*. London: Routledge.
Hofstede, G. 2003. *Culture's Consequences: Comparing Values, Behaviors, Institutions and Organizations Across Nations*. London: Sage.
Hoppe, R. 2002. Cultures of public policy problems. *Journal of Comparative Policy Analysis: Research and Practice*, 4, 305–326.
In 't Veld, R. J., Töpfer, K., Meuleman, L., et al. 2011. *Transgovernance: The Quest for Governance of Sustainable Development*. Potsdam: Institute for Advanced Sustainability Studies (IASS).
Jessop, R. D. 1971. Civility and traditionalism in English political culture. *British Journal of Political Science*, 1, 1–24.
Kao, G. Y. 2011. *Grounding Human Rights in a Pluralist World*. Washington, D.C.: Georgetown University Press.
Kaplan, R. 2013. *The Revenge of Geography. What the Map Tell Us About Future Conflicts and the Battle Against Faith*. New York: Random House Trade Paperbacks.
Kickert, W. J. M. 2003. Beneath consensual corporatism: Traditions of governance in the Netherlands. *Public Administration*, 81, 119–140.
Malinoski, M. 2012. On culture and income inequality: Regression analysis of Hofstede's international cultural dimensions and the Gini coefficient. *Xavier Journal of Politics*, III, 32–48.
Mamadouh, V. 1997. Political culture: A typology grounded on cultural theory. *GeoJournal*, 43, 17–25.
Marshall, T. 2015. *Prisoners of Geography: Ten Maps That Tell You Everything You Need to Know About Global Politics*. London: Elliot and Thompson.
Meuleman, L. 2008. *Public Management and the Metagoverance of Hierarchies, Networks, and Markets*. Heidelberg: Springer.
Meuleman, L. 2010. The cultural dimension of metagovernance: Why governance doctrines may fail. *Public Organization Review*, 10, 49–70.
Meuleman, L. 2013. Cultural diversity and sustainability metagovernance. In: Meuleman, L. (ed.). *Transgovernance - Advancing Sustainability Governance*. Berlin/Heidelberg: Springer Verlag
Meuleman, L. & Niestroy, I. 2015. Common but Differentiated governance: A metagovernance approach to make the SDGs work. *Sustainability*, 12295–12321.
Meyer, E. 2015. *The Culture Map. Decoding How People Think, Lead, and Get Things Done Across Cultures*. New York: Public Affairs.
Pahl-Wostl, C., Tàbara, D., Bouwen, R., et al. 2008. The importance of social learning and culture for sustainable water management. *Ecological Economics*, 64, 484–495.
Painter, M. 2014. Governance reforms in China and Vietnam: Marketisation, leapfrogging and retro-fitting. *Journal of Contemporary Asia*, 44, 204–220.
Thijs, N., Hammerschmid, G. & Palaric, E. 2018. *A Comparative Overview of Public Administration Characteristics and Performance in EU28*. Brussels: European Commission.
Thompson, M., Ellis, R. & Wildavsky, A. 1990. *Cultural Theory*. Boulder, CO, US: Westview Press.
Throsby, D. 2017. Culturally sustainable development: Theoretical concept or practical policy instrument? *International Journal of Cultural Policy*, 23, 133–147.

Tidwell Jr, C. H. 2001. Trinidad and Tobago: Customs and issues affecting international business. Reaching the World: International, Intercultural, and Ethical issues: A Conference for SDA Business Teachers, Andrews University, USA.

United Nations. 2017a. Report on the sixteenth session (24–28 April 2017) of the United Nations Committee of Experts on Public Administration. United Nations. 2017. New York: United Nations.

United Nations. 2017b. 2017 Synthesis of Voluntary National Reviews Synthesis of Voluntary National Reviews. New York: United Nations.

United Nations. 2018. Elaborating principles of effective governance for sustainable development. Committee of Experts on Public Administration. New York: United Nations.

Verweel, P. & De Ruijter, A. 2003. Managing cultural diversity. *J Today*, 2, 1–20.

Wackernagel, M., Hanscom, L. & Lin, D. 2017. Making the sustainable development goals consistent with sustainability. *Frontiers in Energy Research*, 5, 18–18.

7 Mind-sets and mental silos
Rise and fall of simple switches

> A conservative is someone who believes in reform. But not now.
> – Mort Sahl, Canadian comedian

One of the funniest parts of my workshops on governance is the moment that I present a video clip from the BBC series *Yes, Minister* of the 1980s. The series shows the tensions between an ambitious Minister of Administrative Affairs and his – very hierarchical – permanent secretary Sir Humphrey. The clip I always show is the one where Minister Jim Hacker visits a brand-new hospital with 500 staff, which received the 'Florence Nightingale Award' for being the most hygienic hospital in England.[1] It is also the most efficiently run hospital. The reason is that the hospital does not accept patients: they would disturb the smooth operation of the organization.

The episode of the satirical television series was recorded more than 30 years ago, and is still relevant. In the 1980s, it was an early critical comment on the enthusiasm with which the New Public Management (NPM) movement, that prioritized efficiency over effectiveness, was received in Margaret Thatcher's Britain. Meanwhile, after so many years, most people I have shown the video could easily name similar real-life examples. Apparently, NPM has not only delivered 'goods' but also 'bads'.

On the positive side of the balance, NPM has resulted in shaking up inert bureaucratic organizations, making them more efficient and service-oriented. However, this chapter illustrates governance failures triggered by the blind faith many politicians and policy makers have had in NPM as a solution to all problems. It starts with a short introduction on the relevance of holistic thinking, mindfulness and certain mind-sets. A rich catalogue of popular and simple solutions can be linked to the New Public Management movement. They have become part of our 'software': simple programmes or mental 'apps' which are running in the background all the time. They have in common that their usefulness is much more known than their failures or even perversions. They also have in common that they are often believed to be universally applicable.

With NPM, a managerialist perspective on governance (and metagovernance) has entered the public service, which has resulted in depolitization. I agree with Sørensen and Torfing (2017) on the importance of realizing that it is a choice which can be made consciously: to continue on this pathway or to find a new balance with the political (and political science) perspective on (meta)governance.

7.1 Mind-sets and mindfulness: the end of one-size-fits-all?

Complex policy challenges like the implementation of the SDGs require systemic thinking, a comprehensive approach (taking into account all relevant aspects) and, in addition, a holistic view. The latter means trying to keep in mind the importance of the whole and the interdependence of its parts, both horizontally and vertically (downstream as well as upstream). Horizontal coordination is essential because progress in one goal area could generate positive spill-over effects in others, which may not be recognized in the absence of such coordination (Meuleman and Niestroy, 2015). Without vertical coordination policy actions at different levels of administration may undermine each other.

Metagovernance capacity has some similarity with the concept of mindfulness, which was based on experiences in highly responsive organizations (HROs) in the USA (Weick and Sutcliffe, 2009). "Mindfulness" is used here with the connotation of combining ongoing scrutiny of existing expectations, bringing in new experiences, the willingness and capability to make sense of unprecedented events, dealing with context and improving foresight. It is also close to the capabilities and capacities (Termeer et al., 2015) that are considered to be essential for dealing with 'wicked problems'.

Changing the mind-sets of public officials may be needed in order to facilitate public sector reform to perform better on sustainability targets, for example. In this context, the European Commission's Quality of Public Administration Toolbox mentions metagovernance as an approach which helps policy makers to be open-minded about how they interact with each other and with external stakeholders (European Commission, 2017b).

Dealing effectively with complexity suggests that one is able to think with different perspectives in mind. Personal development and organizational development research suggest that there is no clear correlation between the personal capacity to think in multiple perspectives and the hierarchical position people with this capacity have in organizational hierarchy (Laske, 2006). This question is relevant for metagovernance because it illustrates that 'natural' metagovernors might be found at any level in (public sector and other) organizations.

7.2 Best practices or inspiring examples?

Learning from each other is one of the most frequently prescribed remedies when things don't work out successfully. At the same time the impact of attempts to improve (inter)organizational learning tends to be overestimated. It is easier to write about learning than to learn, just as I can talk for an hour about the importance of listening. In the 1980s, the concept 'best practice' became a buzzword, together with the term 'benchmark'. In order to 'measure' whether a policy or governance practice is good, it should be compared with the 'best practice' among similar cases.

The term 'best practice' has nested itself between our ears and still is extremely popular. The term has a score of 47.3 million hits on the search engine Google: a lot more than 'good practice' (18.0 million) or 'inspiring example' (0.4 million). The main problem I have with the term is that it promises universality, while universally applicable solutions are very rare in social systems and therefore also in governance. Empirical research has shown that copying 'best practices' from one to another country or situation may result in serious failure. A successful approach to community policing in difficult neighbourhoods in the United Kingdom completely failed when it was applied in South Africa (Collier, 2004). Merging municipalities to make them save money because of economy of scale may result in better service to citizens in some, and worse services in other situations in the same country.

When and if a governance approach becomes a success depends on many direct and indirect factors. An approach can be successful because it was smart, timely, cost-effective, top-down or bottom-up, or it might have been sheer coincidence. The critical success factor (if any – because also this is a term that suggests a static hierarchy of factors) may be taking into account the cultural dimension or the history of an issue. A successful project can be the tip of an iceberg of work: it might be the result of years of trial and error, leading to a high-level of public and political awareness, followed by acceptance of measures, which had been perceived as unpleasant in the past. A recent example is the sudden increase in political attention to plastic waste. The problem is not new, neither are solutions. What is new is the sudden 'chemistry' between many and very different factors that created a 'window of opportunity' – to use another classic term for such situations.

The expression 'best practice' is not value-neutral. It is related to the market governance value of competition ("who is the winner, who is the best?"). Alternative wording is available, such as 'good practice', 'successful practice', or 'inspiring example'; the latter is systematically used by the European Commission in its Toolbox Quality of Public Administration (European Commission, 2017b).

7.3 Less is more – or maybe also less?

The assertion in the context of New Public Management that a small government is per definition better than a large government is expressed with the slogan 'less is more'. On the positive side, it should be said that 'cleaning up' an organization now and then has its merits, as each public or private organization tends to develop over time redundant features or becomes obsolete altogether, due to changing circumstances and challenges. The agency responsible for collecting charges for having a radio or TV set in the Netherlands was abolished in 2000, when a new government realized that the original need to have such an agency had vaporized. Since everybody meanwhile had a radio and television set, there was no need any more to target charges to those who possessed these articles, based on a registration and by having radio & television charge inspectors.

Smaller government means, among other things, having less human resources. This has resulted in a series of weaknesses: less quality assurance, reliability, accountability and internal and external checks and balances. Creating smaller ministries means less democratic control. I have observed cases where ministers are worrying, from the beginning of their term, about things going wrong for which they will be blamed. In a way they are right because this will indeed happen when they inform the parliament insufficiently due to lack of expertise, capacity or transparency among the minister's staff. The real expertise – if not completely outsourced to the market – is meanwhile often at 'arm's-length' in agencies, where no direct political steering happens and where political sensitivity is usually at a low level. Having fewer rules means unfairer application of legislation because different cases need to be treated equally.

'Less is more' is inspired by efficiency thinking, which would normally be a secondary objective of an organization. As soon as the primary objective suffers from efficiency measures, something is wrong. Therefore, it is not a universally sound principle. 'Less is more' is often used to suggest that governmental organizations should be smaller. The full meaning of the slogan actually is 'Less state is more market'. Indeed, 'less is more' implies that more public tasks will be given to market parties whose objective is not to 'serve the public' but to make profit.

Related to the 'less is more' mantra is the idea that governments should be 'steering, not rowing' (Peters, 2011): they should make policies but utilize other sectors to deliver public services.

7.4 Evidence-based policy making – or cognitive dissonance?

Another NPM slogan with a disguised normative message is 'evidence-based policy making' (Meuleman, 2012). It claims that an indisputable, 'true' knowledge base for policies is necessary, and fortunately also available, including for sustainability challenges. This claim is in sharp contrast with the low level of certainty that social sciences consider realistic with regard to the politics of complex, disputed, so-called 'wicked' problems (Rittel and Webber, 1973) of the sustainability agenda. Many sustainable development challenges are of the 'wicked' type. For wicked problems, 'hard' evidence is only to a limited extent useful. In addition, some sustainability scientists have started to accommodate the needs of political marketing by claiming that their research produces more certainty than is scientifically sound, in order to get policy makers to act.

Still, many governments seem to favour 'evidence-based' policy making for sustainability challenges although they know that this can lead to ignoring uncertainty and unpredictability. Similarly, predictions of economists are often used as evidence although there is no empirical research behind it. What are the reasons for such collective form of 'cognitive dissonance'? Is it because the costs of unwise decisions will be often later and elsewhere?

The notion of 'evidence-based policy making' is borrowed from natural sciences (especially medicine) and law practice. It suggests that policies are based on factual, undisputable knowledge and based on rational models of problem solving, assuming that knowledge is collected, evaluated and then translated straightforwardly into 'better policies', an assumption that is far from the 'messy' reality (Hertin et al., 2009). It seems therefore that the use of the term 'evidence' in the context of sustainable development is more metaphoric – it emulates confidence and authority – than that it expresses a satisfactory quality of knowledge.

The fashion of evidence-based policy making is not only favoured by natural scientists but also by positivist social scientists like neoclassical economists. They have successfully promoted cost-benefit analysis (CBA) as an 'objective' tool to assess policy options, without being transparent about the fact that their models are being fed with assumptions such as the existence of rational human behaviour, a choice for a discount rate, and monetization of intangibles like the environment. These are all normative choices which should be made in the political realm (De Zeeuw et al., 2008). Monetizing is popular because it is believed to produce objective information, and because it can be done. With CBA, quantification of information became *en vogue*, regardless of the accuracy of the produced numbers (Niestroy, 2008) and despite arguments that cost-effectiveness analysis (CEA) would probably produce more relevant information for environmental policy making than CBA (Ackerman and Heinzerling, 2004). The big difference between economic and environmental/sustainability science is therefore not that the first have 'better' evidence, but that politicians tend to use predictions of economists as good evidence although they often turn out to be wrong.

Factual knowledge does exist but is relatively rare in relation to social systems and human behaviour, and this we have known for a long time (Lindblom, 1959, Berger and Luckmann, 1966), be it with regard to economic, social or environmental behaviour. Real 'evidence' is also rare with regard to complex natural systems. The concept of 'planetary boundaries' is usually presented as being completely within the natural science domain. However, such boundaries are influenced by the interactions between nature and humankind. Schmidt (2013) therefore argues that planetary boundaries cannot be described exclusively by scientific knowledge-claims. They have to be identified by science-society or transdisciplinary deliberations.

Students of political science know that evidence/factual political decision making is a contradiction in terms. Already in 1947, Barnard and Simon (1947) argued that political decisions perform an imperative function, based on ethical terms like 'ought', 'good' or 'preferable', and are per definition neither true or false, correct or incorrect.

Is the term 'evidence' then completely useless? Certainly not. The use of a typology of evidence (from 'beyond all reasonable doubt' to 'insignificant') is useful for assessing scientific research, and acknowledges the costs of being wrong, in both directions of a decision (i.e. acting and not acting) and, critically, on their distribution between groups and generations (Gee, 2008). Gee also emphasizes the importance of the precautionary principle. This principle is fundamental in environmental policy (and for example laid down in Article 191 of the Treaty on the Functioning of the EU). It received a prominent place in a proposal for EU legislation on sustainable finance (European Commission, 2018): "An economic activity should not be considered environmentally sustainable if it causes more harm to the environment than the benefits it brings. . . . Where scientific evaluation does not allow for the risk to be determined with sufficient certainty, the precautionary principle should apply, in line with Article 191 TFEU".

The question is then why many governments seem to favour 'evidence-based' policy making for sustainability challenges, while they know that this can lead to decisions which do not take into account uncertainty and unpredictability. Such contradictory behaviour is known in social psychology as 'cognitive dissonance' (Festinger, 1957), and is also observed in the form of the "human inertia that overrides sound logic and reason" with regard to global environmental problems (Salingaros, 2014).

The popularity of the metaphor 'evidence' for knowledge is a function of the culture and traditions of administrative organizations and their political leaders. The culture and traditions are expressed in the predominant application of a governance approach with a specific appreciation of what usable knowledge or 'evidence' is. Therefore, some of the mechanisms behind the cognitive dissonance can be understood through governance theory. The three basic governance styles have different assumptions about what constitutes useful knowledge for decision making. This goes back to the epistemological foundations of the styles. They stem from different cognition theories and have "incompatible contentions about what is knowable in the social world and what does or can exist – the nature of being – in the social world. . . . They derive their governance 'certainties' from propositions drawn from specific methodological families, which reflect particular configurations of epistemological and ontological perspectives" (Dixon and Dogan, 2002, p. 191). Politicians and public managers who are committed to hierarchical governance see the social world though a naturalist-structuralist lens, those committed to network governance see the social world through a hermeneutic-structuralist lens, and those committed to market governance see the social world through a naturalist-agency lens (Dixon and Dogan, 2002, pp. 184–185).

To conclude, a successful use of knowledge to underpin sustainability policies requires an awareness of the normative character of governance approaches with regard to what usable knowledge is and an approach to dealing with normative tensions, such as metagovernance.

There are advantages to have a more or less independent, not politically dominated phase during the process from collecting data to using knowledge. This might make a long-term view and the study of cumulative and indirect effects more feasible than in the short-term world of immediate results. The challenge, therefore, is to find ways to be more transparent about the values, traditions and scientific assumptions of knowledge production and to create mechanisms that give early warnings for the emergence of a state of cognitive dissonance with regard to 'sustainability evidence'. Elements of such institutional innovation could include:

- An ex-ante (integrated/sustainability) impact assessment procedure, like the European Commission has in place, provides a reliable structure for admitting and discussing research results in the political discussions before a decision is made, and is often considered as a good practice.
- We could follow the suggestion of Rayner (2012) that 'clumsy' arrangements may need to be constructed to ensure that uncomfortable knowledge is not excluded from policy debates, especially when dealing with 'wicked problems' where the accepted version excludes knowledge that is crucial for making sense of and addressing the problem. Such arrangements can be placed inside or outside administrative organizations. The internal solution can be easily silenced by the powers that be, but has the advantage that it is more visible. The external solution – a function sometimes fulfilled by sustainability advisory councils – can be easily ignored, but has the advantage that external influence can be developed.
- Those involved in complex sustainability challenges could learn from the research by Weick and Sutcliffe (2009) on how certain organizations succeed in managing unexpected threats with the help of a collective state of 'mindfulness' that "produces an enhanced ability to discover and correct errors before they escalate into a crisis".
- The development of a configuration approach as proposed by Jungcurt (2013) aims to overcome the weaknesses of boundary work between science and society, by allowing the positioning of boundary work institutions with regard to their degree of politicization and mode of representation. Such an approach could yield a more systemic understanding of boundary work for international sustainable development decision making.
- We need to invest more in capacity building at all levels of government, in order to better deal with complexity, reflexivity and transparency (Niestroy, 2000) and to raise awareness of problems like organizational cognitive dissonance with regard to the quality of 'evidence' for sustainability governance.
- It is important to create more 'meeting places' such as communities of practice where researchers and practitioners can meet and exchange ideas.

7.5 Better regulation?

The NPM concept 'better regulation' has often been used as a euphemism for 'less regulation'. The downside of this approach can be less reliability, legitimacy and steering power of government. However, if it is used in a more literal sense, the impact can be positive in terms of effective governance. The European Commission's Better Regulation initiative (European Commission, 2015) includes, besides a 'REFIT' exercise of reviewing existing EU legislation, a philosophy that is close to metagovernance. First Vice-President Timmermans of the European Commission formulated this as follows during a panel discussion in Brussels in 2015:

> What I would challenge is the way we operate in the Brussels system: to think that if there is a problem, we create a law to solve that problem. I am not sure that the modern economy, the sustainable economy is best served by the premise that for every problem there should be a legal solution. That's the only thing I challenge. I don't challenge the goals. . . . I urge people to convince me and others, that going down the path of making a Directive that interdicts or prescribes, is the best way of attaining this. At least, [we should be] looking into the possibility whether the same goal could also be attained by a combination of different methods.[2]

The 'Environmental Implementation Review' initiative of the European Commission (European Commission, 2017a) is based on the latter approach, offering a dialogue tool in addition to legal and financial tools.

7.6 Public-private partnerships – private on top?

New Public Management also introduced public-private partnerships – with benefits and flaws (Meuleman et al., 2016). A public-private partnership (PPP) is a contractual collaboration between public and private actors, generally to provide what traditionally are public sector services. The World Bank has promoted PPPs for more than 30 years. PPPs foster innovation and fill financing gaps for public infrastructure projects. In a PPP, public and private actors are seen as complementing each other, and leading to cost-effective ways to deliver public services. Closely related to PPPs are multi-stakeholder partnerships (MSPs), promoted since the 2002 Johannesburg Summit. While PPPs are contracts between a government and a company, MSPs are voluntary agreements between different stakeholders (Hemmati and Dodds, 2016). The PPP concept became very popular in the slipstream of the New Public Management ideology from the 1980s onwards. The focus on cost-efficiency is important but this is not the main objective of the 2030 Agenda's vision of partnership, which emphasizes effectiveness (i.e. reaching the objectives) and inclusiveness. Over the years, PPPs have had successes and failures (Beisheim et al., 2014). Just "copy-pasting" PPP practice to the 2030 Agenda is hence no guarantee for success. Also MSPs do not ensure inclusiveness and result-orientation.

Furthermore, several PPPs have left a legacy of large disasters, and these examples may serve to show what should not be repeated. One of the worst PPPs is perhaps the water project in Cochabamba in Bolivia (Schiffler, 2015), where the Bolivian government and a private company worked together for several years around 2000 to develop infrastructure for water supplies for the public. Partly funded by the World Bank and implemented by a private company more interested in profit than in serving the public, and with little serious support from the authorities, the project inspired large-scale riots followed by police brutality that left several people injured or killed.

Proposals have already been made to alter the use of PPPs in health. Mininberg (2016) analysed whether stronger accountability measures could make PPPs more in line with the objectives of SDG 17. A case study on infant feeding indicated that intimate involvement of the private sector is detrimental from the point of view of public health. Based on such examples, the study concluded that a traditional PPP approach to partnerships for the SDGs could undermine the 2030 Agenda. While not overgeneralizing from this conclusion, it is clear that the PPP approach has deficiencies and should not serve as a blueprint going forward (Dunn-Cavelty and Suter, 2009). A key part of the way forward will be to ensure accountability and transparency of all SDG partnerships. Beisheim and Simon (2015) have already drafted detailed proposals on this.

To conclude, the downside of "PPP" as role model for partnerships between governments and societal partners is that (1) PPP is mainly about cost-saving mainly; (2) the business partner in the partnership is often dominant; (3) PPP is not designed to include civil society organizations as partner on equal footing. In Section 11.5 an alternative, more inclusive type of partnership will be discussed.

7.7 Breaking down the silos – or teaching silos to dance?

"Breaking down the silos" has become a vivid call, even a mantra, in debates on governance for sustainable development and the 2030 Agenda, in recognition of its comprehensive, holistic and systemic approach. However, what do people imagine when they call for breaking down the silos? This section is based on an earlier publication (Niestroy and Meuleman, 2016).

While the SDGs require breaking down 'mental silos' to allow for change, the common call to break down institutional silos poses risks. Institutions provide the necessary structure, reliability, transparency and communication points. Instead of breaking them down, we need to teach silos to dance. Conceptually, 'breaking down the silos' reflects the long-standing call for policy integration and policy coherence: the former coined more in the legacy of environmental policy integration, the latter more specifically in the context of development policies, but also used as a wider term. In sustainable development and gender, for example, 'mainstreaming' has become common terminology, underlining the need for broader integration, not only integrating environment into individual sectoral policies. Policy coherence seems to be an end, while integrating and mainstreaming are more the means to it.

Three types of silos should be distinguished: political, mental and institutional. In democracies, politicians need to win majorities. This comes with different degrees of competition and power struggles. Individual politicians tend to focus on their file and defend it, in order to raise their own profiles. This can lead to political silos. As political silos are almost inherent to the democratic system, there are limits to tackling them, which also depends on the political culture of the country. Some have constitutional arrangements that reduce or eliminate the decision-making power of individual ministers (as in Sweden), or give a relatively strong (but still limited) "steering power" to the prime minister (as in Germany). In some countries, governments have experimented with so-called project ministers for cross-cutting issues that involve more than one ministry (e.g. the Netherlands).

In addition and often related, there are also mental silos: people have a firm belief that their problem definition and solution are not only the best, but even the only way forward. Different policy sectors like agriculture, transport and environment have their own world view and tend to operate in isolation. There are cultural, political, power- and career-related, cognitive and other reasons why people have 'tunnel views' and argue against change. However, for the comprehensive SDGs, for moving towards sustainable development with a need for policy integration and coherence, we need to step out of our comfort zones.

We also know that most governmental organizations (and in fact most large organizations) work as classical bureaucracies. They organize their work by dividing complex problems into more simple, partial problems, which are dealt with by separate sectoral or functional bureaucratic entities, which we tend to call "silos". You become a civil servant in one silo and typically stay within it. Exceptions apply in administrations where civil servants rotate across silos, as this is beneficial for careers (e.g. England; European Commission). Such institutional silos give people the room to work undisturbed, but may effectively prevent them from working with others, both within government and with stakeholders.

Most people probably think about these institutional barriers when they use or hear "breaking down the silos". But what would this mean? Merging ministries? Putting everybody in one 'super ministry'? And within one ministry: an organigram with all names in one box? Introducing flexible or matrix organizations, as experimented widely under the New Public Management banner, which is all about becoming more efficient and saving costs? Effectiveness, reliability and accountability are often lost on the way. Mergers of ministries have so far mainly taken place for short-term efficiency reasons in public sector reform, and not with the aim of improving policy coherence. Merging ministries makes sense if you have 40 or 50 ministries in place to run the country. But if you are already down to around 20 (China), 15 (Georgia) or even 11 (Finland), further mergers can make governments much less effective.

Hierarchical governance typically wants to maintain organizational structures that assign tasks and responsibilities to dedicated units. This gives structure, accountability, transparency and makes clear who should be contacted. In the market governance logic, silos are an inefficient complication of (public) organizations, makes them too big and therefore they should be 'broken down' so work can be done more efficiently. Network governance believes more in building bridges: connecting different 'silos' and establish alliances.

Breaking down silos has resulted in small government organizations that lack expertise (which is outsourced to agencies or to private companies), and in which the boundaries of units have become fuzzy. In some countries, ministries have implemented this approach to the extent that they have become more efficient but less effective.

However, without silos there is no focus, no structure, no accountability and no transparency. Institutional mergers may create new governance failures and threaten SDG implementation. Three benefits of silos are:

- Silos represent positive features of government organizations such as clear lines of command, responsibility, focus on a given target and having internal "hotspots", where expertise, memory and learning are concentrated. With respect to the SDGs, an accountable "silo" is needed in each country for, inter alia, reporting progress at the national level. Breaking them down can result in institutional 'deserts'.
- Silos have a different function and meaning in different administrative cultures. In *Rechtsstaat* cultures like Germany and hierarchical ones in general, opening up silos has turned out to be more difficult than in consensus cultures like the Netherlands and Denmark, or in public interest models of government like in Australia, New Zealand and the UK (Pollitt and Bouckaert, 2011, pp. 62–63). Hence, a closer look is required into how they operate, and a general verdict that silos are bad is culturally insensitive.
- Silos provide clear, reliable and stable contact points within a ministry for partners and stakeholders. Without them, it can be difficult to develop enough trust that is needed to make a network approach work. Partnerships and participation increase the challenge of coordination, and thus require clear anchor points. Hence the common assumption that silos prevent stakeholder participation is questionable.

We should also keep our silos because there is no perfect alternative, and there may never be one. As Ulrich Beck developed in his "second modernity": our time is so complex that we need to move from thinking in "best" tools, i.e. "either A or B", to the approach of "A and B" (Beck, 1992). Such redundancies are also good for institutional resilience: if one tool doesn't work, it is easier to switch to another.

If silos need to be broken down, it should be mental silos, not institutional silos, although exceptions apply. Tearing down institutional silos is rather a last resort, if necessary at all. Instead of breaking down institutional silos, we should 'teach them to dance' (Niestroy and Meuleman, 2016). The alternative to breaking down institutional silos is striving to make them more flexible, permeable, interactive and transparent, while keeping their typical strengths and their specific functions in different administrative cultures. It remains a key approach for better policy integration to reinvigorate and improve horizontal coordination. Examples of such horizontal coordination arrangements are widespread, e.g. 'inter-service steering groups' (European Commission), state secretaries' commissions or similar bodies for national sustainable development strategies (Germany, Finland), interdepartmental "dossier teams" (the Netherlands) or cross-sectoral project teams. The SDGs reinvigorate the need to better bridge domestic and external policies – a coordination task that has so far hardly been tackled.

While experience has been mixed, a key lesson is that coordination arrangements do not do the trick by sheer existence, but more efforts need to be put into how to run them. This requires a clever use of procedural tools for facilitating dialogue, with the aim to enable learning about others' assumptions and objectives, building respect for other parties' expertise, recognizing different professional competences and skills of other disciplines (including soft skills such as creativity, accuracy and analytical capacity). All of these can bring to bear the potential for cross-fertilization and new solutions.

A focus on facilitating dialogue, interaction and learning is at the core of opening mental silos. It will likely also require different leadership styles, e.g. switching from commanding to coaching, and capacity building for such adaptive leadership. Leaders will also need to allow, and even stimulate mistakes, as making mistakes is a normal feature of innovation. Important for opening both mental and political silos are new narratives, such as sustainable production and consumption as presenting business and investment opportunities. New narratives as well as mutual gains approaches widen the perspective and enable awareness of synergies.

The SDGs as a system of goals provide a great opportunity to think along interlinkages, to identify trade-offs, but also synergies and knock-on effects. The nexus approach, with a focus on groups of connected goals, makes this manageable. Also impact assessments need to serve this better: On the one hand it is positive that in the European Commission's comprehensive impact assessment of new policies, the economic, social and environmental impacts remain visible up to the very end. They are not 'integrated away' into oblivion, but the individual assessments remain transparent. I recommend, however, to also develop assessments that show trade-offs and synergies, and with that provide a knowledge basis for better-integrated decisions. Translating the universal SDGs into national contexts requires "common but differentiated governance", as different cultures and institutional traditions need to be taken into account (Meuleman and Niestroy, 2015). Similarly, there is no single 'best' way to make silos more collaborative: each culture has unwritten rules about how people can work together.

Approaches to make silos 'dance' include reframing issues to make them attractive challenges for other sectors or actors; establishing cross-departmental projects with shared responsibility; building trust to transform 'enemies' into 'opponents' and 'coalition partners' into 'friends'; creating a peer support culture to share experiences across sectors/actors; creating informal meeting places, where colleagues from different silos get to know each other.

To conclude: Silos are features of our institutions, mental comfort zones and political rationales, and they can be very change-resistant. Teaching elephants to dance (Belasco, 1990) has not been easy, and neither is teaching silos to dance. But we need to go for it. This is a capacity-building programme in itself, related to peer-to-peer learning, and will be beneficial for new global partnerships as well.

7.8 Focus or unfocus?

There is a famous psychological experiment in which the observer has to focus on the number of throws of a ball by a group of people. The best 'focusers' tend to miss out completely the gorilla – or in another version a woman with a parasol – passing through the same room.[3] The lesson from this experiment is that *not* being able to focus has the advantage of seeing unexpected things. And this is exactly a key feature of innovative thinking. Of course it is not all black and white – people who can focus are not per definition uncreative. But it is a point to keep in mind when designing novel approaches for the systemic transitions towards sustainability to which the UN member states have committed in 2015 with the adoption of the 17 universal Sustainable Development Goals (SDGs). The problem is that striving for focus has become one of the many mantras of the liberal market view that dominates global and national politics, and has also influenced management literature – just like 'less is more' and 'break down the silos'.

Many theories of sustainability management argue that more focus is desired, which contradicts the need for accepting complexity when dealing with sustainable development challenges (Starik and Kanashiro, 2013). Focusing on one aspect, without recognizing the importance of linkages between issues (which requires a nexus approach), or on short-, medium- or long-term aspects only, is not the way to achieve progress. To be clear, the nexus approach also has limitations because its focus on a small number of interlinkages while leaving others outside the scope. A counter-intuitive example is the fact that risk assessment also prevents accidents that could have become solutions – the discovery of penicillin is a case in point.

Although the mere adoption of the SDGs is a gigantic achievement, they have some built-in deficiencies. For example, there is a contradiction between sustainability (a transformation challenge) and the goal of achieving economic growth (the latter being of the business as usual type) (Stewart, 2015). This is an example of the ultimate focusing problem: focusing on two contradictory goals at the same time without addressing the tensions.

The point is therefore to have a good balance in any team responsible for dealing with implementing the SDG, and in any mix of policy tools, of focused and 'unfocused' approaches. This is comparable with the balance between project management (controlling time, resources and money) and process management (dealing with unpredictability and uncertainty in a complex and dynamic environment). And this in turn is a question of how to organize the governance of the SDGs.

On the level of staffing (teams, networks, division of responsibilities) it is important to have a diversity of personalities. This does not come automatically. When I was involved in the selection of 50 trainees (out of 1500 candidates) in the Dutch Environment Ministry, I was surprised when I met the candidates who had passed the selection. They were all the same: intelligent,

186 *Features of metagovernance*

extrovert, thinking fast (which is not always the best thing, see Kahneman, 2011) and very focused. The opposite is also fascinating: I have seen several times informal groupings of people interested in sustainability innovation who were all of the opposite type – very creative, thinking out-of-the-box, but also out-of-context and not taking into account the constraints that sometimes are just there.

On the level of how to compose a governance framework as a combination of legal, market-based and network tools, the 'gorilla' lies in the acceptance of redundancy. It may be cost-efficient to steer only with marked-based instruments or with a soft network approach, but it may not be effective in the end when for some reason there is no back-up mechanism (regulatory or other) to prevent failure when the preferred tools do not deliver. When we accept that the systemic challenges of implementing the SDGs are highly complex and the behaviour of the 'object' we want to change – e.g. an energy, food production or transport system – is resilient in the sense that it is highly creative in finding ways *not* to transform, we must not focus on one policy tool, on one type of solution only, but should design a pluralist approach. This may be more expensive in the short term, but pay out on the long run.

To conclude: Any implementation process for the SDGs should have an embedded readiness to switch between *zooming out* to see the broader picture and maybe a gorilla or another unexpected animal, and *zooming in* to concentrate on delivering. Especially at the start of a process, zooming out is important. This is built into the 'mutual gains approach' to multi-stakeholder negotiations. It can lead to agreement on packages that address the original policy question but also include other actions in order to create added value (a 'win-win' situation) for all participants.

7.9 Conclusions

This chapter gave an overview of some of the shadowy sides of New Public Management, not because NPM is overall a bad idea, but because some of its mantras have resulted in governance failure. Seven of them were discussed: best practices, less is more, evidence-based policy making, better regulation, public-private partnerships, breaking down the silos, and the focus on 'focus'.

Armed with these insights, the next chapter will introduce some of the main governance and metagovernance challenges of the 17 SDGs.

Notes

1 This video clip was retrieved from the Internet on 08.01.2018: www.youtube.com/watch?v=x-5zEb1oS9A
2 European Commission First Vice-President Timmermans on 27 May 2015 in Brussels during DG Environment's 'Beaulieu Café' on 'Better Regulation and the circular economy'. Transcript by the author.
3 www.youtube.com/watch?v=vJG698U2Mvo

References

Ackerman, F. & Heinzerling, L. 2004. *Priceless: On Knowing the Price of Everything and the Value of Nothing*. New York: The New York Press.
Barnard, C. & Simon, H. A. 1947. *Administrative Behavior. A Study of Decision-Making Processes in Administrative Organization*. New York: Free Press.
Beck, U. 1992. *Risk Society: Towards a New Modernity*. London: Sage.
Beisheim, M., Liese, A., Janetschek, H., et al. 2014. Transnational partnerships: Conditions for successful service provision in areas of limited statehood. *Governance*, 27, 655–673.
Beisheim, M. & Simon, N. 2015. Meta-Governance of Partnerships for Sustainable Development. Actors' Perspectives on How the UN Could Improve Partnerships' Governance Services in Areas of Limited Statehood.
Belasco, J. 1990. *Teaching the Elephant to Dance: The Manager's Guide to Empowering Change*. New York: Penguin.
Berger, P. & Luckmann, T. 1966. *The Social Construction of Knowledge: A Treatise in the Sociology of Knowledge*. New York: Doubleday.
Collier, P. M. 2004. Policing in South Africa: Replication and resistance to new public management reforms. *Public Management Review*, 6, 1–20.
De Zeeuw, A., In 't Veld, R.J., Van Soest, D., Meuleman, L. & Hoogewoning, P. 2008. *Social Cost Benefit Analysis for Environmental Policy-Making*. The Hague: RMNO.
Dixon, J. & Dogan, R. 2002. Hierarchies, networks and markets: Responses to societal governance failure. *Administrative Theory & Praxis*, 24, 175–196.
Dunn-Cavelty, M. & Suter, M. 2009. Public – private partnerships are no silver bullet: An expanded governance model for Critical Infrastructure Protection. *International Journal of Critical Infrastructure Protection*, 2, 179–187.
European Commission. 2015. *Better regulation for better results – An EU agenda*. Brussels. COM(2015) 215 final. Brussels: European Commission.
European Commission. 2017a. *The EU environmental implementation review: Common challenges and how to combine efforts to deliver better results*. COM(2017) 063 final. Brussels: European Commission.
European Commission. 2017b. *Quality of Public Administration – A Toolbox for Practitioners*. 2017 edition, Abridged version. Luxembourg: Publications Office of the European Union.
European Commission. 2018. *Proposal for a Regulation on the establishment of a framework to facilitate sustainable investment*. COM(2018) 353 final. Brussels: European Commission.
Festinger, L. 1957. *A Theory of Cognitive Dissonance*. Stanford, CA: Stanford University Press.
Gee, D. 2008. Emerging E & H issues: Implications for prevention and research SDRN. *Presentation at the Workshop on Health and Environment, Feb 20/21, 08*. Edinburgh.
Hartley, J. 2005. Innovation in governance and public services: Past and present. *Public Money & Management*, 25, 27–34.
Hemmati, M. & Dodds, F. 2016. High-quality multi-stakeholder partnerships for implementing the SDGs. Available: http://blog.felixdodds.net/2016/08/high-quality-multi-stakeholder.html.
Hertin, J., Turnpenny, J., Jordan, A., et al. 2009. Rationalising the policy mess? Ex ante policy assessment and the utilisation of knowledge in the policy process. *Environment and Planning A*, 41, 1185–1200.

Jungcurt, S. 2013. Taking boundary work seriously: Towards a systemic approach to the analysis of interactions between knowledge production and decision-making on sustainable development. In: Meuleman, L. (ed.). *Transgovernance*. Heidelberg: Springer.

Kahneman, D. 2011. *Thinking, Fast and Slow*. Basingstoke: Palgrave Macmillan.

Laske, O. 2006. *Measuring Hidden Dimensions. The Art and Science of Fully Engaging Adults*. Medford, MA: Interdevelopment Institute Press.

Lindblom, C. E. 1959. The science of 'Muddling Through'. *Public Administration Review*, 19, 79–88.

Meuleman, L. 2012. Cognitive dissonance in evidence-based sustainability policy? Reflections based on governance theory. Paper presented at the 2012 Berlin Conference on Evidence for Sustainable Development, 5–6 October 2012.

Meuleman, L. & Niestroy, I. 2015. Common but differentiated governance: A metagovernance approach to make the SDGs work. *Sustainability*, 12295–12321.

Meuleman, L., Strandenaes, J.-G. & Niestroy, I. 2016. From PPP to ABC: A New Partnership Approach for the SDGs. *IISD SD Policy & Practice*. Guest Article, posted on 11 October 2016 [Online]. Available: http://sdg.iisd.org/commentary/guest-articles/from-ppp-to-abc-a-new-partnership-approach-for-the-sdgs/

Mininberg, L. 2016. A precarious paradigm: Seeking alternatives to public-private partnerships in health, a case for the code. *Independent Study Project (ISP) Collection*. 2357. Available: http://digitalcollections.sit.edu/isp_collection/2357

Niestroy, I. 2000. *Die strategische UVP als Instrument zur Integration von Umweltbelangen in andere Politikbereiche. Fallstudien im Bereich Wasserstrassenplanung an Elbe und San Francisco Bay*. Berlin: VWF Verlag.

Niestroy, I. 2008. Sustainability impact assessment and regulatory impact assessment. In: OECD (ed.) *Conducting Sustainability Assessments*. Paris: OECD Publishing,

Niestroy, I. & Meuleman, L. 2016. Teaching silos to dance: A condition to implement the SDGs. *IISD SD Policy & Practice. Guest Article, posted on 21 July 2016* [Online]. Available: http://sd.iisd.org/guest-articles/teaching-silos-to-dance-a-condition-to-implement-the-sdgs/.

Peters, B. G. 2011. Steering, rowing, drifting, or sinking? Changing patterns of governance. *Urban Research & Practice*, 4, 5–12.

Pollitt, C. & Bouckaert, G. 2011. *Public Management Reform: A Comparative Analysis: New Public Management, Governance, and the Neo-Weberian State Third Edition*. Oxford: Oxford University Press.

Rayner, S. 2012. Uncomfortable knowledge: The social construction of ignorance in science and environmental policy discourses. *Economy and Society*, 41, 107–125.

Rittel, H. W. J. & Webber, M. M. 1973. Dilemmas in a general theory of planning. *Policy Sciences*, 4, 155–169.

Salingaros, N. A. 2014. Cognitive dissonance and non-adaptive architecture: Seven tactics for denying the truth. *Doxa*, 11 January 2014, 100–117. Istanbul: Norgunk Publishing House.

Schiffler, M. 2015. Bolivia: The Cochabamba water war and its aftermath. In: Schiffler, M. (ed.) *Water, Politics and Money: A Reality Check on Privatization*. Cham: Springer.

Schmidt, F. 2013. Governing planetary boundaries: Limiting or enabling conditions for transitions towards sustainability? In: Meuleman, L. (ed.) *Transgovernance*. Heidelberg: Springer.

Sørensen, E. & Torfing, J. 2017. The Janus face of governance theory. *Anti-Politics, Depoliticization, and Governance*, 28.

Starik, M. & Kanashiro, P. 2013. Toward a theory of sustainability management. *Organization & Environment*, 26, 7–30.

Stewart, F. 2015. One flaw in the sustainable development goals may make the difference between success and failure. *The Elders: Independent Global Leaders Working Together for Pease and Human Rights* [Online]. Available: https://theelders.org/article/one-flaw-sustainable-development-goals-may-make-difference-between-success-and-failure.

Termeer, C. J. a. M., Dewulf, A., Breeman, G., et al. 2015. Governance capabilities for dealing wisely with wicked problems. *Administration & Society*, 47, 680–710.

Weick, K. E. & Sutcliffe, K. M. 2009. *Managing the Unexpected: Resilient Performance in an Age of Uncertainty.* 2nd Edition. Hoboken, NJ: John Wiley & Sons.

Part 3
Metagovernance for sustainability

> We are the first generation that can end poverty, and the last generation that can take steps to avoid the worst impacts of climate change. Future generations will judge us harshly if we fail in upholding our moral and historical responsibilities.
> – Former UN Secretary-General Ban-Ki-Moon in his address to the World Economic Forum in Davos, Switzerland on 23 January 2015; www.un.org/sg/en/content/sg/statement/2015-01-23/secretary-generals-remarks-world-economic-forum-plenary-session

Chapter 8. Metagovernance challenges for the 17 Sustainable Development Goals
Chapter 9. Metagovernance: sketching a method
Chapter 10. Metagovernance, public sector reform, coherence promotion and capacity building
Chapter 11. Conclusions: metagovernance as framework for SDG implementation

8 Metagovernance challenges for the 17 Sustainable Development Goals

> "These goals", I mumbled, almost embarrassed to reveal a pivotal detail, "there are 17 of them". My mom's response caught me off guard, to say the least: "Seventeen is a great number", she enthused. In mild shock, I blurted, "You're the first person to say that!" Then I asked, with likely the sardonic tone of an unpersuaded teenager, "Why is 17 a great number, mom?" What she said next simply struck me to my core: "It sounds like they didn't fake it. The world is complicated". She later added, "If they had come back with some Letterman-style top 10 then I probably wouldn't have believed them".
> – From a blog by John McArthur, 28 September 2015[1]
> (quoted by Kamau et al., 2018)

The 17 SDGs are interlinked, even by design 'indivisible' and some of the related governance challenges are common, but they also have their own typical challenges. The growing literature on sustainability governance presents examples of design, management, failures and solutions for each of the Goals and combinations of the goals (nexus themes). The objective of this chapter is to hint at possible avenues for developing such governance frameworks for each of the SDGs.

Each SDG stands for one or more policy sectors or challenges. In each of them, also depending on the country's culture and history, a certain governance style or combination may be dominating current governance frameworks. Areas such as waste management may be heavily regulated (in the European Union, for example), which suggests that the main governance style is hierarchy, and further improvement could mean adding elements of the other governance styles. Other areas are so complex (e.g. sustainable urban development) or relatively new (such as circular economy) that there is little legislation and governance frameworks therefore tend to focus on tools inspired by network and/or market governance. If the existing governance approach does not work well in a given situation, the cause of failures should be investigated. In Chapter 3, general governance failures were presented and Chapter 6 discussed governance flaws related to New Public Management. As a rule of thumb, if one approach dominates and the results are not satisfactory, it should be considered to switch to another style or to add elements from other governance styles to the framework.

194 *Metagovernance for sustainability*

The sections in this chapter are short, but they hopefully stimulate interest to dive deeper into how metagovernance can be used to analyse, design and manage the governance frameworks that should help implementing the related Goals. For each Goal, specific 'governance' targets are mentioned. These are the targets numbered with letters (a, b, c . . .), focusing on the 'means of implementation'. They are meant to be seen in relation to SDG 17 but also in the context of SDG 16.

In this chapter, the SDGs are clustered according to a framework based on Waage et al. (2015) and further elaborated by Lucas et al. (2016) and Niestroy (2016) (Figure 8.1), who clustered the SDGs in a first group on well-being, a second group on production and a third group on the natural environment. As explained by Niestroy (2016, p. 10):

- 'People-centred' goals are found in the inner circle: well-being through ending poverty; improved health and education; reduced inequality (including gender) within and between countries (SDGs 1, 3, 4, 5, 10).
- These goals are embedded in the 'middle circle' of 'production, distribution and delivery of goods and services' and their achievement relies on the realization of these SDGs: delivery of food, water, energy (SDGs 2, 6, 7), as well as economic growth and employment, infrastructure, resources and waste management (SDGs 8, 9, 11, 12).
- This is again embedded in and depends on the conditions of the natural environment that represent the basis for life and all human activities. This outer circle hence comprises the three SDGs relating to natural resources and ecosystems: climate, oceans, biodiversity and land (SDGs 13, 14, 15).

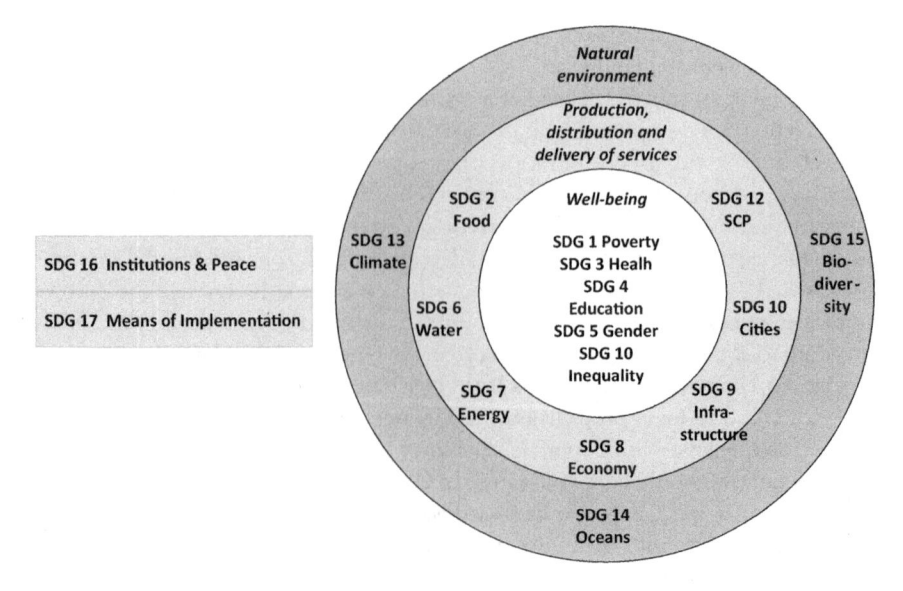

Figure 8.1 Framework for clustering the SDGs
(After Niestroy/Lucas/Waage)

- SDG 17 is placed outside the circle as underlying goal for Means of Implementation and other governance-related targets. For the same reason, SDG 16 on governance and peaceful societies is also placed outside the circle in this depiction of the model.

The first Goals to be discussed are the two dedicated governance SDGs 16 and 17.

8.1 Goals on governance (SDGs 16, 17)

SDG 16: *peace, justice and strong institutions*

SDG 16 was one of the most debated Goals during the negotiations of the Agenda 2030, among others because some countries argued that there is no intergovernmentally agreed definition of 'rule of law' (Target 16.3) and that they did not want to be subject to a 'Western' definition in their own sovereign states (Kamau et al., 2018, p. 202). Besides rule of law, SDG 16 covers many more dimensions of statehood, such as peace, corruption, institutions and decision making, access to information, capacity building on combating crime, and promoting legislation to implement the SDGs. It has 12 targets:

SDG 16 Targets:
16.1 Significantly reduce all forms of violence and related death rates everywhere.
16.2 End abuse, exploitation, trafficking and all forms of violence against and torture of children.
16.3 Promote the rule of law at the national and international levels and ensure equal access to justice for all.
16.4 By 2030, significantly reduce illicit financial and arms flows, strengthen the recovery and return of stolen assets and combat all forms of organized crime.
16.5 Substantially reduce corruption and bribery in all their forms.
16.6 Develop effective, accountable and transparent institutions at all levels.
16.7 Ensure responsive, inclusive, participatory and representative decision making at all levels.
16.8 Broaden and strengthen the participation of developing countries in the institutions of global governance.
16.9 By 2030, provide legal identity for all, including birth registration.
16.10 Ensure public access to information and protect fundamental freedoms, in accordance with national legislation and international agreements.
16.a Strengthen relevant national institutions, including through international cooperation, for building capacity at all levels, in particular in developing countries, to prevent violence and combat terrorism and crime.
16.b Promote and enforce non-discriminatory laws and policies for sustainable development.

The 2017 Progress Report on the implementation of SDGs was as regards SDG 16 overall not optimistic. Violent conflicts had increased, and on the other targets, the message is that "Progress promoting peace and justice, together with effective, accountable and inclusive institutions, remains uneven across and within regions" (United Nations, 2017b). Indeed, in their Voluntary National

Reviews 2017, many countries emphasized the continued need for the development of effective and inclusive governance institutions and processes, consolidation of the rule of law, strengthening of the justice sector and the evolution of an informed civil society (United Nations, 2017a).

Achieving SDG 16 is in the first place political. The Goal covers the nexus between politics and institutions (Whaites, 2016), where important normative decisions have to be made, such as how to ensure rule of law while at the same time stimulating governance innovations allowing stakeholders to take part in implementation of the SDGs. SDG 16 is not only about creating effective institutions, but also about making them accountable and their impact to support the other SDGs measurable.

This is extremely challenging in fragile or 'failed' states. According to the annual ranking of a think tank, only 54 (29%) of the 187 states that are currently monitored can be considered as very sustainable, sustainable or stable (Fund for Peace, 2017). This means that around 70% of the countries vary between the categories 'warning' and 'very high alert'. The last two categories (high alert and very high alert) constitute 35 countries (19%). Also the OECD and the World Bank rank states according to their fragility. Prominent indicators constituting the 'fragility' index are: state legitimacy, quality of public services, human rights and rule of law.

Confidence building is crucial as one of the first steps. The World Bank's Development Report 2011 gives the good example of South Africa, where after the 1994 elections measures were given priority meant to maintain confidence in the new government, such as including maternal and infant health care and using community structures to improve water supply (World Bank, 2011). The report recognizes that the choice of confidence-building measures and institution-building approaches needs to be adapted to each country. In fragile situations, expectations and the credibility of governments are often low, making collaborative action impossible. However, the dominance in governance literature and in the Agenda 2030 on network governance creates a problem in fragile or 'failed' states. Even in a more stable state, the United Kingdom, network governance based on partnerships could not overcome power asymmetries in the partnership arena (Davies, 2007). In fragile states, governments lack the capacity or the legal or actual opportunities to engage civil society and business as partners in public governance.

This situation has a huge impact on effective governance for the SDGs. Institutional capacity building is acknowledged as a key approach, but this involves large funds. Even within the European Union, where the member states are all but one (Cyprus) ranking among the best 54 of the Fund for Peace ranking, over 4 billion Euros has been earmarked (2014–2020) for capacity building. To support public administration capacity building, the European Commission has published extensive guidance on what constitutes a good public administration (European Commission, 2017c). This is general(ized) guidance, which, as the Commission notes, should always (also within the EU) be adapted to national circumstances.

Corruption is a major challenge. Because it is usually systemic, small incremental actions may be without impact – e.g. because they will be encapsulated by the system, and bold measures are needed. Systemic corruption may not even be tackled by legal measures if the judiciary itself is weak. In such situations, it could be an advantage to have strong hierarchical leadership – which is something many fragile states have in common. In 2003, Georgia was one of the most corrupt countries in the world. The then-government abolished the whole traffic police force in July 2004 and fired around 30,000 traffic police officers in order to create a new corruption-free force. In addition, approximately 15,000 policemen were fired within a day, and the organization was completely rebuilt (Centre for Public Impact, 2016). Nowadays, Georgia is a quite safe country to live in and travel through, and it ranks (2017) almost in the middle of the fragility scale of the Fund for Peace and is among the most improved countries in 2017.

Development cooperation funds have, in the past, sometimes tried to circumvent corrupt national governments by funding civil society organizations instead. This has been criticized as undermining what is left of statehood. For example, it has been argued that the World Bank uses metagovernance to 'colonize' civil society, and while doing this, "depoliticizes social and economic life by distancing the allocation of social goods from the centres of political decision making" (Jayasuriya, 2003).

Implementation of SDG 16 generates many trade-offs. It is the most 'hierarchical' goal but it requires values related to the other governance styles to become effective. Reform of institutions is usually a slow process – examples from relatively fast-reforming fragile states indicate that countries who came from a fragile state situation needed decades to reach a reasonable institutional quality (World Bank, 2011), and improving accountability and access to justice and information requires the support and involvement of stakeholders and citizens – as well as the possibility of citizens to organize themselves in civil society groups.

To keep an overview of the possible impact of decisions based on such trade-offs, and to design measures to prevent, mitigate or remediate them, metagovernance capacity is important. SDG 16 requires 'orchestrating' reforms, not only by means of multi-level and decentralized (polycentric) governance but also by metagovernance through situationally appropriate governance frameworks (Roth and Ulbert, 2018).

For example, a metagovernance approach to tackling systemic corruption could try to reframe the issue. Where corruption is an essential source of income for public employees, resistance to change will have an existential character. An analysis of the governance environment could result in reframing corruption from a criminal to an income policy issue. This could create new leverage for reform. Secondly, by mindfully combining the three basic governance styles, the predominant hierarchical style of fragile states could be enriched by network and market governance elements.

Another illustration of metagovernance is a case in the field of inner security and crisis management in Sweden (Larsson, 2017b). Crisis management is on the one hand a typical command and control approach but cannot work without collaboration among decentralized actors. Direct involvement and increased deliberation between the central metagovernors and the other actors helped to develop a shared understanding of crisis management. A similar metagovernance approach can be observed in the collaboration in the Netherlands at the subnational level in the 'safety regions' established between municipalities, police, health organizations and fire brigades. In such a metagovernance approach, one public actor should be able to take the lead. A coordinated response to a disaster is essential because large-scale crises cut across governmental boundaries and can quickly exceed the capacity of local responders (Moynihan, 2013):

> Crisis-response networks of different public, non-profit and private organizations coordinate more effectively if responders speak the same language and follow the same general principles. Only a national government can enforce common standards, and so avert the confusion when multiple approaches conflict in their attempts to manage hazards.

Depending on how metagovernance is defined, different conclusions can be drawn. For example, Beisheim and Simon (2015) define metagovernance as a top-down approach and then found, not surprisingly, that metagovernance ideas "were rooted in perceptions of what may be institutionally feasible and politically possible" – which is almost a pleonasm – and that bottom-up, adaptive, context-specific or tailored approaches were needed. However, taking the wider definition that metagovernance is about combining all different governance styles – which I do in this book – it would be a priority to find synergies between top-down and bottom-up.

SDG 17: *partnerships for the goals*

The second governance Goal is about the 'means of implementation' for the SDGs: financial and other resources, including development assistance commitments, debt relief, investments, science and technology transfer, capacity building for SDG implementation, and trade systems. A whole cluster of targets addresses systemic issues such as policy and institutional coherence, multi-stakeholder partnerships, data availability, monitoring and accountability. The 19 targets of this Goal are therefore very diverse:

SDG 17 Targets (abridged):
Finance
17.1 Strengthen domestic resource mobilization . . . to improve domestic capacity for tax and other revenue collection.
17.2 Developed countries to implement fully their official development assistance commitments.
17.3 Mobilize additional financial resources for developing countries from multiple sources.
17.4 Assist developing countries in attaining long-term debt sustainability through coordinated policies.
17.5 Adopt and implement investment promotion regimes for least developed countries.

Technology
17.6 Enhance North-South, South-South and triangular regional and international cooperation on and access to science, technology and innovation and enhance knowledge sharing on mutually agreed terms.
17.7 Promote the development, transfer, dissemination and diffusion of environmentally sound technologies to developing countries on favourable terms.
17.8 Fully operationalize the technology bank and science, technology and innovation capacity building mechanism for least developed countries by 2017.

Capacity building
17.9 Enhance international support for implementing effective and targeted capacity building in developing countries to support national plans to implement all the Sustainable Development Goals.

Trade
17.10 Promote a universal, rules-based, open, non-discriminatory and equitable multilateral trading system under the World Trade Organization.
17.11 Significantly increase the exports of developing countries, in particular with a view to doubling the least developed countries' share of global exports by 2020.
17.12 Realize timely implementation of duty-free and quota-free market access on a lasting basis for all least developed countries.

Policy and institutional coherence
17.13 Enhance global macroeconomic stability, including through policy coordination and policy coherence.
17.14 Enhance policy coherence for sustainable development.
17.15 Respect each country's policy space and leadership to establish and implement policies for poverty eradication and sustainable development.
17.16 Enhance the Global Partnership for Sustainable Development, complemented by multi-stakeholder partnerships that mobilize and share knowledge, expertise, technology and financial resources.
17.17 Encourage and promote effective public, public-private and civil society partnerships, building on the experience and resourcing strategies of partnerships.

Data, monitoring and accountability
17.18 By 2020, enhance capacity building support to developing countries, including for least developed countries and small island developing States, to increase significantly the availability of high-quality, timely and reliable data.
17.19 By 2030, build on existing initiatives to develop measurements of progress on sustainable development that complement gross domestic product, and support statistical capacity building in developing countries.

Bowen et al. (2017) argue that three governance challenges are central to implementing the SDGs, namely cultivating collective action by creating inclusive decision spaces for stakeholder interaction across multiple sectors and scales; making difficult trade-offs focusing on equity, justice and fairness; and ensuring mechanisms exist to hold societal actors to account regarding decision making, investment, action and outcomes. These are all covered under SDG 17. In 2017, the UN called for a stronger commitment to partnership and cooperation achieve the Sustainable Development Goals, and more coherent policies (United Nations, 2017b).

Financial targets and the role of taxation

The first five targets concern financial resources, including domestic revenue generation through taxation. Taxes are market-based instruments from the market governance toolbox. In addition to generating revenue, they can be very successful in regulating the behaviour of citizens or economic operators. This 'double sword' is used in many countries and it has a great potential to support the SDGs. EU countries have over the past decades developed a wide repertoire of regulatory taxes, and at the joint level of the EU, there is consensus that taxes should shift, where possible, from taxing 'goods' to taxing 'bads'. Prominent in this area is the shift from taxing labour (a 'good') to taxing environmental pressure and pollution ('bads'). Taxes, but also charges or fees, on landfilling municipal waste have contributed strongly to the increase of recycling and reuse of waste. Taxes on ground water extraction, and generally, the legal obligation to price water, are limiting water scarcity problems. The European Commission generally considers tax on environmental pollution, on property and on consumption (VAT) as the types of taxes which are the least detrimental to economic growth (European Commission, 2017b, p. 7). A recent example of a metagovernance mix that includes taxation is the announcement of a proposal for an EU 'tax' on single-use plastics, which adds another market-based tool to the 2018 Plastics Strategy which is essentially already a combination of hierarchical, network and market-based means (European Commission, 2018).

Technology

The three targets on technology focus on international cooperation on access to science and technology and an innovation capacity-building mechanism for least developed countries. International technology transfer is never an easy topic because investment money and vested interests are involved. Governance for this target can be expected to be slowly delivering results, and not primarily by means of legal agreements but through other tools from the broad governance toolbox, such as incentives and voluntary cooperation.

Capacity building

The challenge with enhancing international support for capacity building in developing countries to support national plans to implement all the Sustainable Development Goals (Target 17.9) is to do it in a way which supports and not disrupts already functioning capacity. Overall, I would say that in many developing countries there is already sufficient capacity to make national plans on any subject for which donors in the past gave financial support. The weak point is implementation – and this cannot be done only in and by the capitals. Capacity building for the SDGs should therefore focus on how to get decentralized authorities as well as civil society and private sector motivated and equipped to be part of implementation processes. This may require also a cultural change on the side of donor countries, who are used to apply New Public Management–inspired control, auditing and accountability mechanisms that are top-down.

Trade

Trade governance frameworks need to promote a universal, rules-based, open, non-discriminatory and equitable multilateral trading system and to conclude negotiations under the Doha Round (Target 17.10). Organizations such as the Asian Development Bank (ADB) have taken on metagovernance functions such as providing the framework for financial and economic policy at the national and sub-national levels and supporting regional financial initiatives (Jayasuriya, 2004).

Policy and institutional coherence – partnerships first?

Under the title 'Policy and institutional coherence', Targets 13–17 cover global macroeconomic stability, policy coordination and coherence, leadership, multi-stakeholder partnerships and other forms of partnerships.

Policy and institutional coherence is seen as a precondition for macroeconomic stability (Target 17.13) but also has a self-standing importance for the implementation of the SDGs (Target 17.14). This is a crucial target to make the SDGs work in synergy with each other, horizontally and vertically. It means that governments at all levels need to cooperate better across policy areas and across levels. However, the 2017 Voluntary National Reviews show that policy coherence and multi-sectoral coordination still present a major challenge for many countries (United Nations, 2017a). Section 10.2 contains an elaborate reflection on what coherence implies as governance and metagovernance challenge, based on a paper I wrote as member of the UN Committee of Experts on Public Administration (CEPA) (Meuleman, 2018).

Target 17.15 calls for respecting each country's "policy space and leadership" to implement the SDGs. Many countries have some form of high-level entity that makes political decisions on the country's vision, priorities and development frameworks through which the 2030 Agenda is to be implemented (United Nations, 2017a). In most cases, these entities already existed prior to the adoption of the 2030 Agenda, but in some countries a new arrangement was adopted (e.g. Costa Rica, Nigeria). In Section 5.4 (feature 39), I have argued that situational leadership is a precondition for sustainability governance and metagovernance.

The two partnership Targets 17.16 and 17.17 have attracted so much attention that SDG 17 is usually called the Partnership Goal. As a key term of network governance, 'partnership' has become a buzzword in many circles, including among those involved in sustainability governance. The word is used to describe almost any relationship that pools the resources of different actors to address societal challenges and concerns, and can only be fully understood in relation to its practical context (Stott, 2017).

Further on, I will address this theme separately (Section 11.5). Since the Johannesburg World Summit on Sustainable Development in 2002, so-called multi-stakeholder partnerships (MSPs) have grown very popular within the UN. This has been reinforced with the adoption of the Agenda 2030 in 2015. MSPs can be defined as partnerships in which "non-state actors (such as civil society organizations and businesses) and state actors (such as international organizations and national governments) collaborate across transnational, national and local levels to provide collective goods" (Beisheim et al., 2014).

MSPs are thought to be a solution – in fact they are considered to be a key tool of SDG governance frameworks. However, transnational multi-stakeholder partnerships for sustainable development are still "neither as successful as their proponents claim nor as ineffective as their critics argue" (Meuleman et al., 2016). Several factors have been suggested as responsible for sub-optimal performance of these partnerships. One point is about their design: such partnerships are not always composed of partners which are on equal footing. Especially when they are based on the model of public-private partnerships (PPP), the private actors often dominate the cooperation. In Section 11.5 I will elaborate the argument that such partnerships should stay away from the PPP blueprint and be designed as 'ABC partnerships' based on transparent and fair collaboration between administrations (A), business actors (B) and civil society organizations (C) (Meuleman et al., 2016).

A second problem is the governance of this governance tool, i.e. the metagovernance. Beisheim and Simon (2015) asked involved actors whether lessons learned from experience with partnerships should be followed up with rules and standards for existing and new partnerships. This would be a typical metagovernance approach: strengthening the weaknesses of the predominant governance style (i.e. network governance for partnerships) by using features of hierarchical governance: rules and standards. Inspired by their research topic (governance of MSPs on, in particular, areas of limited statehood) and following the more narrow approach of metagovernance as governing networks (see Section 4.3), these authors characterize metagovernance as a higher level of decision making: higher-level rules that guide transnational partnerships, through e.g. guidance for institutional design and management of partnerships. This higher-level governance can be used by state or non-state actors.

Another challenge is the relation between partnerships and other formal and informal arrangements. The 'democratic anchorage' of partnerships, a challenge researched since more than a decade (Sørensen et al., 2005, Davies, 2007) remains problematic, with power asymmetries in the partnership arena being a normal feature.

Notwithstanding these critical aspects, the adoption of the SDGs has already stimulated many countries to create new institutions for stakeholder engagement, and to ask stakeholders to participate in their SDG-related decision-making processes (United Nations, 2017a).

Data, monitoring and accountability

The two last governance targets of SDG 17 address data, monitoring and accountability.

A big step forwards on indicators to measure progress on the SDGs was the agreement on a global indicator framework for the SDGs in the United Nations Statistical Commission in March 2017, which was subsequently adopted by the General Assembly in July 2017. A Global SDG Indicator Database was launched in July 2017, which presents country-level data as well as global and regional aggregates compiled through the UN system and other international organizations (United Nations, 2017a, p. 32). However, still much needs to be done, because the Goals should be implementation in a contextualized way at the national and sub-national levels; central monitoring even at the national level will not be enough.

Accountability can be organized in many ways (see Section 5.3, feature 31): hierarchical (top-down, 'order and observance'), horizontal (network governance, 'interactive persuasion, participation and co-working'), market-oriented ('market competition') or various combinations. In addition, like monitoring progress, accountability mechanisms should be ready at all levels.

Both monitoring and accountability therefore have a multi-level dimension, which implies that the interfaces connecting the levels are important. When accountability or monitoring systems are different at different levels, the interfaces need to interpret, translate and accommodate the results in order to make them effective. For example, a national accountability system that uses only financial data to monitor how effectively the SDGs are being implemented at lower levels will miss relevant results identified at the local level, for example on skills gaps across the labour force. Existing sustainability indicators of the province of Manitoba (Canada), for example, contained no finance or policy indicators, although they are available in the global SDG indicators (Thrift and Bizikova, 2016).

8.2 People-centred goals (SDGs 1, 3, 4, 5, 10)

SDG 1: *end poverty in all its forms everywhere*

Governance Targets:
1.a Ensure significant mobilization of resources from a variety of sources, including through enhanced development cooperation, in order to provide adequate and predictable means for developing countries, in particular least developed countries, to implement programmes and policies to end poverty in all its dimensions.
1.b Create sound policy frameworks at the national, regional and international levels, based on pro-poor and gender-sensitive development strategies, to support accelerated investment in poverty eradication actions.

SDG 1 aims at ending poverty "in all its forms everywhere". The UN website explains this as a systemic challenge:

> Poverty is more than the lack of income and resources to ensure a sustainable livelihood. Its manifestations include hunger and malnutrition, limited

access to education and other basic services, social discrimination and exclusion as well as the lack of participation in decision making.[2]

In addition, poverty eradication is strongly correlated with Goal 10 on inequality.

The governance targets of SDG 1 are about ensuring that developing countries have the resources and supporting policy frameworks to implement poverty programmes and policies. Building the resilience of the poor and strengthening disaster risk reduction is a core development strategy for ending extreme poverty in the most afflicted countries, as disaster risk is highly concentrated in low- and lower-middle-income countries, and small island developing states are facing a disproportionate impact (United Nations, 2017b).

In the environment-poverty nexus, there are difficult trade-offs such as between environmental conservation and economic development, but often the goals are mutually reinforcing, for example in protecting scarce water resources (Uitto, 2016). As the UNDP-UNEP Poverty Environment Initiative (PEI) has shown, poverty and environmental degradation are often strongly interlinked, and it makes sense to have joint programmes. Cases in Central Asia have shown that national and local development planning and private initiatives should be better linked and that multi-level governance should be more coherent (Challe et al., 2017).

SDG 3: good health and well-being

Governance Targets:
3.a Strengthen the implementation of the World Health Organization Framework Convention on Tobacco Control in all countries, as appropriate.
3.b Support the research and development of vaccines and medicines for the communicable and non-communicable diseases that primarily affect developing countries, provide access to affordable essential medicines and vaccines, in accordance with the Doha Declaration on the TRIPS Agreement and Public Health, which affirms the right of developing countries to use to the full the provisions in the Agreement on Trade-Related Aspects of Intellectual Property Rights regarding flexibilities to protect public health, and, in particular, provide access to medicines for all.
3.c Substantially increase health financing and the recruitment, development, training and retention of the health workforce in developing countries, especially in least developed countries and small island developing States.
3.d Strengthen the capacity of all countries, in particular developing countries, for early warning, risk reduction and management of national and global health risks.

SDG3 aims at ensuring healthy lives and promoting well-being for all at all ages. The governance Targets 3a–3d are relatively detailed and focus more on certain aspects of the quality of health care than on systemic issues. Health metagovernance could have several focal areas. One is the governance of national health systems. In developed countries, the rising costs of health care have resulted in large-scale privatization, which sometimes has reduced costs, but sometimes also to the detriment of health care as primary task. A review of health governance in such systems could include reframing part of the problem from taking care of the sick to preventing sickness and tackle precursors of bad health situations, such as poverty, lack of sanitation, inequality and hunger. It could aim at remediating the negative impacts of programmes mainly focused on short-term efficiency gains.

Global health governance is a fragmented field of international cooperation with a perceived lack of synergy and choreography between international and transnational organizations (Holzscheiter et al., 2016). Nevertheless, global "metagovernance norms" have emerged: international norms that define good global governance as orderly and harmonized global governance. They are formalized in reflexive practices which govern and order the relationships between international organizations (Holzscheiter et al., 2016).

Many developing countries still lack comprehensive health systems. A metagovernance perspective could prevent making some of the same mistakes as many countries already have made under the influence of New Public Management, where health care reforms were primarily targeted at reducing costs and less on improving effective health care. Lessons from health care reforms which focused on market governance have not only been learned in Western countries, but also elsewhere. A comparative research on China, India and Thailand shows a largely disappointing experience with non-hierarchical modes and that all three countries, especially Thailand, are in the process of reverting to a more hierarchical mode of health service delivery (Ramesh et al., 2015). The metagovernance style conclusion the authors draw is that

> non-hierarchical governance is not a substitute for or an improvement upon hierarchical governance in health care due to the many market and government failures that afflict the sector and affect the ability of different governance modes to function effectively. The hierarchical mode of government is also imperfect but less so than the alternatives in delivering health care.

Public, private or personal (family-level) health care is strongly embedded in national values and traditions, and metagovernance for SDG 3 should explicitly take this into account, for example by integrating endogenous knowledge and using partnerships with non-public partners. Finally, health care governance requires intersectoral cooperation as well as cross-border collaboration. An example is the collaboration initiated by Belize with Guatemala and Mexico on combating malaria (United Nations, 2017a).

SDG 4: *quality education*

Governance Targets:
4.a Build and upgrade education facilities that are child, disability and gender sensitive and provide safe, non-violent, inclusive and effective learning environments for all.
4.b By 2020, substantially expand globally the number of scholarships available to developing countries, in particular least developed countries, small island developing States and African countries, for enrolment in higher education, including vocational training and information and communications technology, technical, engineering and scientific programmes, in developed countries and other developing countries.

(*Continued*)

4.c By 2030, substantially increase the supply of qualified teachers, including through international cooperation for teacher training in developing countries, especially least developed countries and small island developing States.

SDG 4 aims to ensure inclusive and equitable quality education and promote lifelong learning opportunities for all. Its governance Targets 4a–c give priority to building up suitable education facilities, expanding the number of scholarships available to developing countries, and substantially increasing the supply of qualified teachers. The targets reflect that these common targets are a greater challenge for developing countries than for developed states, and that the latter should support the former.

Most existing education systems are of a hierarchical nature: quality requirements, accountability and targets are usually formulated at the national level and implemented through legislation. More in detail, the systems reflect the predominant governance style (combination) in each country: relatively more top-down in Germany, more centralist in France, more decentralized/devolved in the UK, and more supported by networks in the Netherlands and Denmark. In some countries, New Public Management has resulted in an overall focus on efficiency to the detriment of education quality – in fact a similar process with similar impacts as in health governance.

Teachers have been described as 'street-level bureaucrats' (Lipsky, 1980) who typically need sufficient discretionary space to be effective. New Public Management has decreased this space substantially, as a UK research project showed (Taylor and Kelly, 2006). A comparative study on local governance of schools in Sweden, Norway and Denmark focused on three types of strategies which could be linked to the three basic governance styles: control-based steering strategies (hierarchy), competitive steering strategies (market governance) and collaborative steering strategies (network governance) (Thøgersen, 2017). It was shown that, although the three countries are relatively similar with regard to administrative culture and tradition, the national context still leads to clear differences in the governance mix. For example, Sweden is the most marketized of the three Scandinavian countries and at the same time has the largest degree of control and regulation over private schools. This is a typical New Public Management combination of market and hierarchical governance.

Magalhães et al. (2013) argue that education requires metagovernance, quoting (Peters, 2010, p. 3) who states that metagovernance is needed "in sectors whose institutions enjoy a substantial degree of autonomy". According to Magalhães et al., EU-level education governance is already a clear case of metagovernance: the European Commission uses a dynamic combination of different governance styles and their tools to (Magalhães et al., 2013).

In the UK, a trend towards stronger network governance has been identified in the education system, which raises questions about how to organize accountability (Ehren and Perryman, 2017). Earlier, De Grauwe (2007, pp. 14–16) suggested different accountability systems in education governance which can be linked to the basic governance styles: a hierarchical public or state control model, a

SDG 5: gender equality

Governance Targets:
5.a Undertake reforms to give women equal rights to economic resources, as well as access to ownership and control over land and other forms of property, financial services, inheritance and natural resources, in accordance with national laws.
5.b Enhance the use of enabling technology, in particular information and communications technology, to promote the empowerment of women.
5.c Adopt and strengthen sound policies and enforceable legislation for the promotion of gender equality and the empowerment of all women and girls at all levels.

SDG 5 aims at achieving gender equality and empowering all women and girls. Gender inequality persists worldwide, depriving women and girls of their basic rights and opportunities. Achieving gender equality and the empowerment of women and girls will require more vigorous efforts, including legal frameworks, to counter deeply rooted gender-based discrimination that often results from patriarchal attitudes and related social norms (United Nations, 2017b). UN Women has called for a "revolution in democratic governance" for women and girls to claim their rights and shape sustainable development, and point at the gap as regards gender statistics (United Nations, 2018). The UN identified gender inequalities in all global regions. In the EU, for example, there is a persistent gender pay gap and pension pay gap, many women in the EU have experienced some form of sexual harassment, and women in the EU work two times as many hours for unpaid care than men (Gabizon, 2018). Boas et al. (2016) have argued for a focus on the 'gender-energy-poverty nexus', because women are much more affected by energy poverty than men.

The governance Targets 5a–c focus on having legislation in place to ensure gender equality. Target 5.a links women's equal rights to "accordance with national law". This can be problematic in the many countries where inequality of women and men is currently enshrined in national law. To fully implement SDG 5, it would, in such cases, be logical to put a high priority on adjusting the relevant national legislation. This could be accompanied by market-based instruments such as incentives and loans, and network governance action such as encouraging the degree of organization of women.

SDG 10: reduced inequalities

Governance Targets:
10.a Implement the principle of special and differential treatment for developing countries, in particular least developed countries, in accordance with World Trade Organization agreements.

(Continued)

10.b Encourage official development assistance and financial flows, including foreign direct investment, to States where the need is greatest, in particular least developed countries, African countries, small island developing States and landlocked developing countries, in accordance with their national plans and programmes.
10.c By 2030, reduce to less than 3 per cent the transaction costs of migrant remittances and eliminate remittance corridors with costs higher than 5 per cent.

Achieving SDG 10, aiming at reducing inequality within and among countries, is a huge task. Like with the other Goals, implementation is complex and will not be easy, and national governance frameworks will need to be closely linked to those on other SDGs such as SDG 1 on poverty and SDG 5 on gender equality. The three governance Targets 10 a–c focus on means of implementation for least developed countries, which makes developed countries co-responsible of implementing SDG 10.

8.3 Goals on production, distribution, delivery, goods and services (SDGs 2, 6, 7, 8, 9, 11, 12)

SDG 2: hunger, food security and sustainable agriculture

Governance Targets:
2.a Increase investment, including through enhanced international cooperation, in rural infrastructure, agricultural research and extension services, technology development and plant and livestock gene banks in order to enhance agricultural productive capacity in developing countries, in particular least developed countries.
2.b Correct and prevent trade restrictions and distortions in world agricultural markets, including through the parallel elimination of all forms of agricultural export subsidies and all export measures with equivalent effect, in accordance with the mandate of the Doha Development Round.
2.c Adopt measures to ensure the proper functioning of food commodity markets and their derivatives and facilitate timely access to market information, including on food reserves, in order to help limit extreme food price volatility.

SDG2 has the objective to ending hunger, food insecurity and malnutrition for all.

More investments in agriculture, including government spending and aid, are needed to increase capacity for agricultural productivity (United Nations, 2017b).

This Goal is interlinked with especially Goal 1 on poverty. For both Goals, UN member states consider supporting local initiatives important, with measures stimulating e.g. small-scale food producers such as family farmers, enabling women to be involved in food production, and stimulating organic farming. SDG 3 is also interlinked with Goal 11 on climate change (United Nations, 2017a).

The governance Targets 2a–c focus on rural infrastructure and research, agricultural markets and food commodity markets. These are very complex and broad themes. Fisheries are part of this complex. For fisheries governance, metagovernance has been proposed as important approach for states to drive and support governance arrangements, based on case studies in Japan, Vietnam and Norway (Thang, 2018). Another case where there is scope for metagovernance is brought forward in a study on organic seed regulation in the USA, the European Union and Mexico, which shows that growth of the organic sector is hindered by regulatory imbalances and trade incompatibilities and the varying capacity for self-organizing governance of the seed sector in relation to the state's regulatory role (Renaud et al., 2016).

SDG 6: clean water and sanitation

Governance Targets:
6.a By 2030, expand international cooperation and capacity building support to developing countries in water- and sanitation-related activities and programmes, including water harvesting, desalination, water efficiency, wastewater treatment, recycling and reuse technologies.
6.b Support and strengthen the participation of local communities in improving water and sanitation management.

SDG 6 (Ensure availability and sustainable management of water and sanitation for all) has two governance targets, on (6.a) international cooperation and capacity-building support to developing countries in water- and sanitation-related programmes, and (6.b) strengthening the participation of local communities in improving water and sanitation management. One of the specific challenges in this area which is traditionally dominated by technical discourse is that institutional dimensions of water management and decision making do not effectively take into account the needs of poor households. The result has been that poor households were not connected by water suppliers, or did not apply for a connection (Bakker et al., 2008). This is an example of the need for more vertical policy and institutional coherence, as well as for the need to encourage informal action by non-governmental actors to support poor households.

Effective water and sanitation management relies on the participation of a range of stakeholders, including local communities. A 2016–2017 survey found that over 80% of 74 responding countries had clearly defined procedures for engaging service users/communities in water and sanitation management (United Nations, 2017b).

Water scarcity is another challenge, even in 'wet' countries such as the UK. A case study used a metagovernance lens to investigate the failure of the government to ensure effective water management, identifying a 'laissez-faire' mentality and strong focus on deregulation and market mechanisms (Walker, 2014). In an earlier study it was concluded that failure of water management reform in Australia was caused by a lack of serious application of metagovernance (Bell and Park, 2006). In the case study, network governance was ultimately a hierarchical process displaying distinct limits to participatory governance. Such an approach to networks which is only rhetoric risks frustrating participants – also in the long run.

SDG 7: affordable and clean energy

Governance Targets:
7.a By 2030, enhance international cooperation to facilitate access to clean energy research and technology, including renewable energy, energy efficiency and advanced and cleaner fossil-fuel technology, and promote investment in energy infrastructure and clean energy technology.
7.b By 2030, expand infrastructure and upgrade technology for supplying modern and sustainable energy services for all in developing countries, in particular least developed countries, small island developing States and landlocked developing countries, in accordance with their respective programmes of support.

SDG 7 aims at ensuring access to affordable, reliable, sustainable and modern energy for all. Progress in every area of sustainable energy falls short of what is needed to achieve energy access for all and to meet targets for renewable energy and energy efficiency. Meaningful improvements will require higher levels of financing and bolder policy commitments, together with the willingness of countries to embrace new technologies on a much wider scale (United Nations, 2017b).

Energy is essential for development and requires bold thinking from both governance and access perspectives. SDG 7 aims to (7.a) enhance access to clean energy research and technology and promoting the relevant investments. Target 7.b aims at expanding infrastructure and upgrading technology for modern and sustainable energy services, in particular for least developed countries, small island developing states and landlocked developing countries.

For the target on universal access to energy it is relevant that energy production in many countries has moved from (local) public service to private enterprises, followed by scaling up into powerful power companies which are almost monopolies. As a counter reaction local initiatives have emerged in many countries to decentralize energy production and making renewable energy again part of the 'commons'. In some countries a governance transition has taken place from state/local via market steering (followed by market failures), back to local network initiatives steered and owned by local authorities and/or citizens. In other countries, energy production has been privatized to monopolist firms from the onset. Energy production and consumption are largely cross-national and scaling up of renewable energy would benefit from increased policy coherence across nations. In institutional terms, universal access to energy might require that centralized energy systems are willing and able to integrate or create alliances with small local initiatives. Sustainable energy programmes in Croatia and Mongolia, where market (private), network (collaborative) and hierarchical (centralized) governance mechanisms were combined, have been identified as successful (Christopoulos et al., 2012).

The situation pictured above also applies to SDG Target 7.b on expanding energy infrastructure and upgrading energy technology. In many cases energy (infrastructure) policy is strongly influenced by market forces and financial and geopolitical priorities. Achieving SDG 7 in the full spirit of Agenda 2030 might require a stronger regulatory influence from governments and at the same time a stronger influence from bottom-up initiatives (leaving no one behind). In addition, policy and institutional coherence debates should be linked to existing global initiatives around energy such as Sustainable Energy for All and the Covenant of Mayors.

SDG 8: decent work and economic growth

Governance Targets:
8.a Increase aid for trade support for developing countries, in particular least developed countries, including through the Enhanced Integrated Framework for Trade-related Technical Assistance to Least Developed Countries.
8.b By 2020, develop and operationalize a global strategy for youth employment and implement the Global Jobs Pact of the International Labour Organization.

SDG 8 aims at promoting sustained, inclusive and sustainable economic growth, full and productive employment and decent work for all. Its two governance targets are about (financial) means of implementation and not really about governance as the ways to reach the Goal. Interestingly, it seems that tourism is the economic sector where governance and metagovernance are being discussed most. Tourism is an often-underestimated economic sector which is much larger than most people think. It is partially unsustainable and on the other hand needs sustainable destinations to survive. Metagovernance addresses some of the perceived failures of traditional governance approaches and associated interventions; examination of how governance is itself governed enables a better understanding of the practices of planning and policy making affecting tourism and destinations (Amore and Hall, 2017).

SDG 9: industry, innovation and infrastructure

Governance Targets:
9.a Facilitate sustainable and resilient infrastructure development in developing countries through enhanced financial, technological and technical support to African countries, least developed countries, landlocked developing countries and small island developing States.
9.b Support domestic technology development, research and innovation in developing countries, including by ensuring a conducive policy environment for, inter alia, industrial diversification and value addition to commodities.
9.c Significantly increase access to information and communications technology and strive to provide universal and affordable access to the Internet in least developed countries by 2020.

SDG 9 aims to build resilient infrastructure, promote inclusive and sustainable industrialization and foster innovation. The governance targets focus on developing countries, but also in more developed countries there are substantial governance challenges: existing industry, technology and innovation approaches may be not sustainable at all. In each country there is a different mix of governance styles which would benefit from a 'sustainability review'. Can change best be achieved with more or with less state involvement, market involvement and active roles of civil society?

SDG 11: sustainable cities and communities

Governance Targets:
11.a Support positive economic, social and environmental links between urban, peri-urban and rural areas by strengthening national and regional development planning.
11.b By 2020, substantially increase the number of cities and human settlements adopting and implementing integrated policies and plans towards inclusion, resource efficiency, mitigation and adaptation to climate change, resilience to disasters, and develop and implement, in line with the Sendai Framework for Disaster Risk Reduction 2015–2030, holistic disaster risk management at all levels.
11.c Support least developed countries, including through financial and technical assistance, in building sustainable and resilient buildings utilizing local materials.

SDG 11 (Make cities and human settlements inclusive, safe, resilient and sustainable) has three governance targets, on (11.a) improving national and regional development planning linking urban and non-urban areas; on (11.b) promoting integrated policies and plans towards inclusion, resource efficiency, mitigation and adaptation to climate change, resilience to disasters, and developing and implementing holistic disaster risk management; and on (11.c) supporting least developed countries, including through financial and technical assistance, in building sustainable and resilient buildings utilizing local materials.

Cities are hotspots of innovation, wealth and at the same time of extreme poverty. In Africa, Asia and Latin America there are fast-growing cities in which sustainability has not yet been a high priority. On the positive side, many cities have become frontrunners on addressing climate change, other environmental issues and social challenges. One of the big challenges is vertical coherence between national and local policies and institutional arrangements and mandates. A general complaint from front-running big cities is that their national governments are blocking progress, and from others it is their bureaucracy.

The general pull towards cities has weakened the sustainability of rural areas and created large peri-urban areas. The relation between urban, peri-urban and rural areas is a challenge of horizontal coherence. A good practice is Cape Town in South Africa, where the mayor organizes monthly meetings with under-resourced municipalities in its peri-urban area to let them benefit from the power of the bigger metropolitan area (see also Section 3.5).

In addition, informality characterizes many of the bottom-up sustainability issues at community level.[3] This supports the hypothesis that 'endogenous' development, based on existing traditions, values and leadership, deserves a prominent place as alternative to 'exogenous'-inspired development.

Over the last decade, the division between sustainable and unsustainable city environments has increased. A growing number of cities have become frontrunners in addressing climate change, environmental issues and social challenges.

The Covenant of Mayors is a strong example, but there are others, like ICLEI as a global organization of sustainable cities and Eurocities for major cities in the EU.

Frontrunners, especially large cities, express themselves confidently, claiming that their national governments are blocking progress. "I don't need slow-functioning and bureaucratic multilateral agreements to work out solutions: when I need advice, I just call a major of another big city in the USA or any other country, and we discuss how to best approach issues".[4] "Cities have been around for thousands of years, national governments are much younger. Who do you think understands better what citizens want?" asked another big city mayor.

Urban governance in the context of sustainable development is receiving increasing attention in the research world. Because of the inherent complexity of the topic, metagovernance has been used (or 'discovered') already quite some time ago. In Section 4.8 the example was already mentioned of a case study showing that governance styles in the field of urban regeneration in England succeeded each other: (1) pre-partnership collaboration followed the principles of *network* governance; (2) partnership creation and consolidation used *hierarchy* to incorporate some organizations, and to formalize authority; (3) partnership programme delivery used *market* governance mechanisms of tendering and contractual agreements, while *hierarchy* ensures regulation and supervision of contractors, and *networking* assists in production of bids and management of expenditure programmes; (4) partnership termination and succession is the last phase, where *networking* between individuals and organizations is used as a means of maintaining agency commitment, community involvement and staff employment. In this case, a management logic had been applied that looks like the 'switching' strategy of metagovernance (see section 4.8). In another study on urban regeneration, also in the UK, it was concluded that governmental hierarchies continued to have a crucial role in coordinating the activities of governance regime (Whitehead, 2003).

Many recent research publications on urban sustainability use the network (meta)governance and transition management approaches, instead of the broader approach that combines hierarchical, network and market governance (e.g. Doberstein, 2013, Frantzeskaki and Loorbach, 2014, Larsson, 2017a, Hölscher, 2018). One reason may be that many publications on urban sustainability governance originate from countries where the network and transition management are popular, such as the Netherlands, Scandinavian countries and also the UK and Canada. Another reason may be that cities may want to have a larger 'free operating space' than is allowed within the rules set by national government and/or supranational organizations, and therefore believe that the 'extra miles' they want to go will be easier when network and market-based instruments will be used.

SDG 12: *responsible consumption and production*

Governance Targets:
12.a Support developing countries to strengthen their scientific and technological capacity to move towards more sustainable patterns of consumption and production.

(*Continued*)

12.b Develop and implement tools to monitor sustainable development impacts for sustainable tourism that creates jobs and promotes local culture and products.

12.c Rationalize inefficient fossil-fuel subsidies that encourage wasteful consumption by removing market distortions, in accordance with national circumstances, including by restructuring taxation and phasing out those harmful subsidies, where they exist, to reflect their environmental impacts, taking fully into account the specific needs and conditions of developing countries and minimizing the possible adverse impacts on their development in a manner that protects the poor and the affected communities.

SDG 12 aims to ensure sustainable consumption and production (SCP) patterns. The word 'patterns' suggests that this requires systemic change. Indeed, SDG 12 is a far-reaching goal in terms of which economic sectors it regards and who should be involved. National governments are expected to take the lead on SCP, promote resource efficiency, halve food waste, achieve sound management of chemicals and wastes, reduce waste, promote sustainable public procurement, and ensure better information availability. Private companies, especially large and transnational companies are expected to adopt sustainable practices and integrate sustainability information into their reporting cycle (United Nations, 2015). In its 2017 progress report on the SDGs, the UN is particularly concerned about the rising use of natural resources worldwide, in particular in Eastern Asia – a trend contrary to the fulfilment of Goal 12, and about weak implementation of international agreements on hazardous waste and on persistent organic pollutants (Lieder and Rashid, 2016).

These policy targets are supported by three governance targets, the first on science and technology, the second on monitoring tools for sustainable tourism, and the third one on removing market distortions, including by restructuring taxation and phasing out environmentally harmful subsidies.

In the 2017 progress report, the UN argues for an additional governance arrangement:

> Achieving Goal 12 requires a *strong national framework* for sustainable consumption and production that is integrated into national and sectoral plans, sustainable business practices and consumer behaviour, together with adherence to international norms on the management of hazardous chemicals and wastes.
>
> (United Nations, 2017b)

A potential governance failure is to consider the transition to the circular economy and to CSR mainly as an informal network governance challenge, with market mechanisms being able to deal with the more difficult challenges. This neglects weaknesses of network governance such as lengthy discussions, lack of rules and lack of democratic accountability. In such a case, support from hierarchical (regulatory) governance mechanisms may be needed. It is a rule of thumb that when a governance framework is malfunctioning, adding something from a 'neglected' governance style may improve the situation. In the case of SDG 12, a stronger role of governments and their legislative power would make the existing voluntary approaches more meaningful (i.e. effective). This means in terms of policy and institutional coherence that voluntary and informal initiatives and agreements between government, private sector and civil society (e.g. 'green deals') should be embedded in a regulatory coherence framework that sets out the rules of engagement between different parties. As regards the role of business, it is becoming clear that businesses who have signed up for corporate social responsibility (CSR) want to be a part of the SDG implementation. Various CSR networks, for example the International CSR Network, which works mainly in African countries, have come to understand that quality of government and governance is fundamental to reach their objectives.

SDG 12 covers three policy areas: consumption, production (including circular economy and resource efficiency) and corporate social responsibility. The first two are included at the UN level in the 10-year framework of programmes on sustainable consumption and production patterns (10YFP) is a global framework of action to enhance international cooperation to accelerate the shift towards sustainable consumption and production (SCP). This framework has a governance structure consisting of a Secretariat, a UN Interagency Coordination Group, a small Board, National and Stakeholders Focal Points, and a Member State body to receive reports. This is only an organizational structure which does not inform about the (meta)governance of the implementation itself.

A number of countries have adopted circular economy plans, usually with a mixture of standards (hierarchical governance), awareness-raising campaigns (network governance) and market incentives, including through public procurement (market governance). However, the governance approach of the plans tends to be static. The idea (of metagovernance) that corrections in the governance mix may be necessary over time, and that to identify the need thereof one should have an early alert mechanism, is still rare.

Promoting sustainable *consumption* is usually based on network governance: voluntary action/measures, and less on hierarchical governance (legislation, standards). However, quality standards for products are on the rise, at least in the European Union. Sustainable *production* is being promoted through series of measures on resource efficiency (water, minerals, metals, etc.) and circular economy (aimed at a zero-waste economy). In 2015 and 2017, the European Commission proposed a reframing of resource efficiency and waste policies into the comprehensive concept of the transition to the 'circular economy'. For implementation of the circular economy, top-down and bottom-up strategies should go together to maintain the interests of all stakeholders, i.e. policy makers, governmental bodies and manufacturing industries (Lieder and Rashid, 2016).

SDG 12 calls for action by governments, consumers as well as producers. On the producer side, the term 'corporate social responsibility' (CSR) expresses the expected roles of the private sector. The United Nations Global Compact promotes concrete business practices, and CSR is also about integrating sustainability in corporate reporting cycles – and about companies helping capacity building on this in developing countries (Mori Junior and Ali, 2016).

It could be argued that business' responsibility for sustainable development also includes a responsibility to support statehood where needed, both in terms of human rights and in terms of ensuring level playing fields and sufficient standards to stimulate innovation and transformation towards sustainability. Weak public administration is not only a nightmare for citizens but also for economic operators (see also SDG 16/17 and Section 11.5 on partnerships). A good example of this broad CSR approach is the International Network for Corporate Social Responsibility (INCSR), which organizes annual conferences on the intersection between CSR, law and ethics.

A recent analysis of CSR policy in China concludes that metagovernance, defined as coordinated governance through via designing and managing sound combinations of hierarchical, market and network governance, is the main approach used by the Chinese government to promote CSR development. The authors see CSR activity as self-governing behaviour, which takes place in a "market environment supervised and coordinated by a hierarchical networked fragmented governance system", which is how they characterize China's governance context (Tang et al., 2018).

Another avenue for CSR is what Glasbergen has coined 'private metagovernance' (Glasbergen, 2011) (see also Section 4.4). With private metagovernance, businesses address fragmentation of private regulations, which emerge through collaboration between private and civil society actors – on standards, for example. Certification schemes are increasingly being used, for example by minerals companies, in order to be able to demonstrate that they are operating responsibly; some schemes can also be used by civil society actors to hold mineral companies to account (Ostrom, 1990).

A research on CSR initiatives from 1994 to 2014 shows a gradual development from independent and fragmented multi-stakeholder networks, collaborative CSR governance, networked CSR governance and integrated networked CSR governance (Ostrom, 1990). The authors conclude that CSR governance can be seen as a whole network, including new interorganizational networks, experimenting with metagovernance. They equal metagovernance with Earth system governance and to governing the commons as framed by Ostrom (1990).

I have already argued that restricting the term 'metagovernance' to one field of application (e.g. the commons, or the earth system), even if it is a huge area, does not fit with the meaning of the prefix 'meta' as 'over and beyond'. Moreover, it excludes areas where new opportunities could be found. For example, governing the commons could go well together with revising labour market policies through education and skills development supporting SCP and CSR.

CSR metagovernance as researched by Albareda and Waddock (Albareda and Waddock, 2016) is a form of metagovernance of network governance as developed by (Sørensen and Torfing, 2009). 'Networked CSR governance' is expected to be able to "better solve the paradoxes and tensions inherent in the processes of multi-stakeholder dialogues" (Davies, 2007). I am not so sure about this, because power relations and trust are crucial – and some interest conflicts are just not reconcilable, not even with a 'mutual gains approach' to conflicts. Another restriction, if not weakness, of the 'whole of network' metagovernance approach on CSR is that makes the *a priori* choice for network governance – which is in my view a political, not a scientific choice to be made. However, the authors themselves take up the role of moral consciousness when they argue that as regards CSR governance "over time, the overall governance type is likely to shift to more centralized (lead organization or NAO)[6] structures" and there is a need "to watch for such developments".

To conclude, metagovernance for SDG 12 would include looking into some of the weaknesses of the current developments. One potential weakness is the trend that the challenge of the transition to the circular economy and to CSR is considered as a network governance challenge mainly, with the market mechanism being able to deal with the more difficult challenges. Metagovernance as engaging in facilitation, mediation and arbitration and managing interactive processes could mitigate this partially, as was observed in a case study in China on conflicts around establishing a waste incineration plant in Guangzhou, China (Fund for Peace, 2017). However, as I have argued before, when a governance framework is not functioning well, adding something from a 'neglected' governance style might do the trick. In the case of SDG 12, it may be that a stronger role of governments and their legislative power would make the existing voluntary approaches more meaningful (i.e. effective).

8.4 Goals on natural resources and ecosystems (SDGs 13,14,15)

SDG 13: *climate action*

Governance Targets:
13.a Implement the commitment undertaken by developed-country parties to the United Nations Framework Convention on Climate Change to a goal of mobilizing jointly $100 billion annually by 2020 from all sources to address the needs of developing countries in the context of meaningful mitigation actions and transparency on implementation and fully operationalize the Green Climate Fund through its capitalization as soon as possible.
13.b Promote mechanisms for raising capacity for effective climate change-related planning and management in least developed countries and small island developing States, including focusing on women, youth and local and marginalized communities.

SDG 13 calls for urgent action to combat climate change and its impacts. Climate change governance has seen various phases in which different governance styles dominated. Around 2009 – the Copenhagen Conference which did not lead to an agreement – there was a strong push to have a legally binding agreement for all UN member states. In order to make this work, consciously or not, the problem definition was reframed from a wicked, multifaceted and diverse problem into a global disaster for which hierarchical governance (command and control) would be the only solution (Meuleman, 2010). Later, a broader approach emerged with a mixture of different governance approaches, which on 12 December 2015 resulted in the Paris Climate Agreement. Climate governance had a steep learning curve, but there is more to learn. For example, it has been argued that basing the negotiations on a pollution model fails to address the underlying cause of climate change, and works against each national interest by emphasizing 'burden sharing' (Moomaw and Papa, 2012).

SDG 14: *life below water*

Governance Targets:
14.a Increase scientific knowledge, develop research capacity and transfer marine technology, taking into account the Intergovernmental Oceanographic Commission Criteria and Guidelines on the Transfer of Marine Technology, in order to improve ocean health and to enhance the contribution of marine biodiversity to the development of developing countries, in particular small island developing States and least developed countries.
14.b Provide access for small-scale artisanal fishers to marine resources and markets.
14.c Enhance the conservation and sustainable use of oceans and their resources by implementing international law as reflected in the United Nations Convention on the Law of the Sea, which provides the legal framework for the conservation and sustainable use of oceans and their resources, as recalled in paragraph 158 of "The future we want".

SDG 14 is about conserving and sustainably using the oceans, seas and marine resources for sustainable development. This combines preserving biodiversity, preventing waste polluting the seas, and developing sustainable fishery practices. On the latter, case studies have shown how metagovernance is and can be used in various countries (Japan, Vietnam and Norway) (Thang, 2018).

SDG 15: *life on land*

Governance Targets:
15.a Mobilize and significantly increase financial resources from all sources to conserve and sustainably use biodiversity and ecosystems.
15.b Mobilize significant resources from all sources and at all levels to finance sustainable forest management and provide adequate incentives to developing countries to advance such management, including for conservation and reforestation.
15.c Enhance global support for efforts to combat poaching and trafficking of protected species, including by increasing the capacity of local communities to pursue sustainable livelihood opportunities.

SDG 15 aims to protect, restore and promote sustainable use of terrestrial ecosystems, sustainably manage forests, combat desertification, and halt and reverse land degradation and halt biodiversity loss. The Goal contains a governance target on (15.a) financial resources to conserve and sustainably use biodiversity and ecosystems, (15.b) to finance sustainable forest management and provide adequate incentives to developing countries to advance such management, and (15.c) to enhance global support for efforts to combat poaching and trafficking of protected species, including by increasing the capacity of local communities.

Some of the other challenges that countries referred to when implementing SDG 15 include: inadequate institutional capacities; lack of participatory coordination frameworks in land and forestry management; illegal logging; and inadequate sewerage coverage and insufficient treatment of effluent. As for next steps, they referred to the development and revision of natural resources and environment-related laws and broad, multi-stakeholder collaboration (United Nations, 2017a).

SDG 15 emphasizes the need to increase finances for the conservation of biodiversity and ecosystems. This, however, can be an ineffective strategy or at least low value for money, when there are at the same time undermining and contradictory policies in other areas such as agriculture, transport and infrastructure, mining and economic development. Policy coherence, supported by institutional coherence, is therefore crucial and strong leadership at central level seems to be a precondition.

8.5 Nexus approach

The interlinkages between the SDGs may be tackled in various nexus approaches. The word nexus stands for a connection or series of connections linking two or more things. The food-energy-water nexus seems the most researched so far (e.g. Venghaus and Hake, 2018). Another important nexus is the climate-conflict nexus, discussed based on case studies in north-western Kenya, the Nile Basin and Israel/Palestine by Ide et al. (2016). From an environmental policy perspective, examples of three nexus areas are (European Commission, 2017a):

- **Air – Mobility:** The impact of transport on air quality and the related environmental, economic and social (including health) costs require environmental authorities, mobility planners, urban planners and economic sectors to work together on a more sustainable concept of mobility, including cleaner vehicles and better transport modality and addressing traffic congestion.
- **Water – Nature – Food:** The way our food is produced and consumed influences water quality and management, the related environmental, economic and social costs, as well as nature and biodiversity. A sustainable food system is therefore needed. At the same time, agriculture needs water of good quality and of sufficient quantity to fulfil its purposes.
- **Nature – Rural land use – Urbanization:** Pressure on nature and biodiversity is caused by both rural and urban land use. On the other hand, nature and biodiversity keep rural areas attractive for various types of land use such as tourism and recreation, ultimately contributing citizens' well-being to the economy.

8.6 Conclusions

In this chapter we have looked at the SDGs from a governance and metagovernance perspective. It turns out that the publications so far are mainly anecdotal and are not sufficient to get to a broad analysis. Another conclusion is that the governance challenges of the SDGs, due to their complexity, the interlinkages, and the differences in governance context in the countries where the Goals should be implemented, cannot be dealt with without a broad governance concept and the intention to use the full metagovernance toolbox. On several SDGs, we have seen literature showing metagovernance analysis and/or the observation that metagovernance is being used by state (or other) actors.

Notes

1 www.brookings.edu/blog/future-development/2015/09/28/why-17-is-a-beautiful-number/
2 www.un.org/sustainabledevelopment/poverty/
3 Oral communication from Ms. Patricia de Lille, Mayor of Cape Town, June 2017, Brussels.
4 A mayor of a large African city speaking at a conference in Brussels, June 2017.
5 Lowndes and Skelcher (1998: 320): *The Dynamics of Multi-Organizational Partnerships: An Analysis of Changing Modes of Governance.*
6 NAO = network administrative organization.

References

Albareda, L., & Waddock, S. 2016. Networked CSR governance: A whole network approach to meta-governance. *Business & Society*, 57 (4), 636–675.
Amore, A. & Hall, C. M. 2017. From governance to meta-governance in tourism? Re-incorporating politics, interests and values in the analysis of tourism governance. *Tourism Recreation Research*, 41(2), 109–122.
Bakker, K., Kooy, M., Shofiani, N. E., et al. 2008. Governance failure: Rethinking the institutional dimensions of urban water supply to poor households. *World Development*, 36, 1891–1915.
Beisheim, M., Liese, A., Janetschek, H., et al. 2014. Transnational partnerships: Conditions for successful service provision in areas of limited statehood. *Governance*, 27, 655–673.
Beisheim, M. & Simon, N. 2015. *Meta-governance of partnerships for sustainable development. Actors' perspectives on how the UN could improve partnerships' governance services in areas of limited statehood*. SFB-Governance Working Paper Series. Berlin: Collaborative Research Center (SFB).
Bell, S. & Park, A. 2006. The problematic metagovernance of networks: Water reform in New South Wales. *Journal of Public Policy*, 26, 63–63.
Boas, I., Biermann, F. & Kanie, N. 2016. Cross-sectoral strategies in global sustainability governance: Towards a nexus approach. *International Environmental Agreements: Politics, Law and Economics*, 16, 449–464.
Bowen, K. J., Cradock-Henry, N. A., Koch, F., et al. 2017. Implementing the "Sustainable Development Goals": Towards addressing three key governance challenges – collective action, trade-offs, and accountability. *Current Opinion in Environmental Sustainability*, 26–27, 90–96.
Centre for Public Impact. 2016. Seizing the moment: Rebuilding Georgia's police. [Online]. Available: https://www.centreforpublicimpact.org/case-study/siezing-moment-rebuilding-georgias-police/.

Challe, S., Christopoulos, S., Kull, M., et al. 2017. Steering the poverty-environment nexus in Central Asia: A metagovernance analysis of the poverty environment initiative (PEI). *Development Policy Review*, 36(4), 409–431.
Christopoulos, S., Horvath, B. & Kull, M. 2012. Advancing the governance of cross-sectoral policies for sustainable development: A metagovernance perspective. *Public Administration and Development*, 32, 305–323.
Davies, J. S. 2007. The limits of partnership: An exit-action strategy for local democratic inclusion. *Political Studies*, 55, 779–800.
De Grauwe, A. 2007. Alternative models in reforming school supervision. *Reforming school supervision for quality improvement, Module 7* [Online].
Doberstein, C. 2013. Metagovernance of urban governance networks in Canada: In pursuit of legitimacy and accountability. *Canadian Public Administration*, 56, 584–609.
Ehren, M. & Perryman, J. 2017. Accountability of school networks: Who is accountable to whom and for what? *Educational Management Administration & Leadership*, 174114321771727.
European Commission. 2017a. The EU environmental implementation review: Common challenges and how to combine efforts to deliver better results. COM(2017) 063 final. Brussels: European Commission.
European Commission. 2017b. *European semester thematic factsheet taxation*. Brussels: European Commission.
European Commission. 2017c. *Quality of Public Administration – A Toolbox for Practitioners*. 2017 edition, Abridged version. Brussels: European Commission.
European Commission. 2018. *A European strategy for plastics in a circular economy*. COM(2018) 028 final. Brussels: European Commission.
Frantzeskaki, N. & Loorbach, D. 2014. *Transition management as a meta-governance experiment to build energy resilience in European cities*. Abstract Submitted to Resilience Conference 2014. Montpellier, France, May 4–8, 2014.
Fund for Peace. 2017. Fragile States Index 2017. [Online]. Available: http://fundforpeace.org/fsi/2017/05/14/fragile-states-index-2017-annual-report/
Gabizon, S. 2018. To achieve Agenda 2030 we must end gender discrimination [Online]. Available: www.sdgwatcheurope.org/to-achieve-agenda-2030-we-must-end-gender-discrimination/.
Glasbergen, P. 2011. Mechanisms of private meta-governance: an analysis of global private governance for sustainable development. *International Journal of Strategic Business Alliances*, 2, 189–189.
Hölscher, K. 2018. So what? Transition management as a transformative approach to support governance capacities in cities. In: *Co-Creating Sustainable Urban Futures*. Heidelberg: Springer.
Holzscheiter, A., Bahr, T. & Pantzerhielm, L. 2016. Emerging governance architectures in global health: Do metagovernance norms explain inter-organisational convergence? *Politics and Governance*, 4.
Ide, T., Michael Link, P., Scheffran, J., et al. 2016. The climate-conflict nexus: Pathways, regional links, and case studies. In: Brauch, H. G., Oswald Spring, Ú., Grin, J. & Scheffran, J. (eds.) *Handbook on Sustainability Transition and Sustainable Peace*. Cham: Springer.
Jayasuriya, K. 2003. *Workfare for the Global Poor: Anti-Politics and the New Governance*. Working Paper No.98. Perth: Murdoch University.
Jayasuriya, K. 2004. The new regulatory state and relational capacity. *Policy and Politics*, 32, 487–501.

Kamau, M., Chasek, P. & O'connor, D. 2018. *Transforming Multilateral Diplomacy: The Inside Story of the Sustainable Development Goals*. London: Routledge.

Larsson, O. L. 2017a. Meta-governance and the segregated city: Analyzing the turn to network governance, knowledge alliances, and democratic reforms in Malmo city, Sweden. Paper presented at the ECPR General Conference, Oslo, 609 September 2017.

Larsson, O. L. 2017b. Meta-governance and collaborative crisis management – competing rationalities in the management of the Swedish security communications system. *Risk, Hazards & Crisis in Public Policy*, 8, 312–334.

Lieder, M., & Rashid, A. 2016. Towards circular economy implementation: a comprehensive review in context of manufacturing industry. *Journal of Cleaner Production*, 115, 36–51.

Lipsky, M. 1980. *Street-Level Bureaucracy: Dilemmas of the Individual in Public Services*. New York: Russell Sage Foundation.

Lowndes, V., & Skelcher, C. (1998). The dynamics of multi-organizational partnerships: An analysis of changing modes of governance. *Public Administration*, 76(2), 313–333.

Lucas, P., Kanie, N. & Weitz, N. 2016. *Translating the SDGs to High-Income Countries: Integration at Last?* New York: IISD [Online]. Available: http://sd.iisd.org/guest-articles/translating-the-sdgs-to-high-income-countries-integrationat-last/. *Sustainable Development Policy and Practice, Guest Article No. 49* [Online].

Magalhães, A., Veiga, A., Ribeiro, F. M., et al. 2013. Creating a common grammar for European higher education governance. *Higher Education*, 65, 95–112.

Meuleman, L. 2010. *Metagovernance of climate politics: Moving towards more variation*. Paper presented at the Unitar/Yale conference 'Strengthening Institutions to Address Climate Change and Advance a Green Economy'. Yale University, New Haven, Connecticut, 17–19 September 2010.

Meuleman, L. 2018. *Promoting Policy and Institutional Coherence for the Sustainable Development Goals*. New York: United Nations, Committee of Experts on Public Administration.

Meuleman, L., Strandenaes, J.-G. & Niestroy, I. 2016. From PPP to ABC: A New Partnership Approach for the SDGs. *IISD SD Policy & Practice*. Guest Article, posted on 11 October 2016 [Online]. Available: http://sdg.iisd.org/commentary/guest-articles/from-ppp-to-abc-a-new-partnership-approach-for-the-sdgs/

Moomaw, W. & Papa, M. 2012. Creating a mutual gains climate regime through universal clean energy services. *Climate Policy*, 12, 505–520.

Mori Junior, R. & Ali, S. H. 2016. *Designing Sustainability Certification for Greater Impact: Perceptions, Expectations and Recommendations in Sustainability Certification Schemes*. Brisbane: Centre for Social Responsibility in Mining (CSRM), The University of Queensland.

Moynihan, D. 2013. A central agency is crucial for disaster response. *Nature*, 495.

Niestroy, I. 2016. *How Are We Getting Ready? The 2030 Agenda for Sustainable Development in the EU and its Member States: Analysis and Action So Far*. Bonn: Deutsches Institut für Entwicklungspolitik.

Ostrom, E. 1990. *Governing the Commons: The Evolution of Institutions for Collective Action*. Cambridge: Cambridge University Press.

Peters, B. G. 2010. Meta-Governance and Public Management. In: Osborne, S.P. (ed.). *The New Public Governance: Emerging Perspectives on the Theory and Practice of Public Governance*. London: Routledge.

Ramesh, M., Wu, X. & Howlet, M. 2015. Governance health care in an imperfect world: Hierachy, markets and networks in China and Thailand. In: Capano, G., Howlett, M. & Ramesh, M. (eds.). *Varieties of Governance: Dynamics, Strategies, Capacities*. Basingstoke: Palgrave Macmillan.

Renaud, E. N. C., Bueren, E. T. L. V. & Jiggins, J. 2016. The meta-governance of organic seed regulation in the USA, European Union and Mexico. *International Journal of Agricultural Resources, Governance and Ecology*, 12, 262–291.

Roth, M. & Ulbert, C. (eds.). 2018. *Global trends analysis. Cooperation in a post-western world: Challenges and future prospects*. Bonn: Development and Peace Foundation (SEF).

Sørensen, E. & Torfing, J. 2009. Making governance networks effective and democratic through metagovernance. *Public Administration*, 87, 234–258.

Sørensen, E., Torfing, J. & Rhodes, R. A. W. 2005. The democratic anchorage of governance networks the new governance: Governing without Government. *Scandinavian Political Studies*, 28, 195–218.

Stott, L. 2017. *Partnership and Social Progress: Multi-Stakeholder Collaboration in Context*. PhD, The University of Edinburgh.

Tang, Y., Ma, Y., Wong, W. C., et al. 2018. Evolution of government policies on guiding corporate social responsibility in China. *Sustainability*, 10.

Taylor, I. & Kelly, J. 2006. Professionals, discretion and public sector reform in the UK: Re-visiting Lipsky. *International Journal of Public Sector Management*, 19, 629–642.

Thang, H. V. 2018. *Rethinking Fisheries Governance. The Role of States and Meta-Governance*. Basingstoke: Palgrave Macmillan.

Thøgersen, M. 2017. Local Governing of Schools in Scandinavia-Between State, Market and Civil Society. In: Sivesind, K.H. and Saglie, J. (eds.). *Promoting Active Citizenship*. Cham: Springer International Publishing.

Thrift, C. & Bizikova, L. 2016. Bottom-up accountability for sustainable development goals (SDGs). Are subnational indicator systems aligned? *IISD Briefing Note* [Online]. Available: www.iisd.org/sites/default/files/publications/bottom-up-accountability-sdgs-briefing-note.pdf.

Uitto, J. I. 2016. The environment-poverty nexus in evaluation: Implications for the sustainable development goals. *Global Policy*, 7, 441–447.

United Nations. 2015. *Transforming Our World: The 2030 Agenda for Sustainable Development*. New York: United Nations.

United Nations. 2017a. *2017 Synthesis of Voluntary National Reviews Synthesis of Voluntary National Reviews*. New York: United Nations.

United Nations. 2017b. Progress towards the sustainable development goals – Report of the Secretary-General. New York: United Nations.

United Nations. 2018. Turning promises into action: Gender equality in the 2030 agenda for sustainable development. New York: United Nations.

Venghaus, S. & Hake, J. F. 2018. Nexus thinking in current EU policies – The interdependencies among food, energy and water resources. *Environmental Science & Policy*. https://doi.org/10.1016/j.envsci.2017.12.014

Waage, J., Yap, C., Bell, S., et al. 2015. Governing sustainable development goals: Interactions, infrastructures, and institutions. In: Waage, J. & Yap, C. (eds.) *Thinking Beyond Sectors for Sustainable Development*. London: Ubiquity Press.

Walker, G. 2014. Water scarcity in England and Wales as a failure of (meta)governance. *Water Alternatives*, 7, 388–413.

Whaites, A. 2016. Achieving the Impossible: Can we be SDG16 believers? Paris: OECD.

Whitehead, M. 2003. 'In the shadow of hierarchy': Meta-governance, policy reform and urban regeneration in the West Midlands. *Area*, 35, 6–14.

World Bank. 2011. *World Development Report 2011: Conflict, Security, and Development*. Washington, DC: World Bank.

9 Metagovernance
Sketching a method

> Today we are all culturally conditioned. We see the word in the way we have learned to see it. Only to a limited extent can we, in our thinking, step out of the boundaries imposed by our cultural conditioning. This applies to the author of a theory as much as it does to the ordinary citizen: Theories reflect the cultural environment in which they were written. . . .
>
> Since most present-day theorists are middle-class intellectuals, their theories reflect a national intellectual middle-class culture background.
>
> – Hofstede, 1980, p. 50

With the wisdom in mind of Hofstede's quote above, I doubt that I am able to present a universal method of applying metagovernance. What this chapter presents is a practical approach with empirical examples, but it is only a sketch; other methods will also be possible, and I realize that the approach I present is more rational than what happens in reality in public governance: applying metagovernance is often first based on intuition, while the analytical reflection comes only later.

9.1 From heuristic to theory and back to practice: metagovernance as method

As someone with natural science as first academic education, I am a little uncomfortable with the apparent ease with which contrasting definitions are being used for the same social science terms. I know that it's all 'in the eye of the beholder' but it makes academic discourse not easy. There are dozens of definitions of governance, and at least three or four of metagovernance. This does not hurt as long as discussions take place within 'schools' which have formed around similar definitions. More difficult I find it when representatives of different schools undermine each other's findings without taking into account that the definitions they are using are different. This happens often in books reviewing the state of play. Fortunately, there are also scholars who succeed in making the relations between the different schools understandable for the student or practitioner of governance. Excellent examples include Mark Bevir in the introduction of his *Handbook of Governance* (Bevir, 2011) and, respectively, Bob Jessop in the chapter on metagovernance and James Meadowcroft in the chapter on sustainable development in the same volume (Jessop, 2011, Meadowcroft, 2011).

Academic concepts such as metagovernance are perceived by different people in very different ways. Practitioners of governance tend to believe that terms like metagovernance have little added value. A public manager responsible for a successful policy process told me: "What you call metagovernance is what I do intuitively and based on my own experience. I don't need such a term" (interview for Meuleman, 2008). Indeed, when I wear my practitioner's hat, I do not really care whether the approach is called metagovernance, modern governance, governance 2.0 or bird's-eye governance or something else. But as a scholar I find the term appropriate and useful. Appropriate because it is governance 'over and beyond' (which is the meaning of the Greek prefix 'meta') governance – and everybody knows the similar relation between physics and metaphysics. Useful, because having an academic term has helped the emergence of quite a body of literature and research projects to help understanding better what the phenomenon metagovernance is and how it can be used.

Using the same 'natural science' type of logic – a concept cannot be at the same time 'over and beyond' another concept and at the same time be part of the other concept – it is clear to me that metagovernance is not a governance style. It is not even governance, but about how to deal with governance challenges. One of the pioneers in governance theory, Rhodes (2017) seems to disagree when he names metagovernance, network governance and decentred governance the three main "waves of governance". His conclusion might be explained from the fact that he implicitly chooses one of the narrow definitions of metagovernance, namely 'managing networks'.

Metagovernance is a structured approach and therefore we should call it a *method*. But methods are seldom value-free. The sociologist Hofstede (1980) rightly claimed that all theories and methods reflect the values of those who designed them. Indeed, behind the 'metagovernance method' there are assumptions and some of them will be linked to the fact that I have a long experience in public sector and – being a network governance–prone born Dutchman – have had for a long time difficulties understanding hierarchical governance and – to a lesser extent – market governance. My assumptions behind metagovernance as method include:

- Governance is less straightforward as proponents of hierarchical and market governance assume;
- Governance is more rational and 'manageable' than proponents of network governance believe is possible;
- It is possible, to a certain extent, to 'rationalize' the governance mess by taking a bird's-eye perspective and keeping an open mind on what's happening in the governance environment.

In Chapter 4, the concept of metagovernance was introduced. In Chapters 5–7 several ingredients for a metagovernance mix were discussed: 50 'shades of governance' explaining the differences between hierarchical, network and market governance (Chapter 5); values and traditions as context of governance (Chapter 6); a series of mind-sets and in particular the ones promoted by New Public Management (Chapter 7). Chapter 8 shows some metagovernance challenges for each of the SDGs and Chapter 10 will review aspects of metagovernance as framework for sustainability governance, in particular for the SDGs.

In this chapter, I propose a sketch of an operational metagovernance method, based on a number of steps (or phases, or rounds). It is an elaboration of the seven-step approach introduced in 2015 (Meuleman and Niestroy, 2015). The seven-step method (Table 9.1) is only in a few but crucial aspects different from traditional policy making, institution building, reform processes, etcetera and their management. It takes into account the governance environment, the applicability of different governance styles, and the use of metagovernance strategies. In addition, like with normal policy making, it is not necessary to execute the steps in the proposed order.

Table 9.1 Sketch of a metagovernance method in seven steps

Step 1. Map the governance environment
Step 2. Evaluate the current situation
Step 3. Define, reframe, refine the problem
Step 4. Formulate context-specific goals and options
Step 5. Design a governance framework
Step 6. Metagovern the governance framework
Step 7. Review the effectiveness of the governance framework

I have not yet found in governance literature well-documented examples of metagovernance applied as method. I will therefore illustrate the suggested method in this chapter in text boxes with reflections on the preparation of a recent initiative of the European Commission: the 2017 Environmental Implementation Review (EIR) which is related to SDGs 16 and 17, from an environmental perspective (European Commission, 2017a). The EIR, of which I coordinated the first ideas, the development of the first package and its evaluation, is an example of an attempt to practice metagovernance, although the term was not used explicitly. I will also use some examples from the European Commission's proposal for a Plastics Strategy, which should contribute to implementing SDG 12 (European Commission, 2018b). Box 9.1 introduces the EIR case.

Box 9.1   The case of the EU Environmental Implementation Review (EIR)

It is estimated that 80% of the existing environmental protection rules in European Union member states are based on European rules. Weak implementation of EU environmental policy and law results in economic, social and environmental costs of around €50 billion per year, notwithstanding the elaborate governance framework the European Commission – the EU's executive – has at its disposition. In 2005 it was concluded that the dual approach of 'stick' (i.e. legal enforcement procedures and fines: a hierarchical governance approach) and 'carrot' (i.e. EU funding of environmental infrastructure on e.g. waste and water: a market governance approach) were not sufficient to close the existing implementation gaps and to create a level playing field across the European Union. Therefore, in 2016 a new tool was announced, the twice-yearly Environmental Implementation Review (EIR), which would add a third approach of analysis, dialogue and collaboration on the ground (i.e. network governance). In February 2017 the first EIR cycle started with the publication of 28 country reports with a 'snapshot' of the state of implementation in each member state. The reports mention the relevant SDGs as part of the goals, identify implementation gaps and suggest actions. The publication of the reports marked the start of a series of national (stakeholder) dialogues on the main implementation challenges, the launch of a peer-to-peer tool to finance exchange of experience between the 28 countries, and political debates at EU level on the common trends and challenges.

The result so far (mid-2018) is not only an increased awareness about the implementation gaps and enhanced mutual learning across countries, but also, in governance terms, that the European Commission now possesses a broader and more flexible toolbox to monitor and stimulate environmental implementation in the member states, which can be used situationally, e.g. by switching from one to another style when appropriate. When legal tools are not effective because they cannot address systemic problems, dialogue and/or funding can be used. When there is no appetite for dialogue, legal procedures may create the necessary pressure. When a national dialogue shows that the problem could be solved with reallocation of funds, this could be a track to explore also at the EU level.

This case can be seen as application of metagovernance, because it started with the observation that the existing governance framework was (a) not functioning in a satisfactory way, because (b) it was lacking an essential part: the 'stick' and 'carrot' were there but the 'dialogue' toolbox was missing. Moreover, (c) it was decided that the EIR would become an addition to legal infringements and EU funding, not replacing these instruments: it would add to and therefore broaden the governance toolbox of the Commission.

9.2 Step 1: map the governance environment

Context is crucial for achieving effective governance. The context in which governance works should therefore be analysed at a very early stage, even before a detailed problem definition is developed. Mapping the governance environment is about questions such as who the relevant actors are and which roles and interests they have. It includes also questions about which values and traditions may play a role, and what is the context in terms of the existing constitutional, legal, political and administrative settings.

Stakeholder mapping

Mapping the relevant internal and external stakeholders, or more generally, actors, is not a one-off exercise: the 'map' needs updating regularly because the context is dynamic and things will change. For example, when I was manager of a land use planning project in the Netherlands it was important that I was aware that three key stakeholders (representing nature protection, tourism and farmers) were preparing a 'green coalition' to actively support the project. This coalition became a game changer among others because I was able to anticipate and support it. In another project my team had to develop guidance for all policy makers in the Ministry of the Environment about how to make an inventory of relevant stakeholders, how to analyse their interests, strengths and weaknesses, and how to reach out to them. The methods we gathered are presented in a book, which is available as open access (Meuleman, 2003). In short:

1 *Factor analysis* – What is the problem? Whose problem is it? Why is it a problem now? Is it a simple or a complex problem? Different stakeholders within and outside government may define societal problems in very different ways, with matching levels of ownership. In addition, "Social problems like the natural environment, minorities, criminality, health care, privacy, mobility, drug abuse and safety can survive under the surface, and arise suddenly in the face of attention from the public and the press" (Bartels, 2013). They become 'issues': "controversial social problems that attract a lot of attention from the public opinion and the media". Issues come and issues go, but not every issue automatically becomes a problem figuring on the political agenda – time, opportunity and power are important factors.

2 *Actor analysis* – Composing a list of the stakeholders that may be interested – or needed, from the perspective of the policy maker, discussing their possible roles, assessing their interests and standpoints, including a possible general preference for one particular governance style, and how to select their representatives from a larger group of candidates. In practice, the tendency will be to concentrate on external stakeholders, but internal actors are equally important: hierarchical actors/layers, other departments, agencies. Among the external actors it is important to consider non-usual suspects such as young business organizations (e.g. young farmers), think tanks and media. Stakeholders may 'reshape' themselves. In this respect, it has been argued that in recent times a new version of classical actors has emerged which exists next to the old ones. There is social (online) media and classical (print) media, monodisciplinary science and transdisciplinary science, and we have among politicians proponents of representative and of deliberative democracy (In 't Veld et al., 2011).

3   *Strength analysis* – Identification of power and positions of actors. Can we trust them? Are their interests opposite or congruent with government's interests? What are the strategies for dealing with 'enemies', 'opponents', 'coalition partners' and 'friends'? Experience shows that the main strategies are about building trust, to convince 'enemies' to become 'opponents', and 'coalition partners' to become 'friends' (Figure 9.1).
4   *Network/relation analysis* – What are the relations between the stakeholders; are they involved in other networks/co-operations? Drawing a network graph is something that could be useful to do periodically, because of the dynamics in networks. Actors can enter or be excluded from a network, for example, or form supportive coalitions or the opposite, enmity. Stakeholders may also have a relevant joint history that influences their positions or their willingness to have an open deliberation.
5   *Argumentation analysis* – What is the present state of the debate? Who argues on an ideological level, a system level, a problem level or an instrumental level (Hoppe and Peterse, 1998)? Take today's newspaper and, sometimes even on the same page, it will show that stakeholders interpret the same issues in a different way. The ideological argumentation puts world views in the spotlight. Corruption, for example, can be seen as the result of inequality or suppression (ideological level), as a lack of detailed regulation and weak institutions (system level), as a wrongly defined problem, namely as a criminal issue and not e.g. an income policy issue (problem level), or as an issue about which not sufficient reliable data are available (instrumental level).
6   *Risk analysis* – What are the risks during the policy-making process or internal project? Are these risks important, and can they be influenced or not? This is a useful tool typically used for project management, but as I have

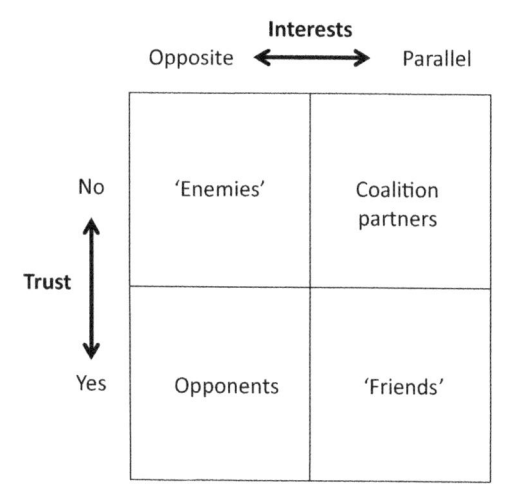

	Interests	
	Opposite ⟷	Parallel
Trust No ↑↓ Yes	'Enemies'	Coalition partners
	Opponents	'Friends'

Figure 9.1 Strength analysis
(After Meuleman, 2003)

argued before, for metagovernance both project and process management are important approaches and should be combined (see Section 5.4, feature 33).

7 *Organizing tailored meetings* – Examples of types of meetings for specific situations include: an internal start-up meeting (internal partners, such as other ministries, also have interests), an external start-up meeting (if this is about creating many ideas, the 'open space' technique or a 'world café' setting could work well), an expert meeting, an opinion-forming meeting, a reflection meeting, a decision-making meeting and an evaluation meeting. Organizing the appropriate type of meetings and communicating about this is a matter of management of expectations and of respect for the target group. Most of us have experienced the confusion (or anger) when attending a meeting that does not fulfil its promises. An open dialogue with local authorities may turn out to be a marketing event to sell a politician's proposal with a flashy presentation. This is one of the best ways to chase citizens away from participating at (and taking co-responsibility for) public challenges.

8 *Negotiation methods* – In the classical method of negotiating, governments and other actors try to create a compromise. However, in a compromise, everybody loses a bit (or more than a bit). A compromise often results from a situation in which there is little trust between parties. This is normal in a hierarchical governance approach: some partners (in this case: governmental actors) are 'more equal' than other partners. In a market approach trust is a delicate issue. Involved parties are, in principle, autonomous and will strive for their own interests. If one chooses a more participatory approach, then building trust is required. Networks rely on mutual understanding of interests, and on the notion that actors are more or less interdependent. This context requires a type of negotiation that concentrates on creating a consensus (everybody may win, to a certain extent).

In the early 1990s Harvard University in the USA developed the so-called Mutual Gains Approach (MGA), which does exactly this (Susskind and Field, 1996). This approach is especially suitable for complex issues in which many stakeholders are involved, such as sustainable development. A core concept of MGA is to make a distinction between positions and interests which are lying behind these positions. Knowing each other's interests opens up a wider range of possible actions than positions/standpoints. Another difference with the classical 'compromise' approach is that one does not start with trying to simplify the issue, but on the contrary, to make the problem more complex. Looking at the interests of stakeholders outside the direct focus of the policy issue may provide more interesting package deals in the end. The strength of this approach is "the acknowledgement of the participants' conflicting interests as a natural fact" (Levin, 2004, p. 80). Negotiating techniques for a multi-stakeholder environment such as the MGA approach requires excellent communication skills of the involved public officials.

9 *Plotting actors on the 'rings of influence'* – After all or some of the above-mentioned methods have been applied it is possible to design the preferred roles for the different actors in the process, for example in a graph similar to

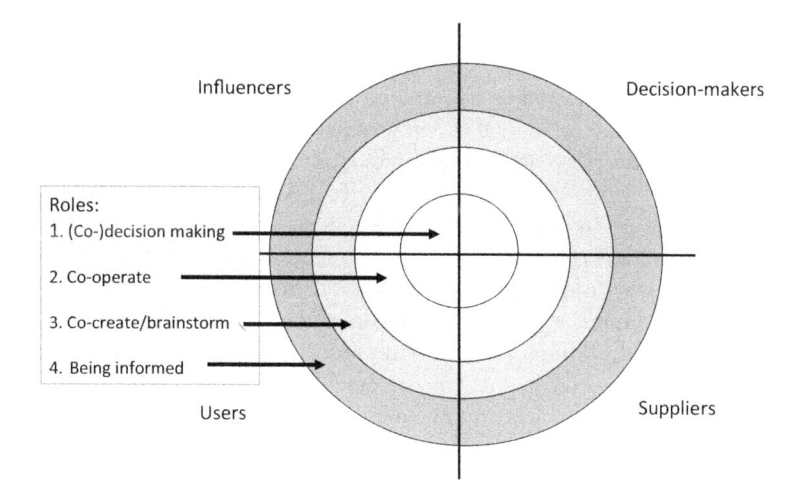

Figure 9.2 Stakeholder roles in the 'rings of influence'

Roles:
1. (Co-)decision making
2. Co-operate
3. Co-create/brainstorm
4. Being informed

Influencers — Decision-makers — Users — Suppliers

the so-called 'rings of influence' (Figure 9.2) which I have used many times in workshops on stakeholder analysis and management.

The centre of the graph belongs to the actors who are directly involved in key decisions. This could be a steering group or board. The second ring is for those who work together within the project or policy process; they are co-operators. Co-creators or brainstormers are not involved on a daily basis but only now and then. They fulfil a sounding board role, for example. The last ring is about being kept informed. These actors do not play an active role in the project but could be in a later phase a co-decision maker (for example members of a parliament. The 'rings of influence' graph is not only useful to plot a stakeholder involvement design on, but also to identify potential conflicts – e.g. when a stakeholder envisions another, more or less active, for itself.

Box 9.2   Stakeholder mapping for the Environmental Implementation Review

As the EIR (adopted in 2017) would become a cycle of analysis, dialogue and collaboration with the aim to improve the implementation of the EU environmental policy and law, connecting well with many stakeholders was essential. Redundancy was seen as a positive principle, in line with Ulrich Beck's second modernity concept (Beck, 1992): it is better inviting too many actors than risking leaving actors out who might have a relevant contribution. Already in an early stage (November 2015), an ad hoc meeting with all national environmental ministries was organized, which was formalized into an official expert group a few months later, and which would meet twice per year. Several civil society and business organizations, as well as relevant other organizations (e.g. OECD, EU Environment Agency) were invited as observers. It was also considered important that other departments of the European Commission would become committed to participate, to promote cross-sectoral thinking and to further policy coherence. Contacts with the EU's official advisory bodies on stakeholder participation (the European Economic and Social Committee – EESC) and sub-national governments (the European Committee of the Regions – CoR) resulted in their close involvement, including by adopting Opinions on the EIR and co-organizing several seminars (CoR). Also specialized networks like IMPEL and Eurocities were contacted. Brussels' offices of environmental NGOs were asked to stimulate their national members to participate in EIR dialogues. Because implementation should not only be a technical but also a political challenge, the package was presented to all EU Environment Ministers at the Environment Council shortly after the adoption. In addition, the European Parliament adopted a Resolution supporting the EIR.

Having worked more than 30 years with stakeholders in fields such as agriculture, transport, economics and finance, some rules of thumb have emerged which could be relevant for working with stakeholders in partnerships in different contexts. In my experience, these rules of thumb do not only apply to working with 'external' stakeholders but also to 'internal' stakeholders within government:

- Always invest time to find out the deeper interests behind stakeholder positions, also – or especially – when these positions are fiercely defended.
- Always approach stakeholders who seem to be highly opposed to the policy objective or the governance approach and who possess little trust in government representatives at a personal level. Often, a large part of the problems is based on unfounded assumptions, on both sides.
- Never exclude stakeholders too light-heartedly; have patience and keep looking for options. There may be an unexpected window of opportunity due to external developments, or they may engage in alliances which make them more prone to cooperate.
- Use tools to find the not-so-usual stakeholders. For example, young farmers' groups may have a different view on the needed long-term policies than the main farmers associations. One of the tools to find 'new' stakeholders is the 'reversed advice' exercise, where participants are asked to list the most

unusual, illogical, controversial actors as potential stakeholders. Following this, it is discussed what each of the top 10 'strange' actors represents (e.g. a policy sector, societal interests), and whether it could have an added value to involve this sector or interest.
- Help making weakly or unorganized interests become vocal, for example by supporting them financially. In the policy development to protect the 'Green Heart' of Western Netherlands in 1995, small recreation enterprises and a group of cultural history experts were both given a small grant to put together better data to underpin their interests (Meuleman, 2003). The latter example can be directly linked with the strong influence of cultural history interests in spatial planning policies in the Netherlands in the late 1990s.
- Be sensitive to a possible imbalance between vocal minorities and silent majorities, especially when many people are not silent because they have no interest or opinion, but because speaking up may result in losing face or other unpleasant reactions.

Mapping values and traditions

Mapping the relevant values and traditions is not an easy task. A useful database about cultural characteristics of countries compared to other countries, based on work of the Dutch sociologist Hofstede is available online[1] and in his publications (e.g. Hofstede and Hofstede, 2005). In addition, the cultural scales formulated by Meyer (Meyer, 2015) are a useful tool, among others, in order to distinguish between 'high context' and 'low context' forms of communication with people from other countries (see Chapter 6).

In addition, it helps to be aware of the impact different cultures can have on what works well and what not, even within one country or organization. Culture has also a historical dimension. The European Commission has in the first place a hierarchical culture, as mentioned earlier "half way between a French Ministry and the German Economics Ministry" (Dimitrakopoulos and Page, 2003); Germany and France were the two largest and most influential countries among the founders of the European Union. But the Commission has at the same time an underlying professional network culture and some aspects of market governance, and much of their external governance is by many authors considered as metagovernance (see Section 4.6).

Box 9.3   Values and traditions and the EU Plastics Strategy

Promoting the circular, no-waste economy cannot succeed without behavioural change among citizens and businesses alike. A Strategy with actions on plastics which should trigger change in all EU member states should be underpinned with knowledge about the different administrative and general cultures in the countries. In 2017, the European Commission started a study project to collect better information about public administration quality in the member states (Thijs et al., 2018). The study confirms what was known from governance and public sector reform studies in the past, namely that public administration everywhere is strongly influenced by hierarchical principles. In addition, North-western European countries are more prone to use network and market governance tools, while the remaining (majority) is traditionally more centralist (e.g. France) or legalist (e.g. Germany, Austria), where hierarchical values are more important. These cultures are mirrored in what citizens accept as interventions from governments. Citizens in North-western EU countries tend to be more individualistic, elsewhere more collectivist. This difference is relevant for the take-up of waste prevention measures, for example: is the decision to separate waste and useless plastics a personal decision, guided by inspiring examples and convincing arguments, or are citizens more likely to follow what the leaders of their in-group ask them to do? As the EU tends to promote the same measures for all countries, stressing the unity among them, the Plastics Strategy (European Commission, 2018a) could not differentiate between different types of countries. The solution was to offer a wide range of actions, some of which might work better in some countries, and others in other countries. As far as legal measures are concerned (on various wastes, not yet on plastics specifically), the EU environmental legislation already offers room for national differentiation: almost all EU environmental laws are so-called Directives which need to be translated into national law, leaving some room for differentiation.

Mapping the institutional environment

In Section 5.3 (feature 17), three different institutional logics were introduced: the logic of hierarchies, the logic of networks and the logic of markets (Meuleman, 2008, 2013). The logic of hierarchies produces centralized institutions, which work on the basis of authority, with rules and regulations and imperatives. The logic of networks tends to produce more informal institutions in which trust and empathy are key values. The logic of markets aims at small, decentralized government, and at using market types of institutions such as contracts, incentives and public-private partnerships.

The report of the EUPACK study comparing 28 EU member states (Thijs et al., 2018) shows how these institutional logics correlate with other aspects, in particular with cultural dimensions of Hofstede (see also Chapter 6). The authors confirm that administrative culture in EU member states is a reflection of the wider societal and political culture and values. They present the example of the high respect of hierarchy, a lack of initiative and a fear of confrontation in Bulgaria, against the background of historical experiences that influence how public sector reforms are conducted. A contrasting example is given from Sweden, which has a much more individualist and egalitarian culture, where it would be expected to consult employees while conducting reforms.

> **Box 9.4 Institutions relevant for the Environmental Implementation Review**
>
> The EU is ruled by three co-legislators: the European Commission, the European Council and the European Parliament. Because the EIR is a political approach to stimulate member states to boost environmental implementation, the Council (i.e. member states) was the most important other institution. However, the Parliament, and two advisory bodies (Economic and Social Committee and the Committee of the Regions) were keen to be also involved. A special challenge is that implementation of EU environmental policy and law in the EU countries is mainly done by subnational governments, with which the European Commission normally does not have direct contacts. This implied that a multi-level approach was necessary, albeit in a relatively informal way as not to 'disturb' the distribution of tasks and responsibilities in the member states.

9.3 Step 2: evaluate the current situation

What does an analysis of the strengths, weaknesses, opportunities and threats (SWOT) of the current policy and governance frameworks say? Are there existing standard solutions which may not be valid anymore? Such a SWOT analysis might take the form as shown in Table 9.2.

Other methods can be found in academic literature and in practical guidance such as the Better Regulation Toolbox (revised 2017) of the European Commission (European Commission, 2017b). It is important that the latter report explicitly refers to the normative dimension of assessment methods and tools – normative, because they contain (often implicit) assumptions about reality, about human behaviour, about what is right and wrong. These assumptions may be related to the values linked to a specific governance styles (Meuleman, 2015). The EU Toolbox argues that assessment methods should be combined and tailored to the needs of a specific situation. Since 2015 ex-ante evaluation is obligatory for the European Commission departments, before a new initiative is allowed to be launched; when it is launched, the first action is starting a full impact assessment process, which combines regulatory and sustainability impact assessment. In Section 5.3 (Feature 29), the relation between governance, metagovernance and impact assessment was discussed.

236 *Metagovernance for sustainability*

Table 9.2 Model for a SWOT analysis of existing policy and governance frameworks

	Existing policy framework		Existing governance framework	
(Internal)	Strengths	Weaknesses	Strengths	Weaknesses
(External)	Opportunities	Threats	Opportunities	Threats

Box 9.5   Ex-ante evaluation, the Environmental Implementation Review and the EU Plastics Strategy

The EIR is a good example that a new initiative does not always start with an evaluation of what was done before. It could also be said the other way around: a thorough evaluation may not have a follow-up until the stars have reached a certain constellation. In the case of the EIR, evaluation reports existed already for years, the EU member states were periodically reporting about their implementation of EU water, waste, air, nature, etcetera policy and law, and the European Commission reacted to this in various ways – but until 2015 there was never a sufficient sense of urgency and of opportunity to do more than sectoral action: it was not possible to approach the issue holistically.

The European Commission's (2018a) Plastic Strategy is a different case. Plastics was included in the existing legislation on waste management and – prevention, but had turned out to be a more serious problem for the environment and human health than was believed in the past. It took the broader scope of the Circular Economy – which added the whole value chain to the standard focus on products – to shed more light on plastics. The fact that China closed its border for import of plastic waste, and that pollution through plastics is close to citizens' experience, also helped developing a higher political priority.

9.4   Step 3: define, reframe, and refine the problem

This step is about what the policy and governance challenges to be addressed in a specific context (e.g. country) are, and if there is agreement on the problem definition. A list of questions such as in Table 9.3 (Meuleman, 2003) could be used to analyse the complexity of the issue at stake.

Problems can be defined in different ways in order to trigger certain types of solutions: hierarchical instruments and disasters are a good match, as well as complex 'wicked' problems and network governance tools, and routine, undisputed problems and market governance methods. Many sustainable development challenges are of the 'wicked' type (Rittel and Webber, 1973). A useful model to distinguish different problem types is depicted in Figure 9.3.

Table 9.3 Model for analyzing the complexity of the problem

Aspects of the complexity of the problem	Yes	No
Is there sufficient 'objective' information available?		
Are there unified standards to weigh different solutions?		
Can the problem be solved without solving other problems?		
Can the problem be solved without cooperation with other parties?		
Is the level of contradictions of the interests low?		
Is the problem expected to stay stable (i.e. not dynamic)?		

(After Meuleman, 2003)

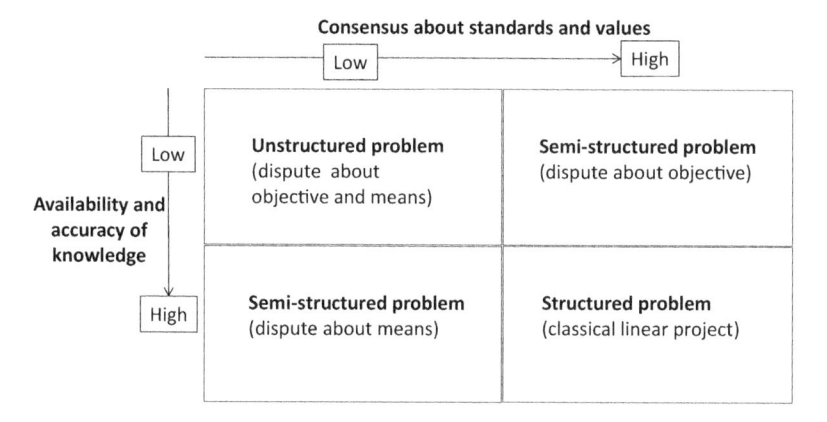

	Consensus about standards and values	
	Low	**High**
Availability and accuracy of knowledge — Low	**Unstructured problem** (dispute about objective and means)	**Semi-structured problem** (dispute about objective)
Availability and accuracy of knowledge — High	**Semi-structured problem** (dispute about means)	**Structured problem** (classical linear project)

Figure 9.3 Problem types
(After Hisschemöller and Hoppe, 1996)

Although implementation of the SDGs is a complex challenge for government organizations, it is good to realize that in daily life at a ministry or other public organization not all problems are complex. Simple, routine and complex problems mingle or succeed each other. At the same time, policy issues are often fuzzy, contested and equivocal and require an 'appropriate' rather than the 'best' answer (Noordegraaf, 2000). Public sector managers/governors should be able to deal with the whole range of problem types. Especially the multi-level, multi-sector and multi-actor character of many contemporary challenges requires design, implementation and tailoring of comprehensive approaches through novel governance practices.

Understanding the dominant or politically given main governance style is important for the analysis of the problem type. Hierarchical governance normally works well with urgent problems and disasters, network governance stimulates addressing complex, multi-faceted and disputed issues, and market governance is at its best when problems are not very complex and have aspects of routine.

Sometimes a problem definition is given by political leaders, which then almost necessitates a specific governance framework. When I started as project manager of a complex, multi-level and multi-actor project in the Netherlands on regulating land purchase for urban development, I had already designed a network-based governance approach when the Minister told me that we should not involve stakeholders at all: in the past, three governments had fallen already on this sensitive issue and he didn't want to be reason for the fourth. So, the approach was redesigned (Meuleman, 2003, p. 86). I like this example because it showed that the political framing power has its limits. The topic was not only sensitive but also hugely important for local authorities and project developers. They started a consensus-oriented network approach among themselves, from which I benefited. The example also shows that metagovernance is not possible under all circumstances.

Framing (see also Section 4.3) can make or break an initiative. In the example of the EU Environmental Implementation Review (Box 9.6), framing and timing went hand in hand: an existing frame (implementation gaps) was reformulated to link it to the political priority of the moment, which was in this case 'better regulation'.

Box 9.6   Problem definition of the Environmental Implementation Review (EIR)

The fact that the implementation of the EU's environmental policy and law was lagging behind in many areas and across countries, with high costs, was not new. Already in 2011, the overall costs of weak implementation were estimated around 50 billion Euros. It was the presentation of the European Commission's Better Regulation package in May 2015 that offered the political opportunity to take additional action. Better regulation included shifting the priority from making new legislation to better implementation of existing law. The EIR was initiated to help closing implementation gaps in order to increase environmental quality, to reduce social and economic costs, and to improve the level playing field of economic operators across the EU. To this frame it was added that existing tools (legal enforcement – hierarchical governance) and EU funding for e.g. waste and water infrastructure (market governance) would remain. In addition, the new tool would not result in additional administrative burden for the member states. This comprehensive problem frame made the EIR initiative politically feasible, even attractive, where earlier attempts had failed. For example, from 2003 to 2010 so-called Environmental Policy Review (EPR) reports were published which had relatively little impact, although they stressed similar themes (namely implementation, integration and involvement (Jordan, 2012, p. 4)) as the EIR does.

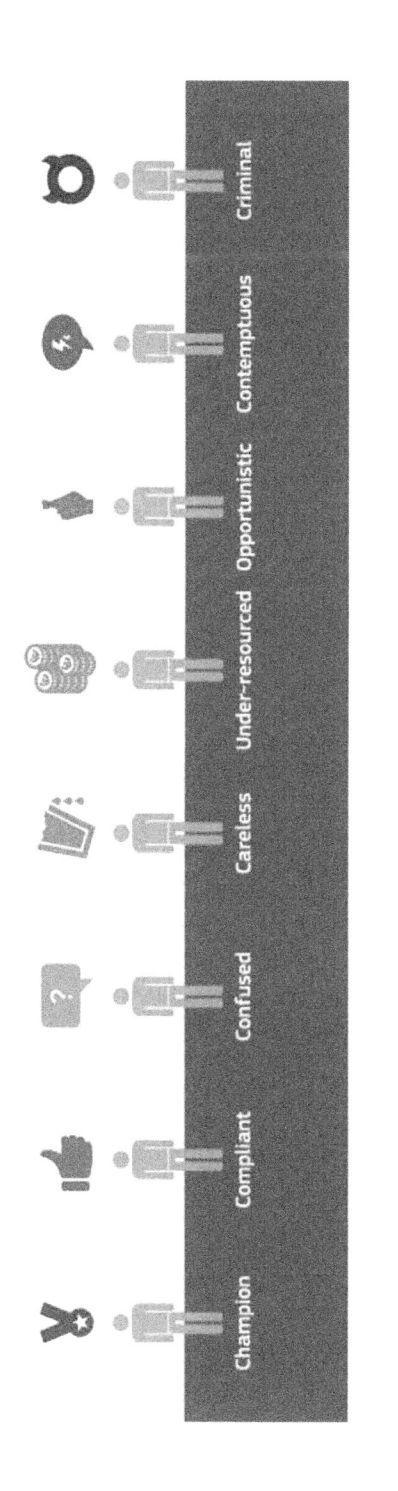

Figure 9.4 Human behaviour and appropriate intervention types
(European Commission, 2018b)

Trying to change human behaviour – when it is harming the common interest – is a general objective of government interventions. This plays an important role in all SDGs, in the social domain (employment, gender balance), economics (poverty, sustainable production) and environmental issues (climate, energy, water). Figure 9.4 gives an example of an analysis of types of behaviour and what could be appropriate responses, prepared during the preparation of a governance framework to promote compliance with EU environmental law.

9.5 Step 4: formulate context-specific goals and options

The fourth step is translating (internationally agreed) goals into national context; formulating country-specific goals and policy options; assessing their benefits and costs on environmental, economic and social parameters; and proposing targets, indicators and time frames. This has the risk of cherry-picking but is also an opportunity to take advantage of a national context to become a regional or global forerunner. A set of principles and priorities as given in Table 9.4 could be the starting point for translation of the SDGs in overall government policies. Box 9.7 gives an example of contextual goal definition.

Box 9.7 Contextual goal definition of the Environmental Implementation Review (EIR)

The political Communication launching the first EIR package in 2017 mentions as a main goal to improve EU environmental implementation through analysis, dialogue and concrete collaboration – three typical network governance concepts. The Communication recognizes that challenges per country are different, but that they still could learn a lot from each other. For the first time, the administrative and governance root causes of weak implementation were addressed, including in the 28 country reports mentioned for each environmental theme the relevant SDG(s). The holistic approach made clear that the same cause, e.g. lack of capacity at the local level, was responsible for a range of different environmental implementation problems which had been dealt with separately. The 2017 Communication ended with three concrete goals: (1) organize inclusive dialogues per country on the main implementation gaps – both on substance and on root causes; (2) stimulate exchange of good practice (peer-to-peer learning); and (3) bring implementation problems to the political level in order to create breakthroughs.

Table 9.4 Principles and priorities for SDG integration

Characteristics of integration	Governance principle	What needs to be coordinated/integrated?	Agenda 2030 principles [and some challenges]
Policy sectors/ areas	Horizontal coordination/ integration	Multiple sectors: economic, social and environmental policies	**Integrative character, for achieving policy coherence** [break down silos]
Policy levels	Vertical coordination/ integration	Multiple levels: local, subnational, national and supranational	Fostering bottom-up and top-down [break down silos]
Actors	Participation	Multiple actors: from politics, business and civil society	**Shared responsibility Leave no one behind** [different levels and styles of consultation and participation]
Knowledge	Reflexivity	Knowledge from different sources ('transdisciplinarity')	**Accountability** Continuous reflection and (peer) learning, evidence-based policy making (e.g. impact assessment)
Time	Intergenerational justice	Long- and short-term thinking	[Election cycles]

(Niestroy, 2014, pp. 154–168, adapted 2017)

9.6 Step 5: design a governance framework

The next step is to design a specific governance framework, based on a selection of elements (institutions, instruments, processes and actor roles) from different governance styles. These elements should not mutually undermine, but enforce and complement each other, with a view on multi-level, multi-actor, cross-cutting and long-term aspects. Such a framework could look as follows (Table 9.5), and could be filled in in a concrete case with the help of the '50 shades of governance' presented as a toolbox in Chapter 5. The column on the left side contains the main general (meta)governance and sustainability principles.

Table 9.6 gives an example of a table which could be used to support the design of a governance framework.

Table 9.5 Design of a governance framework: principles

Focus theme and SD governance principles (*)	Type of questions to be addressed
Institutions	Who is in the lead, who has (legal) responsibilities and duties, what should be multi-level relations, etc.? What role for formal and for informal institutions? Who will manage the governance framework and with which discretionary room?
* Rule of law	How can be ensured that the governance framework and its application are following the rule of law?
* Equity	How is equity anchored in the governance framework?
* Accountability	Which type(s) of accountability (e.g. central, shared, bottom-up) will be appropriate and how will this be organized?
* Transparency	How will pro-active transparency be organized?
* Resilience	How can sufficient resilience against internal or external disruptions be embedded in the framework?
Instruments	Which instruments will be needed to design a new policy or implementation strategy, or organizational change?
* Context-specificity	Can instruments be selected to match with the type of problem and with the governance environment, including the generally preferred/dominating governance style? If they are needed to influence behaviour, do they match the challenge?
* Intergenerational justice (long-term orientation)	How will the long-term dimension be integration – and not as an after-thought – in the governance framework?
Processes	How will decision making, policy and institutional coherence and stakeholder participation (internal and external) be organized and managed?
* Horizontal coherence (coordination/integration)	How are relations between different sectoral departments defined and organized in practice? Balance between formal/legal and informal collaboration? Are there relevant differences in organizational cultures? Need for a change or additional processes?
* Vertical coherence (coordination/integration) (multi-level)	How are the relations between the different layers of government defined and organized in practice? Balance between formal/legal and informal collaboration? Need for a change or additional processes?
* Reflexivity	How can reflexive learning be embedded in the process/project/programme/policy? Sensitivity for early warning signals? Mindfulness of the organization/team?
* Flexibility	How can institutions, instruments and processes be organized that they can be relatively easily adapted to new circumstances?
* Knowledge-based	How can be ensured that the best available 'usable' knowledge will be available when it is needed?
Roles of actors	Which position are stakeholders supposed to have in the management and implementation of the governance framework? (See e.g. the 'rings of influence' method, Section 9.2)
* Inclusiveness	How to make sure that 'no one is left behind'?
* Participation	Which levels of participation are foreseen, for each process phase, and with which purpose?
* Collaboration	How to engage partners in collaboration based on mutual interest and trust?

Table 9.6 Possible template for the design of a governance framework: combining governance styles

Sustainability (meta) governance principles (*)	Features of hierarchical governance	Features of network governance	Features of market governance
Institutions * Rule of law * Equity * Accountability * Transparency * Resilience			
Instruments * Context-specificity * Intergenerational justice (long-term orientation)			
Processes * Horizontal coherence (coordination/ integration) * Vertical coherence (coordination/ integration) (multi-level) * Reflexivity * Flexibility * Knowledge-based			
Roles of actors * Inclusiveness * Participation * Collaboration			

Box 9.8 EU Environmental Implementation Review (EIR) governance

The design of the EIR in relation to already existing governance tools made it possible to start using all governance styles in situational mixtures to promote environmental implementation. Because all policy Units of DG Environment of the European Commission are involved, they could all make use of the new tool in their own way. Some discussed implementation gaps in the context of their annual bilateral meetings about legal enforcement of EU law; others used the new momentum of national dialogues on implementation problems to organize specialized dialogues with member states on their themes, such as air quality or nature protection. Again others focused on promoting the EIR Peer 2 Peer tool for their own themes, such as waste or water management, and the transition to a circular economy. Developments as regards the EIR and the Peer 2 Peer tool are updated weekly online.[2]

9.7 Step 6: metagoverning the governance framework

The metagovernance of a governance framework is more than about implementation. It requires application of principles like reflexivity, resilience, flexibility and allowing redundancy ("and" rather than "or"). It is about using the metagovernance strategies (see Chapter 4): (1) reframing the problem/challenge, (2) switching to another style (or features of another style), (3) recombining style features from different styles, and (4) maintenance of the governance framework.

When a governance framework has a broad toolbox with elements from different governance styles, the 'management' of the framework is on the one hand about implementing these targeted tools, but on the other hand it could be necessary to reshuffle the toolbox and switch to another tool, from the same or another governance style.

9.8 Step 7: review the effectiveness of the governance framework

Reviewing the effectiveness of the governance is usually part of a cyclical policy process, but I would say that it is also a permanent activity during the metagovernance/management phase (step 6).

9.9 Conclusions

The method presented in this chapter is only a rough framework for practitioners. It has to be developed further by practitioners and academic reflection. But it needs more than that. The metagovernance method should become part of training programmes of public sector organizations. Moreover, it is not overly ambitious to argue that, if governments are able to establish special units or teams for approaches like 'nudging' (such as in the UK and USA), it should also be possible to establish a metagovernance 'cockpit' at high-level, where expertise exists to support prevention of governance failures.

Notes

1 www.hofstede-insights.com/product/compare-countries/
2 http://ec.europa.eu/environment/eir/index_en.htm

References

Bartels, G. 2013. *Issues en maatschappelijke problemen. Wat zijn issues en hoe ontstaan ze?* Tilburg: Tilburg University.
Beck, U. 1992. *Risk Society: Towards a New Modernity.* London: Sage.
Bevir, M. 2011. Governance as Theory, Practice, and Dilemma. In: Bevir, M. (ed.) *The SAGE Handbook of Governance.* London: Sage.

Dimitrakopoulos, D. & Page, E. 2003. Paradoxes in EU administration. In: Hesse, J. J., Hood, C. & Peters, B. G. (eds.) *Paradoxes in Public Sector Reform: An International Comparison.* Berlin, Germany: Duncker and Humblot.
European Commission. 2017a. The EU Environmental Implementation Review: Common challenges and how to combine efforts to deliver better results. COM(2017) 063 final. Brussels; European Commission.
European Commission. 2017b. Better regulation toolbox European Commission. Available: http://ec.europa.eu/smart-regulation/guidelines/docs/br_toolbox_en.pdf. Brussels: European Commission.
European Commission. 2018a. A European strategy for plastics in a circular economy. COM(2018) 028 final. Brussels: European Commission.
European Commission. 2018b. Environmental compliance assurance – Scope, concept and need for EU actions. COM(2018) 10 final. Brussels: European Commission.
Hisschemöller, M. & Hoppe, R. 1996. Coping with intractable controversies: The case for problem structuring in policy design and analysis. *Knowledge and Policy: The International Journal of Knowledge Transfer and Utilization,* 8, 40–60.
Hofstede, G. 1980. Motivation, leadership, and organization: Do American theories apply abroad? *Organizational Dynamics,* 9, 42–63.
Hofstede, G. & Hofstede, G. J. 2005. *Cultures and Organizations: Software of the Mind.* New York: McGraw-Hill.
Hoppe, R. & Peterse, A. 1998. *Bouwstenen voor argumentatieve beleidsanalyse.* Den Haag: Elsevier, VUGA.
In 't Veld, R. J., Töpfer, K., Meuleman, L., et al. 2011. *Transgovernance: The Quest for Governance of Sustainable Development.* Potsdam: Institute for Advanced Sustainability Studies (IASS).
Jessop, B. 2011. Metagovernance. In: Bevir, M. (ed.). *The Sage Handbook of Governance.* London: Sage
Jordan, A. 2012. *Environmental Policy in the European Union: Actors, Institutions, and Processes.* London: Earthscan.
Levin, M. 2004. Organising change processes. Cornerstones, methods, and strategies. In: Boonstra, J. (ed.) *Dynamics of Organisational Change and Learning.* Chichester: John Wiley and Sons Ltd.
Meadowcroft, J. 2011. Sustainable development. In: Bevir, M. (ed.). *The Sage handbook of governance.* London: Sage.
Meuleman, L. 2003. *The Pegasus Principle – Reinventing a Credible Public Sector.* Available: http://www.ps4sd.eu/wp-content/uploads/2017/12/2003-The_Pegasus_Principle-book.pdf. Utrecht: Lemma.
Meuleman, L. 2008. *Public Management and the Metagoverance of Hierarchies, Networks, and Markets.* Heidelberg: Springer.
Meuleman, L. 2013. Cultural diversity and sustainability metagovernance. In: Meuleman, L. (ed.). *Transgovernance – Advancing Sustainability Governance.* Berlin/Heidelberg: Springer Verlag.
Meuleman, L. 2015. Owl meets beehive: How impact assessment and governance relate. *Impact Assessment and Project Appraisal,* 33, 4–15.
Meuleman, L. & Niestroy, I. 2015. Common but differentiated governance: A metagovernance approach to make the SDGs work. *Sustainability,* 12295–12321.
Meyer, E. 2015. *The Culture Map. Decoding How People Think, Lead, and Get Things Done Across Cultures.* New York: Public Affairs.

Niestroy, I. 2014. Governance for sustainable development: How to support the implementation of SDGs? In: (ASEF), A.-E. F. (ed.) *ASEF Outlook Report 2014/2015 – Facts & Perspectives. Volume II: Perspectives on Sustainable Development.* Singapore: Asia-Europe Foundation (ASEF).

Noordegraaf, M. 2000. Professional sense-makers: Managerial competencies amidst ambiguity. *International Journal of Public Sector Management*, 13, 319–332.

Rhodes, R. W. 2017. Understanding governance: 20 years on. *Organization studies*, 28(8), 1243–1264.

Rittel, H. W. J. & Webber, M. M. 1973. Dilemmas in a general theory of planning. *Policy Sciences*, 4, 155–169.

Susskind, L. & Field, P. 1996. *Dealing with an Angry Public: The Mutual Gains Approach to Resolving Disputes.* New York: Free Press.

Thijs, N., Hammerschmid, G. & Palaric, E. 2018. *A Comparative Overview of Public Administration Characteristics and Performance in EU28.* Brussels: European Commission.

10 Metagovernance, public sector reform, coherence promotion and capacity building

> The world is littered with examples of innovations that led either to few, if any, improvements, or which had unintended consequences (for example high-rise housing and out-of-town supermarkets).
> – Hartley, 2005

In this chapter we discuss four themes. Firstly, why is public sector modernization is so often without goal or even reference related to major policy goals? How can public sector reform in so many countries and for so long have been executed without much clarity about the relations with strategic goals other than efficiency? The most elaborate part of this chapter is about how to ensure effective policy and institutional coherence, in particular for the implementation of the SDGs (Section 10.2). Lack of coherence is one of the most important governance failures. This section is based on the paper I wrote as member of the UN Committee of Experts on Public Administration, which was published in 2018. The last parts of the chapter are about capacity building for metagovernance (Section 10.3) and measuring progress (Section 10.4).

10.1 Reform to perform: a new public sector reform agenda

In the early 2000s, most public sector reform programmes (of the United Kingdom, the Netherlands, Germany, the European Commission and the OECD, for example) promoted market governance and network governance, and aimed also at restoring elements of hierarchical governance. These were considered as having become too weak and needed to be stimulated, according to the New Public Management doctrine, such as control and accountability procedures. Although conflicts and potential synergies between hierarchical, network and market governance are bound to emerge in public sector organizations and their work, the reform programmes contained no proposals about how to deal with such conflicts and synergies. They did not address governance style interactions. Consequently, the question how to deal with these interactions (the question of metagovernance) was also not addressed. A metagovernance perspective would have implied an awareness of the conflict potential of governance mixtures. Such reform programmes would have discussed the requirements for a metagovernance approach in the implementation phase of the programmes (such as willingness, discretion and capability – see Section 10.3). Finally it would have added an awareness of the limitations of reform programmes. This might have led to explicitly addressing the stumbling blocks for public sector reform, including those resulting from governance style interactions.

A second problem I see with many public sector reform programmes is that they are mainly inspired by New Public Management type of thinking, primarily based upon literature derived from the private sector (Hartley, 2005). Such NMP reforms have been carried globally, starting in Anglo-Saxon countries, spreading across Europe, and inspiring many Asian countries. Exceptions exist, for example in Central Eastern European countries, where post-communist countries chose the classical Weberian model of centralized hierarchy rather than adopting business-like practices when they organized reforms to consolidate the democratic process and enhance economic development (Neshkova and Kostadinova, 2012). Pollit and Bouckaert (2011, p. 10) mention the resistance to NPM in for example France, Germany and the Mediterranean countries because it was considered as not matching with their cultural, ethical and political features. They distinguish besides New Public Management (NPM) two other drivers of reform, namely the Neo-Weberian State (NWS), which is hierarchical and New Public Governance (NPG), which is based on network theory (2011, p. 19).

Hartley argues that in the private sector, successful innovation is often seen to be a virtue in itself, as a means to ensure competitiveness in new markets or to revive flagging markets. In public sector organizations, however,

> innovation is justifiable only where it increases public value in the quality, efficiency or fitness for purpose of governance or services. Moreover, in the public sector at least, innovation and improvement need to be seen as conceptually distinct and not blurred into one policy phrase. Unfortunately, this is not always the case in UK practice where public organizations feel almost obliged to provide evidence and arguments that they are 'modernizing' and 'improving'.
>
> (Hartley, 2005; see also quote at the beginning of this chapter)

Public sector reform often starts with key NPM objectives, but often diverges in a later phase to take into account national specificities. This was observed in multi-country comparative research on 12 OECD countries (Pollitt and Bouckaert, 2011) and in a case study on Bhutan (Ugyel, 2016), which concludes that context and culture matter greatly, especially in relation to transformational public sector reforms.

A third and maybe the most important flaw of many public sector reform programmes is that they lack a sense of direction other than promoting efficiency. Today's public administrations face numerous challenges which are increasingly intertwined, cross-jurisdictional and less predictable; globalization, new technologies and demographic and societal changes challenge public administrations to respond to the ever-changing diverse needs of the populations they serve (OECD, 2015). This fast-changing world requires public sector organizations to innovate. The OECD paper continues about the new leadership that is needed to guide the necessary transformations, and what this implies for human resource management.

Although I value the overall quality of OECD paper, it strikes me that also here a sense of direction is missing, and this is by far not the only example I could name. Looking at the aims and concrete programmes of public sector quality or reform conferences and seminars, one might get the impression that this is all without linkage to the main tasks governments and their organizations have to fulfil in these times. I am questioning whether all change, and every innovation in whatever direction should be strived for. The objective should not be merely readiness to survive in this changing world. Public sector organizations also have a role to play to guide, support, regulate or even block such changes. For this a set of goals should be the framework, and with the SDGs such a set was adopted in the same year (2015) as the OECD paper. This is just one example of how public sector reform and modernization of public administration have become buzzwords without direction.

This absence of goal-orientation is also reflected at the OECD's webpage, which offers a 'framework for public sector innovation'.[1] The 'new landscape' that drives the need for modernization consists of technology, which has changed how citizens interact with government, the higher information level of citizens, increased public expectations, the need for 'responsive government', and the need for governments to come up with new ideas, according to the OECD.

I would argue that governance (form) should follow policy priorities (function) and not the other way around. Modernization of public administration is a means to an end, not an end in itself. This implies that the principles and objectives of public sector modernization in any given country should be derived from the SDGs, both as regards policy priorities and concerning the requisite policy and institutional coherence. Improving the quality of public administration and governance is essential to deliver on the SDGs by 2030. In all policy sectors, systemic and other changes are necessary, and institutions, instruments and work processes need to be (re)directed towards SDG implementation.

Considering that all UN member states in 2015 have adopted the 2030 Agenda, it seems logical that the objectives, strategic direction and operational targets of public sector reform at all levels should be determined by the SDGs. For this to happen, in all countries an evaluation of existing public sector reform and innovation programmes is urgent. Many modernization programmes from the (even recent) past focus on making public administration efficient and smoothly operating machines delivering services to citizens. However, when efficiency is the main driver, effectiveness is bound to be the victim. What the SDGs require is in the first place *effective and tailor-made governance* instead of *efficient and standardized* modernization recipes.

In Section 10.2 the example of policy and institutional coherence will be given. The wider challenge is that in order to get public sector modernization geared to support the SDGs it might be recommendable to allocate a substantial percentage of funds for governance including administrative reform and capacity building to priorities based on the SDGs. An example of what is in a similar case considered to be 'substantial' is that at least 60% of the EU's Horizon 2020 Research and Innovation programme budget is expected to contribute to sustainable development and in particular to implementation of the SDGs.[2]

SDGs 16 and 17 – and the governance targets (a, b, . . .) in the other SDGs – are meant to stimulate targeted public sector modernization. This requires that public sector reform and management are guided by principles essential for the SDGs, such as:

- The indivisibility of the SDGs: focus on promoting policy and institutional coherence, including horizontal and vertical coordination and integration;
- Leadership at all levels should be geared to deal with multiple challenges of various kinds; a good basis for leadership development is still the 'situational leadership' approach developed by Hersey and Blanchard (1988) to be able to switch between directing, coaching, supporting and delegating;
- Reflexivity, resilience, flexibility should be developed in a comprehensive way, because these principles should support each other;
- Accountability: there are various ways accountability can be organized (top-down, and/or participatory);
- Inclusiveness and participation: How can the private sector and civil society become part of SDG implementation?
- Long-term orientation (intergenerational justice): Is there a foresight unit at high-level who stimulates long-term thinking and identifies early warnings about what is coming up?
- Knowledge-based: What kind of mechanisms are in place to ensure that SDG implementation is informed by the best available knowledge?

It also means that, in each country, reflection should take place on the balance between these new requirements and current drivers of public administration and governance modernization, which mainly come from New Public Management ideas such as striving for efficiency above all, 'less is more', 'small government', etc. Ongoing reform programmes should also be adjusted to facilitate the SDGs. Moreover, public administration quality includes the quality of the judiciary, to apply the rule of law.

The points above imply that there is, generally, a need to look at multiple perspectives on governance and to combine different governance styles in smart ways (metagovernance): where hierarchical steering is strong, further improvement comes from integrating tools from network and market thinking, and the other way around. The principle of 'common but differentiated governance' for the SDGs fully applies to the modernization of public administration and governance: there is no one-size-fits-all but there is a need to exchange good practices and adapt them where relevant, using for instance peer-to-peer mechanisms.

Metagovernance and public sector reform 251

The Voluntary National Reviews on progress on the SDGs show that in most countries there are many policies and laws in place that could promote implementation of the SDGs, but their implementation is weak. In all policy fields, quality of public administration and governance are root causes of weak implementation. For EU environmental policy and law, the main root causes identified in 2017 are all in the field of public administration and governance quality: (1) ineffective coordination among local, regional and national authorities; (2) lack of administrative capacity and insufficient financing; (3) lack of knowledge and data; (4) insufficient compliance assurance mechanisms; and (5) lack of integration and policy coherence (European Commission, 2017a).

Public sector reform is particularly important – and difficult – in countries that have been deeply affected by fragility and conflict. Specific guidance to direct such reforms to enable such countries to implement the SDGs is extremely important and it is therefore laudable that the UN Development Programme issued such guidance recently (United Nations, 2018).

This section could only touch upon some features of public sector reform, and is not aiming to be comprehensive in any way. Nevertheless, for the topic of this book, sustainability metagovernance, I think three conclusions can be drawn. Public sector reform should be 'tested' before it starts, on (a) the sensitivity for governance style interactions and metagovernance, (b) on the appropriateness of the normative assumptions – which model or mixture? – on (c) having a sense of direction that should support the implementation of Agenda 2030, and be in any case not detrimental to it.

10.2 Promoting institutional and policy coherence: a metagovernance perspective on the indivisibility of the SDGs

This section analyses how governments could implement the call in SDG 17 (Targets 13–15) to improve "policy and institutional coherence", which the UN considers as a systemic challenge. It identifies interventions to improve coherence and presents some good practice examples of overcoming policy contradictions and improving political steering and administrative quality across policy sectors and across different levels of government. The text below elaborates on a paper I drafted as member of the UN Committee of Experts on Public Administration (CEPA), with input from several other CEPA members.[3]

In the 2030 Agenda for Sustainable Development, the integrated, indivisible and universal nature of the Sustainable Development Goals is stressed. It is essential to build synergies across all dimensions of sustainable development for the effective implementation of the Goals. There is therefore a need for integrated policies that address the relationships among the economic, social and environmental dimensions of sustainable development and among different sectors. In 2017, the High-Level Political Forum on Sustainable Development, as the main platform for reviewing implementation of the Sustainable Development Goals, acknowledged that many countries had already established mechanisms to improve coordination for better implementation of the Goals. Examples include cross-sectoral government working groups, multi-stakeholder committees and high-level coordinators, and some countries are striving for a broader whole-of-government approach.

Incoherence between sectoral policies, between institutional 'silos' and between levels of administration, belongs to the root causes of weak implementation of sustainable development. Lack of policy coherence can be traced back to governance failures such as the lack of dedicated incentives and arrangements to support working together across policy sectors, and between levels of government. Absence of adequate horizontal and vertical coordination is a serious problem. Incoherence linked to political and administrative fragmentation and silo-thinking is prominent and results in huge environmental, social and economic costs.

Although many countries have established mechanisms to improve coordination for better implementation of the SDGs such as cross-sectoral government working groups, multi-stakeholder committees and high-level coordinators, many countries continue to grapple with the challenge of developing and implementing policies that integrate the three dimensions of sustainable development and build on the synergies between the various Goals and targets. Implementation depends on the way institutions are organized and work, and make, deliver and review policies: institutions and institutional infrastructure are crucial for promoting sustainable development. Another precondition for improving policy and institutional coherence for the SDGs is effective leadership with the vision and ownership to build the necessary institutions and policies for domestic resource mobilization, accountability and transparency. Such leadership cannot be outsourced to external experts; it must be in-house capacity. Leadership is needed at the highest level of government and at all levels of public administration.

The synthesis reports of the Voluntary National Reviews (VNRs) of countries at the HLPF and other recent publications have brought together analysis of success and failure on policy and institutional coherence in many countries (United Nations, 2016, United Nations, 2017). However, notwithstanding the positive intentions and promising measures, across the board, sustainability governance by the UN member states is still dominated by centralism, neglect of the complexity and the 'wickedness' of the challenges. This goes together with constructing central, simplified problems to which classical hierarchical governance approaches can be applied (Ziekow and Bethel, 2017). Hierarchical thinking promotes specialization which results in fragmentation and silo-thinking. In addition, the cultural dimension of governance, including its coherence challenges, is often neglected (Meuleman, 2013). The lack of policy coherence can be traced back to governance failures, such as the lack of dedicated incentives and arrangements to support working across policy sectors and among levels of government. The absence of adequate horizontal and vertical coordination is a serious problem.

Definitions of policy and institutional coherence

Targets 13 to 15 of Sustainable Development Goal 17 call for addressing the systemic challenges of policy and institutional coherence for sustainable development. Although the terms "coherence" and "integration" are often used as synonyms, there is a small but significant difference between them which is relevant in the context of the present paper. Policy integration emphasizes taking the objectives of other policy sectors into account (e.g. environmental integration in energy policy) or even merging objectives. The promotion of policy coherence implies ensuring logic and consistency among policies and preventing them from undermining each other. This requires having a kind of coherence "watchdog" function in place when new policies are designed and when policies are being implemented.

Policy and institutional coherence suggests logic and consistency, but the term is subjective and culturally coloured, and no objective measure for coherence exists. Accordingly, the term 'coherence' should always be used in context. On the other hand, the development of logical and consistent policies and functioning institutions is widely recognized as necessary. Political and organizational cultures can hamper coherence, horizontally within or between government departments and vertically between levels of administration. There are often large cultural differences between spending departments, such as those dealing with infrastructure, and regulatory departments, such as departments of justice, the environment or finance. Internationally, this happens between organizations with similar tasks but different national cultural backgrounds. For example, the coherence of energy policies across national borders can be difficult because in some countries such policies are largely privatized, whereas in others they are not.

Policy coherence thus entails achieving consistency between different policies within and across sectors and at different levels of government. Policy coherence for sustainable development, and in particular for the Sustainable Development Goals, builds on the long experience of policy coherence for development in the field of development cooperation, which aims at achieving consistency between foreign aid and other, sometimes contradictory, development-related policy areas, such as agriculture, trade, investment, technology and migration (O'Connor et al., 2016). The objective of both policy coherence for development and policy coherence for sustainable development is to ensure that policy instruments are aligned to support the same objectives. However, many important global agreements lack this requisite coherence during their implementation, which makes them underperform in terms of the desired impact and scale of their outcomes. Moreover, political leaders are not usually held accountable for policy coherence.

Incoherence among policies has a tremendous impact on the implementation of the Sustainable Development Goals. The tackling of climate change (Goal 13) is hampered by the still-existing massive subsidies of fossil fuels, although Goal 7 promotes affordable and clean energy. Hydropower is renewable energy, under the terms of Goal 7, but undermines biodiversity and the protection of nature on land (Goal 15). The shift towards renewable energy may hamper the priority given by some countries to ensuring that people have access to electricity. Transport policies allow the pollution of cities, which is inconsistent with Goal 3, on good health and well-being, and Goal 11, on sustainable cities.

Another example of this is the incoherence between sustainable and inclusive economic development, as promoted in Goal 8, and the fact that national economic policies are usually designed based on the growth of gross domestic product (GDP). Africa's high economic growth over the last 15 to 20 years has been considered good news. However, this growth has not ended the vicious cycle of poverty or ensured inclusive prosperity. Focusing economic growth (Goal 8) on GDP parameters contradicts, among others, Goals 1 (the eradication of poverty) and 10 (reduction of inequalities). Italy is the first country in Europe to adopt a set of development indicators that complement the Sustainable Development Goals and focus on equitable and sustainable well-being, thus implementing Target 17.19. These indicators are now being used to monitor and validate government budgetary policies.

Institutional coherence can be defined as normative integration of institutional arrangements (Scott, 1987). In each society there are political/normative disputes about how institutions should relate to each other, and importantly, "by which institutional logic different activities should be regulated and to which categories of persons they apply" (Friedland and Alford, 1987, pp. 32–33, quoted in Scott 1987). The examples these authors gave three decades ago are still topical, also in the context of implementation of the SDGs: "Are access to housing and health to be regulated by the market or by the state? Are families, churches or states to control education? Should reproduction be regulated by state, family or church?"

Institutional coherence is a means to achieve policy coherence, which is a means to achieve better policy outcomes. Institutional constraints to policy coherence typically include overly hierarchical structures, the lack of a common strategic policy direction and sectoral self-interest. These structural challenges can be compounded by inadequate mechanisms for allocating resources for cross-cutting issues and ensuring shared accountability for shared responsibilities (United Nations, 2015 para 55). There are also often tensions between national policy developers and local policy implementers. These challenges exist, to differing degrees, even in countries where there are clear regulatory mechanisms for the budget across the different levels of government.

Flaws in institutional coherence are responsible for governance failures, including lack of policy coherence, fragmentation of organizations responsible for complex policy challenges, and competition and undermining actions by different administrative organizations. Appropriate institutional coherence requires formal or informal arrangements. In order to prevent policies from undermining each other, leadership is required to establish appropriate reporting lines and guidance on competition for the budget. Good policy coherence may still emerge, even when institutional conditions are not supportive, but the benefits may not last long when different institutions (e.g. sectoral ministries) do not cooperate. This can be even more problematic when there are different political parties in government at the local, regional and national/federal levels and institutions are being used for political purposes.

The policy and institutional dimensions of coherence are aspects of sustainability governance and are highly interrelated. To some extent, they are interdependent. Some degree of institutional coherence is a precondition for policy coherence. Policy officers from transport and environment ministries should be stimulated to work together on traffic congestion, for example. However, policy coherence may also be needed to promote institutional coherence. When ministers from different policy fields agree on a common policy approach, the administrative organization must facilitate its implementation institutionally. One of the inherent problems is that both policies and institutions tend to lose effectiveness over time. The logic they are based on (policy theories or institutional logic) may not apply anymore to changed circumstances after 10 or 20 years. For example, building dikes is a good option to protect against water, unless water levels continue to rise, in which case policy theory should change to work with, instead of against, water, creating "room for the rivers" (the Netherlands). An institutional logic that produces clearly defined silos – which are beneficial for accountability – may need to change to facilitate cross-sectoral programmes. The promotion of coherence should therefore be a dynamic challenge. Similarly, coherence issues take different forms during the policy cycle. A coherent national policy and institutional framework to address climate change may face incoherence during implementation at the sub-national level; conversely, local initiatives may be hampered by lack of coordination between national ministries and agencies.

In order to improve policy effectiveness, institutional coherence may require improvement, for example, by creating interdepartmental project or programme teams or a matrix type of organization and/or using a cluster approach. Quite often, the merging of departments is considered a quick fix to promote coherence. It is clear that climate and energy policies should be integrated as much as possible and, in several countries, this has resulted in the merging of those themes under one ministry. However, it is not at all clear if, and under which conditions, such a merger leads to better handling of the nexus of climate and energy, and if this approach is appropriate for countries in which there is a scarcity of energy.

Experience shows that efforts to promote policy and institutional coherence should focus on: (a) horizontal challenges across sectors, by, among other things, overcoming silo-thinking; (b) vertical challenges across levels of administration; and (c) involving civil society and the private sector in all stages, from policy design to implementation and evaluation (see Figure 10.1).

In order to address the lack of policy and institutional coherence, strategic approaches and tools that can cover both challenges simultaneously are needed. In the present paper, nine approaches or intervention types are suggested as

256 *Metagovernance for sustainability*

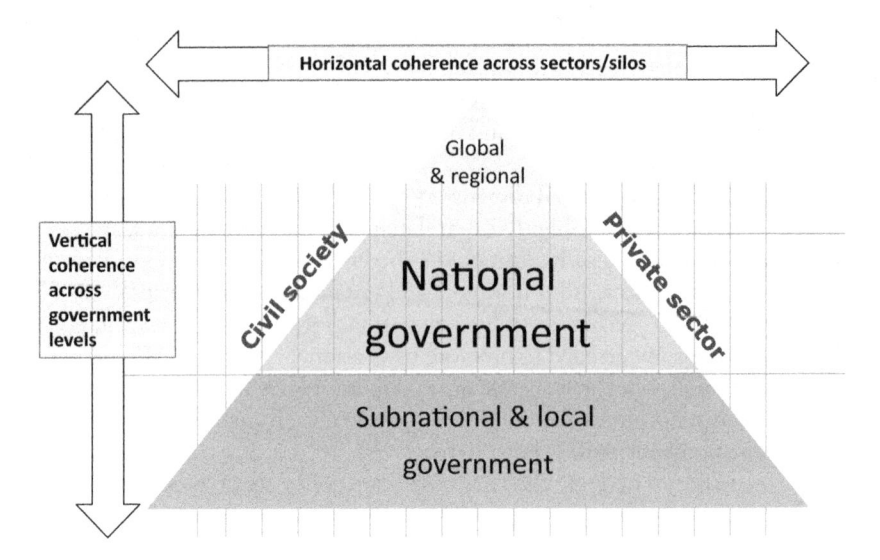

Figure 10.1 Policy and institutional coherence for the SDGs: horizontal, vertical and inclusive

(Meuleman 2018)

potentially helpful, as they have proven to be useful in practice: coordination, integration, alignment, multi-level governance, compatibility, reconciliation, reform, capacity building and empowerment. Those approaches are discussed below.

Approaches for promoting policy and institutional coherence for the SDGs

Table 10.1 gives an overview of nine intervention types to improve coherence, with a selection of examples. The nine interventions are described below in some more detail.

Coordination

Coordination or structured cooperation guided by principles/rules is the best-known approach to promote coherence. It may be more effective, efficient and faster to create working arrangements between institutions representing policy sectors to coordinate policies and institutions than to start a formal reorganization process to merge them. Reorganizations are typically accompanied by a long period of tension and confusion. A number of countries have created interagency/ministerial (high-level) committees to deal with nexus issues and better integrate policy making. In Bhutan, the Gross National Happiness Commission, which

Table 10.1 Intervention types for promoting SDG coherence: examples of successful practice

Intervention types of coherence promotion	Good practice examples of Policy (P) and Institutional (I) coherence promotion
Interventions inspired by hierarchical governance	
Coordination: Structured cooperation guided by principles/rules	(I) Structured involvement of parliament (Argentina, Ethiopia, Germany, India, Trinidad and Tobago)
	(I) High-level coordination arrangements in government
	(P) National SDG implementation strategy (many countries)
	(P) Voluntary National Reviews on the SDGs
Integration: Taking into account another policy or merging policies or institutions	(P) Green public procurement policy (integration environment & economy) (e.g. Netherlands, European Commission)
Alignment: Mutual adaptation of policies/institutions, through formal or informal collaboration	(P&I) Introduction of policy clusters across departments and with non-governmental actors (Cabo Verde)
	(P) Periodical meetings of mayors of a metropolitan city and surrounding communities (South Africa)
Multi-level governance: Structured collaboration between administrative layers	(P) Mainstreaming SDGs at sub-national level (Denmark, Maldives, Nepal, Netherlands, etc.)
	(P&I) National SD Commission including all levels of government (Belgium, Brazil)
Interventions inspired by network governance	
Compatibility: Making contrasting policies/institutions work together while maintaining their character	(I) 'Green Deals' between government, business and civil society (Netherlands)
Reconciliation: Resolving conflicts while achieving better collaboration	(P) Bridging tensions in conflict areas through environmental management (wastewater treatment in Cyprus)
Capacity building: Coaching/ training and creating ownership for policy and institutional coherence	(P&I) Capacity-building activities offered to stakeholders (Indonesia)
	(P) SDG Lab: Joint problem solving via coproduction (Brazil)
	(I) Strengthen local public finance management systems (Honduras)
Interventions inspired by market governance	
Public sector reform: Changing form, structure and/or culture of public sector organizations	(I) Sustainable standards at national Stock Exchange (Botswana, Indonesia, Japan, Nigeria)
Empowerment: Mandating people to work together across or beyond departments and levels	(I) Interdepartmental project teams or directorates (many countries)
	(I) Interdepartmental 'dossier teams' to increase policy coherence (Netherlands)

(Meuleman, 2018)

is chaired by the prime minister, oversees policy as well as institutional coherence with a view to sustainable development. In other countries, the lead is with various ministries, such as planning (Togo), foreign affairs (China and Egypt), finance (Brazil and Liberia), energy (Maldives) regional development (Ukraine) and environment/sustainable development (Belgium). The synthesis reports of the Voluntary National Reviews at the High-Level Political Forum on Sustainable Development (United Nations, 2016, United Nations, 2017) provide a rich sample of high-level coordinating structures, which are sometimes anchored in the constitution (Belgium and Bhutan) or an act (Luxembourg). Some countries have a high-level coordinating committee chaired by the Prime Minister (Costa Rica). Some have appointed a high-level coordinator with an oversight role on coherence (Bangladesh and Nigeria).

Integration

Another popular approach to achieve coherence is integration. Integration implies taking into account another policy or completely merging policies or institutions. This can be a means to improve coherence and consistency, but it is not the only means by far. Horizontal policy integration is best suited to deliver the coherence requirements of the SDGs at the national, regional or metropolitan strategic planning levels (Ziekow and Bethel, 2017, p. 19). They add that this calls for horizontal coordination mechanisms

> able to overcome the fragmentation of content-related perspectives that result from the silo organization of government, which in turn calls for a mix of arrangements at both the strategic and operational levels, which include organizational measures as well as the creation of budgetary incentives and the training of civil servants.

Policy integration may be needed to tackle complex sustainable development challenges, such as the nexus of water and agriculture or the nexus of energy and transport.

Institutional integration usually refers to scaling up, as the dominant mantra in public administration practice is that larger entities are more efficient and effective than smaller ones. Short-term financial gains are indeed often the result of scaling up. Increased effectiveness may also occur, like with any organizational 'shock' which makes people concentrate better – at least for a while. Scaling up and down (or centralizing and decentralizing) are phenomena comparable with the pig cycle in economics. Scaling up creates economies of scale and with this better expertise on systemic issues; scaling down brings organizations closer to citizens and stimulates having better street-level knowledge.

Sustainable development is itself an integrated policy concept with economic, social and environmental dimensions. The constitutions of Bhutan, Belgium and other countries call for the integration of sustainable development in all policies. The Treaty on the Functioning of the European Union contains a key article on environmental integration in all sectors, with a view to sustainable development. Institutional integration typically has the connotation of the merging of departments. The merging of environment and infrastructure policy into one ministry, for example, may help solve traffic congestion and air pollution problems, but it is no guarantee of success. The merging of agriculture and environment into one ministry has resulted in more policy coherence in some countries and in undermining environmental policies in others. The integration of the monitoring function of the policy areas, including statistical or data collection, is a potentially powerful approach. This could render correlations between two policies more visible, which is especially relevant when they are counter-productive.

Alignment

Policy alignment is a lighter approach, in institutional terms, to promoting coherence. It entails the mutual adaptation of policies/institutions in order to create synergies or prevent them from undermining each other, by creating partnerships or alliances between key governmental actors and between governmental and non-governmental actors, for example. A precondition for this approach is to overcome fragmentation by breaking down mental silos within the government and in the relations between the government and stakeholders, by organizing informal meetings, building mutual understanding and trust, and thereby creating a platform for fruitful collaboration (mutual gains approach) (Niestroy and Meuleman, 2016). This could also allow for "ambassadors" or multipliers that are the first groups of stakeholders to be on board.

Alignment should not be confused with the wider call for breaking down institutional silos: without institutional silos there is less focus, structure, accountability and transparency. Civil servants ought to be encouraged and mandated to discuss sustainability challenges more openly with other actors, including non-governmental stakeholders. Policy alignment can be an efficient way to introduce simple measures, when there is no need for large interventions, and it can pave the way and create support for larger transitions. Examples of such interventions are knowledge sharing, experience exchange and championing. New developments, such as block chain technology, could require or force alignment, leading to a need for capacity building for new technologies.

A good practice example of policy and institutional alignment is the introduction of a cluster-based approach within government and in the relations with the private sector and civil society. A cluster is, in Michael E. Porter's well-known definition, a "geographically proximate group of interconnected companies and associated institutions in a particular field, linked by commonalities and complementarities" (Porter, 2000). Clustering triggers monitoring and reporting about policy impact beyond the existing silo structures. It offers a framework through which policy objectives and incentives for the coordinated area can be aligned and different interests associated with different stakeholders (public and private) can be aggregated. Practical tools duly integrated on a unified platform where all departments/sectors are linked can provide the basis for a higher-quality decision-making process and consequently for policy and institutional coherence anchored in an efficient mechanism of resource allocation (budgeting-programming).

From a public policy and institutional standpoint, a cluster-based approach is a powerful tool to identify and manage institutional hurdles to competitiveness and innovation through dialogue among all stakeholders. The cluster-based approach is a good basis for forging partnerships in various areas, such as infrastructure, research, training and regulation, making possible an integrated and coordinated approach. In the European Union, the declared objective of launching an integrated European maritime policy has led to the creation of national clusters within the European Union as a mean to assure policy and institutional coherence. In Cabo Verde, the national medium- to long-term development strategy has been structured in clusters such as the sea, aero-business, information and communications technology and tourism. The aggregation factor in the case of the sea cluster was the country's geostrategic position as an element of competitive and comparative advantage. The sea cluster functioned as a platform, during the planning-budgeting exercise, involving all stakeholders in defining the policies, which contributed to some coherence.

Multi-level governance

Multi-level governance, or structured collaboration between administrative layers, is a special form of policy alignment which is relevant for all of the Sustainable Development Goals. For multi-level governance to function well, the responsibilities of sub-national authorities need to be clearly defined and their resources and skills need to be in line with their responsibilities. In addition, the quality of the interaction between different levels of government highly influences their effectiveness.

Compatibility

Because policy or institutional incoherence is rooted in cultural values or traditions, and in many countries the composition of the population is far from homogenous, ensuring compatibility can be a good approach. Compatibility entails making different/contrasting policies/institutions work together, while keeping their basic differences (e.g. underlying values and objectives) intact. The existing (and growing) cultural pluralism in most countries is often seen as a threat to sustainable development, especially social sustainability. The dominant attitude therefore has been that assimilation of cultural and ethnic views (often euphemized as integration) should be promoted. This ignores the fact that sustainability governance is grounded in cultural values as drivers for social transformation. An alternative approach could be to focus, not on communality or commonly shared values, but on compatibility (De Ruijter, 1995). The compatibility approach recognizes that there are (in principle, valuable) differences which may cause tensions and incompatibilities. These differences should not be removed, but rather regulated. This requires that the government safeguard consistently the values of empathy, tolerance and appreciation of pluralism.

Reconciliation

The reconciliation approach is related to the accommodation approach. When policy or institutional incoherence is accompanied by long-standing disputes between policy sectors and departments, the reconciliation approach can be helpful. Conflict remediation and training in mutual gains approaches can be applied (Susskind and Field, 1996). Leadership is needed to identify the moment for intervention and to manage those approaches and processes.

Capacity building

There is a huge need for investment in capacity building to create understanding and ownership for the promotion of policy and institutional coherence. This includes coaching and training in having a more holistic view, in understanding the full scope of the SDGs, in the diplomatic skills and mutual gains negotiation skills needed to overcome conflicts of interest which prevent policy coherence, and in modern administrative principles and tools. New kinds of policy instruments need to be developed and tested in addition to the classical rules, taxes, incentives and funding, among other things. Public administration schools and training organizations should take the lead on this. Peer coaching programmes could be developed among governments from different countries.

Public sector reform

Public sector reform (see also Section 11.1), or changing the form, structure and/or culture of policies or institutions, is the most drastic approach. Recent reforms have focused on outsourcing, efficiency gains and productivity gains, as in the private sector. Such reforms often include mergers/integration of departments or outsourcing tasks to agencies, which can backfire in terms of promoting coherence. Public sector reforms guided by cost-saving may lead to the dissolution of arrangements established to involve stakeholders and the wider citizenry, which is contrary to the 2030 Agenda principle of leaving no one behind. There is a wide body of academic literature on public sector reform from a comparative perspective from which lessons could be drawn (Pollitt and Bouckaert, 2011). One of the lessons is that public sector reform for promotion of coherence should focus less on efficiency and more on effectiveness. This includes developing new partnerships and other organizational structures that better connect internal silos and link internal and external actors. Information and communications technology is contributing to this shift.

Moreover, public sector reforms should be focused on delivering the SDGs. An efficient, effective and innovative public sector administration does not automatically produce more sustainable results. Implementation of the Goals and the requisite levels of coherence should be part of any reform programme. As such programmes may span several years, consideration should be given to redirecting the ongoing reforms to better deliver on the Goals.

A good example of how coherence can be promoted through public sector reform is the introduction in Cabo Verde of a structured planning-budgeting system. The aim of Target 17.13 of the SDGs is to enhance global macroeconomic stability, including through policy coordination and policy coherence. The achievement of macroeconomic stability requires policy and institutional coherence on a consistent and long-term basis. Planning-budgeting systems are indispensable frameworks for achieving such a goal, taking into account that such systems, once in place, positively pressure organizations to adopt new procedures and processes (organizational reengineering) for delivery. Coordination, integration, alignment and other types of intervention can be made available within a specific planning-budgeting system to manage the decision-making process for efficient and effective delivery. The achievement of policy and institutional coherence through planning-budgeting systems requires the adoption of the following tools: (a) a definition of a medium- to long-term vision/plan on a participatory basis adopting a cluster approach; (b) a medium-term debt strategy (the sustainable financing strategy that guarantees macroeconomic stability within the vision/plan); (c) a medium-term fiscal framework; (d) a medium-term expenditure framework; (e) a plurennial budget/programming; (f) monitoring and evaluation systems; and (g) a results-based management approach.

Empowerment

Empowerment entails mandating people to work together across departments and giving people in the lower ranks of the hierarchy responsibilities and discretionary power to take appropriate decisions. This type of intervention follows the logic that people working close to the policy and institutional challenges are the best placed to assess what should be done in many non-standard situations. These "street-level bureaucrats" include police officers, teachers and health workers. Policy officers implementing the SDGs at the various levels of administration could be given similar discretion.

Underlying principles and practical tools for promoting policy and institutional coherence

There is no general blueprint for promoting policy and institutional coherence for the SDGs, as the circumstances of the member states of the United Nations are too different. Successful practices from elsewhere should be handled with care, as what is successful in one country can be a failure in another. Therefore, it is more accurate to call such successes inspiring examples or good practice, rather than best practices.

The various approaches to policy and institutional coherence should be implemented through mechanisms and tools that have proven to work well in a specific country. Tools to promote coherence are no exception to the rule that tools are never value-neutral: they are based on assumptions about the role of government and about when legal or non-legal instruments should be used, among other things. The values and assumptions behind effective instruments and tools often align with the organizational culture of government departments, which often reflects the views, values and assumptions basic to the national culture and tradition. This makes understanding the cultural identity and diversity within and between countries a requisite for effective governance for coherence with respect to the SDGs. At the same time, global governance structures are ever more interlinked, pushed by digital technologies and the data they generate and the fact that the data can flow easily. Therefore, striving for the common goal of policy and institutional coherence requires differentiated governance on the ground (Meuleman and Niestroy, 2015).

Current insights in administrative effectiveness and sustainability governance, in particular, suggest that the problem of incoherence may not only be caused by the existence of overly hierarchical structures (Hill and Lynn, 2005, Meuleman, 2008, Rayner, 2015). Hierarchy can be the problem, but it can also be a solution. In non-hierarchical countries, rules may be needed to steer the discussions between stakeholders and the government, just as a centralist government could profit from the knowledge and the acceptance resulting from the involvement of stakeholders. A market-liberal country may promote highly efficient but ineffective mechanisms for coherence, following the motto that "less is more". When, for example, a ministerial national cadastral office with an annual turnover of €300 million was turned into an agency, the politically important interface function between ministry and agency was reduced to one staff member for cost-saving reasons, which created an institutionalized risk of coherence problems.

It is possible to cluster the mechanisms and tools for policy and institutional coherence into three groups with more or less similar underlying values, namely, hierarchical, network and market governance tools governance (Meuleman, 2008). Hierarchical governance tools usable for promoting coherence include rules, coordination procedures, monitoring of compliance and provisions to ensure accountability. Examples of network governance tools are establishing or supporting partnerships and informal alliances and organizing peer support and peer review. Market governance entails the use within and by the government of tools typically used by economic operators. These tools include: efficiency measures to eliminate red tape hindering coherence; the empowerment of policy makers; allowing better cooperation; creating agencies; privatization; financial incentives or taxation; and public procurement.

The promotion of coherence requires having all of these tools or mechanisms available in one "toolbox" and using them in ways that are compatible with national/organizational cultures and are, at the same time, as innovative as possible. This can be challenging when one specific governance style is predominant in a country or administrative organization; the skills to operate the other tool families may need to be developed.

Tools for coherence may need to be applied in combination. A partnership approach across policy sectors may require a formal agreement or rules. Keeping a legal obligation or reorganization as a last – but visible – resort may motivate relevant actors to work together on a voluntary basis. Combining governance tools into a dynamic framework requires thinking from multiple perspectives. In addition to being aware about all the available tools, responsible administrative leaders need to have open minds and clear mandates to select and combine tools from different governance styles into an operational governance framework for policy coherence and appropriate institutional arrangements.

In the future, there will be better ways to follow the real-time impact of policy decisions and flexibly adjust them. One example of this kind of adjustment that has already emerged is the redirection or restriction of traffic when pollution levels are too high. Policy simulation tools will allow policy makers to change certain parameters in their policy toolbox mix and then see what the impact may be. What could be the impact, for example, of a 1% tax increase? More specifically, it could be worthwhile to consider using an impact assessment tool to determine the potential benefits and costs of measures aimed at improving policy or institutional coherence. Currently, legislation and policies undergo such (regulatory or other) impact assessments, but for measures aimed at the achievement of coherence this is still the exception, despite the fact that the societal and other costs of failed attempts to promote coherence can be huge.

The promotion of policy and institutional coherence requires the involvement of all stakeholders: the government, public administration entities, universities and research entities, the private sector, civil society organizations and international partners, among others. By adopting institutional arrangements for policy coherence, such as the aforementioned cluster-based approach within a planning-budgeting system, policy areas can be created, in which, through a common denominator, all stakeholders' visions and interests can be integrated or aligned. All stakeholders would have access to a platform dedicated to dialogue and the identification of innovative and disruptive solutions. This, however, requires a strong information system. In the past five years, information systems have been challenged with the big data concept. Big data is already a reality in the developed world. Societies have become information societies in the sense that billions of bytes are produced and captured by different platforms on a daily basis. The challenge is to create information systems that capture this information for the general good. For example, in the case of Africa, big data is providing the opportunity to leap-frog some of the intermediate development phases by providing farmers with greater access to timely, cost-effective and personally relevant information on best practices, markets and prices.

The promotion of coherence is a particular challenge for so-called fragile states (see also Section 8.1). Fragile states face problems in a wide range of domains, such as physical security, legitimate political institutions, sound economic management and the delivery of social services (OECD, 2006). Policy coherence for sustainable development in such countries is linked to challenges in the areas of security, crime and the legitimacy of government interventions.

What has not shown to be the best way to achieve long-lasting change is the outsourcing of policy and institutional innovation. Outsourcing prevents internal learning and the creation of ownership. The traditional call for external (consultancy) support to prepare and propose strategic innovation should be replaced, where possible, by coaching on the job. That way, the implementation of interventions to promote coherence will contribute directly to capacity building. Learning from difficult tasks should be in-house, while less complicated work could be outsourced.

For the small island developing states, in many cases, the scale is too small to develop or hire the necessary expertise to introduce tools to promote coherence. Measures to create economies of scale through close collaboration, with the support of information and communications technology, may be needed. The small island developing states are among the most vulnerable developing countries, and their vulnerabilities are unique and particular. Smallness can be an advantage for policy coherence; greater personal contact, for example, can facilitate closer communication among public servants in different ministries.

Finally, in a number of countries, political administrations change frequently, with or without elections. The specific challenge in these situations is the absence of long-term consistency. Where it is the tradition that many civil servants are replaced after a new government comes into power, it is difficult to maintain the quality of the administration at a high and constant level. These changing situations happen in both developed and developing states, particularly where the democratic system is of the majoritarian type found in many countries. As regards both policy coherence and institutional coherence for the SDGs, this is an unfortunate systemic issue, for which, however, solutions have emerged. For example, investing in the whole of parliament may be a good approach to prevent disruption after a change of government, as has been shown in Jamaica. In Trinidad and Tobago, the parliament has established a Joint Select Committee on the Environment and Sustainable Development and has committed to help implement the SDGs through all acts of legislation, including budget allocation.

Any work programme for promoting policy and institutional coherence should: (a) be light, clear and not overly complicated; (b) be based on an analysis of how to combine several approaches, principles and tools for policy and institutional coherence which are synergetic and do not undermine each other, in a particular case (country, sub-national area, city); and (c) include a mechanism for learning from successful and unsuccessful practices in both different and similar countries.

The promotion of coherence will not work without learning from each other. Implementation of the SDGs by 2030 is so urgent and demanding that we should try to avoid re-inventing the wheel. It is therefore recommendable to redirect existing peer review, peer learning and twinning projects and programmes related to the Goals in order to dedicate a substantial percentage (for example, 10%) of the resources to learning and exchange for the promotion of coherence.

Peer-to-peer learning is a cost-effective example of such learning tools. The introduction of a global peer-to-peer learning tool for the promotion of coherence should be considered. The tool would finance the travel and accommodation costs of experts from one country who are willing to advise another country, at its request, through expert missions, study visits or small workshops. The peer-to-peer tools established by the European Commission for implementation of EU environmental and regional development policies represent a good practice, including on promoting coherence (e.g. European Commission, 2017b).

In addition, it would be recommendable to establish a global network of national coherence promotion coordinators to enable discussion and exchange information on successes and failures. Several countries (such as Nigeria) have already appointed coordinators for the SDGs at a high level, with coherence as part of their remit.

Recommendations on promoting policy and institutional coherence

The following 10 recommendations are proposed to promote policy and institutional coherence for the Sustainable Development Goals. They focus on national governments, but are also relevant at other levels. The implementation of the recommendations requires sponsors, early adopters and supporters among United Nations bodies and member states, as well as other stakeholders.

Principles for better coherence

1. Promote policy coherence always in synergy with the promotion of institutional coherence. To do this, strategies and concrete tools are needed that both cover challenges and prevent contradictions. Public sector organizations should have these tools within reach.
2. Adapt measures for the promotion of coherence to the specificities of Sustainable Development Goals and to the context in which they will be implemented. The principle of "common but differentiated governance" (Meuleman & Niestroy, 2015) fully applies to policy and institutional coherence: coherent and differentiated practices should go hand in hand.
3. Involve the private sector, civil society and the academic world in concrete action for the promotion of coherence. This will bring in indispensable partners with essential knowledge about what works where and when. This is all the more important because private-private (business and civil society) partnerships across sectors are emerging.

Planning, design and implementation for better coherence

4. Develop national work programmes for the promotion of coherence. Such work programmes could contain objectives and tools for the short, medium and long terms and be informed by an assessment of how to combine various

Metagovernance and public sector reform 267

strategies and tools for policy and institutional coherence which are synergetic and do not undermine each other. The work programmes should include a mechanism to monitor their effectiveness and should not create additional administrative burden.
5 Combine multiple approaches to the promotion of coherence. There are various ways to promote coherence. There are nine approaches mentioned in the present paper: coordination, integration, alignment, multi-level governance, compatibility, reconciliation, capacity building, reform and empowerment. They should be considered in a synergistic way.
6 Combine complementary coherence tools. Select and combine coherence-promotion tools from hierarchical (regulatory), network (collaboration) and market (efficiency/incentives) governance. The three families of tools express different and sometimes contrasting, but in principle complementary, cultural values, traditions and assumptions. The promotion of policy coherence and related institutional arrangements requires having all these tools available and the skills to use them.
7 Redirect public sector reform to deliver on the SDGs. Reforms are currently mostly directed at improving efficiency and effectiveness in general; they need to be redirected to promote policy and institutional coherence to advance the implementation of the 2030 Agenda.

Learning for better coherence

8 Introduce a global peer-to-peer learning tool for the promotion of coherence. This would finance the travel and accommodation costs of experts from one country who are willing to advise another country, at its request, through expert missions, study visits or small workshops. It could be based on existing peer-to-peer tools.
9 Establish a global network (community of practice) of national coherence promotion coordinators. This would accelerate mutual learning and the exchange of good practices and failed attempts, among those who are responsible at the national level for progress on coherence. Peer coaching programmes could be developed among governments from different countries. Coaching, also by professional advisors and consultants, could become the new standard to accelerate policy and institutional coherence. Existing networks could be involved, such as the Centre of Excellence for the Sustainable Development of Small Island Developing States.
10 Training is the basis: national public administration schools should integrate the promotion of coherence for the SDGs as a priority in their curricula. Other schools and universities should join this effort.

10.3 Development of metagovernance capacity

In the previous chapters it was argued that public leadership and management with regard to complex and comprehensive challenges such as the Agenda 2030 require tailored approaches. There is not one size fits all. At the same time, all public sector organizations should be ready to implement the SDGs in their context and in their own way, as long as that way is effective. Although case studies have illustrated that metagovernance may be used intuitively, it seems plausible that the metagovernance capacity of an organization or a person can be developed in terms of institutional setting and skills. Therefore, if metagovernance thinking is an asset for (public) governance, the next question should be how to develop the capacity to make this concept work.

Metagovernance capacity can be defined as the ability to analyse, design and manage comprehensive governance frameworks for specific situations. It requires a set of principles and includes skills and capabilities linked to multi-perspective thinking, reflexive learning and understanding stakeholder interests, the strengths and weaknesses of different types of institutional arrangements, single and multi-level settings, policy instruments, policy processes, and actor roles and constellations, in different cultural settings and in different power constellations.

The addresses of metagovernance capacity building are numerous: all kinds of public policy leaders – ranging from project coordinators to politicians holding an office. In addition, actual and potential 'metagovernors' exist in non-governmental and business organizations (Meuleman, 2011, Derkx and Glasbergen, 2014). Different strategies may be applied to enhance metagovernance capacity, involving concepts from different governance styles. Examples are establishing a central 'metagovernance proxy arrangement'; developing guidance showing the benefits of metagovernance capacity; organizing capacity development bottom-up through peer learning in networks; and stimulating competition with incentives among government organizations.

Increasing the capability to apply metagovernance is a challenge for management development: public managers should learn to manage all three basic governance styles. Personal development models may help to reach a level in which managers are able to take such multiple perspectives.

Readiness to implement the SDGs requires the capacity and willingness to look at concrete challenges from various perspectives, and to aim to design and manage governance frameworks composed of elements, where appropriate, from different governance styles. In other words: to apply metagovernance. When metagovernance is a requisite approach to deal with complex governance challenges and in particular with sustainability governance issues, what kind of capacities should then be developed and how? Capacity is a function of governability, which is "the overall capacity for governance of any societal entity or system" (Kooiman, 2008, p. 173): "Governors, the governed and the interactions among governors and the governed all contribute to governability, as do all kinds of external influences".

Metagovernance is applied intuitively by experienced public managers. Laske (2006) has suggested that on average only 10% of people in professional organizations such as public sector organizations have the multi-perspective attitude that is required for metagovernance. A 'metagovernor' accepts and understands the complexity and dynamics of the governance environment and strives for dynamic instead of static quality in terms of Pirsig (1991).

Earlier (Meuleman, 2008, p. 260 ff) I have suggested that metagovernance capacity requires developing three qualifications: willingness, discretionary power and capability:

- Willingness is about being highly motivated and believing that individual people can make a difference, a strong drive to achieve objectives, and the willingness to apply unorthodox measures.
- Having enough discretionary space to prevent and solve most conflicts; discretion is a condition that Lipsky reported as essential for 'street-level bureaucrats' (Lipsky, 1980), but is at least as important public managers and policy makers: the complexity of their environment makes it impossible to ask permission for every action they deem necessary. Discretion levels have been reported to decrease because of the introduction of New Public Management measures such as detailed reporting and accountability systems (Taylor and Kelly, 2006).
- The capability to apply a metagovernance approach is also related to the ability to take multiple perspectives, to stand 'above' the three governance styles and combine them in a way that does not conflict with the own values of public managers. Capability is linked to personal development stages and is an enabler of competences (Laske, 2006). Jessop (2003) formulated the following essential capabilities of metagovernors: a reflexive orientation, the recognition of complexity and variation, self-referentiality and 'requisite irony'.

Leadership for metagovernance

Leadership for metagovernance starts with the recognition that organizing and policy making are about combining various approaches into dynamic mixtures:

> Today, it is important to understand why organizations are pressured to mix different organizational principles, when they give in (or when they do not) such pressure, how they mix, how the mix evolves, and not least how leaders handle the mixing.
>
> (Aagaard, 2016)

In a Danish case study on the Crime Prevention Council, it was investigated how leaders in praxis mix different and partly dichotomous institutional logics in processes of organizational change, using the lens of 'post-transformative leadership' which is pro-active, creative and based on inspirational motivation. Such studies are important in order to modernize public sector training. However, like with governance approaches, public management approaches are contextual; in the Danish case, for example, direct control (classical hierarchical steering) was not an option in the horizontal, network tradition of the Council (Aagaard, 2016).

A survey of 365 senior public managers in Copenhagen, Rotterdam and Barcelona concludes that it is important to broaden the perspective of leadership research in the public sector beyond the transformational type and to look for other important qualities such as risk, motivation and network governance skills rather than visionary leadership alone (Ricard et al., 2017). The survey observes many common characteristics in the three municipalities of Copenhagen, Rotterdam and Barcelona, situated respectively in northern, central and southern Europe. One reason may be that they are all Western administrations, but the researchers assume that another reason is that leadership literature spreads particular visions of leadership across the public (and private) sector.

Applying metagovernance does not have to be confined to one actor but several 'metagovernors' can exist at the same time, while internal and external actors may be dominant metagovernors, possibly with differing roles, as was observed in urban governance (Kroemer, 2010). Consequently, in discussions during presentations on theory and practice of metagovernance, one question comes up frequently: Who is in charge when metagovernance is applied to analyse, design, structure, manage and evaluate policy development of implementation? Is it the metagovernor? And who is in charge of the metagovernor? These questions underlie several assumptions, such as:

- 'There is always someone in charge' (which is not always the case, especially when responsibility is shared: co-governance or other forms of network governance).
- 'The person who has the bird's-eye perspective (which is typical for a metagovernors) should be the one in charge' (however, it could also be an advisor to a leader who has this broad view).
- 'It must be a government official/manager who is in charge of public policy making, so there cannot be non-public metagovernance' (again, shared governance could include rotating leadership between actors, or even leadership from a central stakeholder).

The fact that public metagovernors often have a central responsibility for a policy process and may be the ones whose career depends on success or failure of the policy development or implementation may distract from the fact that applying a metagovernance approach is not necessarily connected to having (hierarchical) (public) power. Anybody, at each level, has some tasks, responsibilities and discretionary power and may choose (or not) to prioritize his or her actions based on an analysis of the governance environment, the governance style most suitable for the challenge at hand, and of his or her relative position in the governance 'game'.

Peer learning

Implementation of the universal Sustainable Development Goals (SDGs) requires mutual learning and support between countries. Other than with the

Millennium Development Goals this is not one-way traffic: developed and developing countries can show each other ways forward. I already mentioned (Section 10.2) that the peer-to-peer tools established by the European Commission for implementation of EU environmental and regional development policies represent a good practice (e.g. European Commission, 2017b). These tools are based on the much longer experience the Commission has with peer learning and twinning, in its TAIEX programme between its member states and countries neighbouring the EU.

'Peer support' or more extensive 'peer reviews' seem to be indispensable tools. Like every governance tool, peer tools are normative: they have hidden assumptions and norms about reality and about what is right or wrong, which may or may not be recognized or acknowledged. The first assumption behind peer tools is that there are people 'out there' are of equal standing and experience. This is a cultural issue. In societies with strong class differences, only peers are acceptable which stem from the same class as those who should benefit from peer support. India is a classic example – but also the UK, although this is a political taboo, has the signature of a class society. In egalitarian countries like the Scandinavian nations and the Netherlands, the ruling dogma is that everybody is equal and therefore could be a peer. In the Kingdom of the Netherlands, for example, it is said that there are 17 million kings and queens, who together pay one couple to do the royal job. The second assumption is that people are always open to learning, and to change. This also differs across countries. Exposure to criticism as well as change can be dangerous – you might lose face, or power, or even worse, both. In many Asian countries it is a risk to have a provocative opinion – not only could this result in losing face but it also could destroy harmony. When you are a dictator everybody who calls him/herself a peer is a direct threat and should be taken care of immediately.

This is why I would generally suggest as a first step to invest in very simple peer tools such as expert visits, small expert workshops or study visits – and consider more extensive peer review only when the short version is successful.

10.4 Metagovernance of measuring progress

Measuring progress on the implementation of goals requires the availability of meaningful and widely accepted indicators. Such indicators are not just technical, but always also political. "If indicators for the Sustainable Development Goals are to make a difference, they need to compete for attention and gain relevance in the same theatre of decision making as widely used economic indicators" (Pintér et al., 2017, p. 125). In their review of indicators for the SDGs, these authors therefore use a governance lens to look at such indicators and analyse relevant concepts, actors, mechanisms, institutions and instruments. Moreover, I would argue that the selection of indicators should be in line with the purpose of measuring progress. Measuring only what can be measured quantitatively and basing decisions on them is a clear predictor of governance failure.

Various sets of indicators have been developed to measure progress on implementation of the SDGs. The think tanks SDSN and Bertelsmann Foundation developed a country-level SDG Index and SDG Dashboards that covered 149 of the 193 UN member countries and 77 indicators (Sachs et al., 2016). The UN Statistics Division has published a Global SDG Indicators Database with 232 indicators, adopted at the 48th session of the United Nations Statistical Commission in March 2017.[4] The European Commission's statistical office published a list of 100 indicators in 2017 (European Commission, 2017c).

All indicator sets are compromises and never 100% scientific: they contain choices such as trade-offs (which indicators could be a proxy for the very broad targets in SDG 17, for example?), selection of sub-optimal quantitative indicators because they are all there is (and qualitative indicators do not count in an index), and weighing decisions. For example, the decision to give all SDGs an equal weight in the SDG Index leads to the counter-intuitive conclusion that huge spill-over effects such as the footprint of rich countries in developing countries does not count in the overall picture. Therefore, many EU member states score very high on the Index, despite their high footprint. Measuring innovation is a dubious indicator when it is not focused on sustainable innovation.

It has been argued that the countries performing highest on the Bertelsmann/SDSN Sustainability Index for the SDGs (Sachs et al., 2016) are by far not the ones closest to the 'sustainable development quadrant' in the well-known graph that plots countries according to their development achievements (using the UN's Human Development Index) against their resource use (using Global Footprint Network's Ecological Footprint (Wackernagel et al., 2017)). The SDG index ranking mimics the conventional development pattern that links higher development achievements with higher ecological footprints. This raises the question if implementation of the SDGs is a good measure of sustainable development. At least it points at the dispute about how to best measure SDG implementation.

Measuring progress depends on having adequate indicators and data, and this again depends on the quality of statistical offices. To receive their data such agencies need to coordinate with many other agencies and authorities. A case study of Australian and British statistical offices showed a transition from traditional top-down interagency coordination to metagovernance, driven by failures and learning. They use combinations of coordination instruments to steer dispersed actors toward common goals (Howard and Bakvis, 2015). In many countries, there is a lack of (historical) data to base progress on. A UN project establishing an indicator-based assessment for the Arab Sustainable Development Report (ASDR) tried to overcome this by producing a 'snapshot' of trends and progress on 56 key indicators, graphic visualization with 'weather' and 'traffic' symbols (Allen et al., 2017). This approach helped clarify the relations between progress in some areas (e.g. water, sanitation, electricity and health) and the costs thereof, e.g. in terms of sustaining the natural resource base, with consumption of water far outstripping availability.

Metagovernance and public sector reform 273

Indicators are a means, not an end in itself. Therefore, a 'plausibility check' would be recommendable as part of decision making about using them for policy decisions. Are the results of applying an indicator supported by observations on the ground, or not? Are the same indicators equally relevant for very different national circumstances?

A review of SDG indicators selected by nine countries (Finland, Germany, Georgia, Mexico, Sierra Leone, South Korea, Switzerland, Venezuela, Uganda) to illustrate progress on SDG implementation in their Voluntary National Reviews (Bizikova and Pinter, 2017) shows a wide variety which can partially be explained by different national contexts (including availability of data), but is probably more related to different political priorities. The SDGs that are most included in the reports are about the social and economic dimensions of sustainable development, while the least-reported indicators are related to environment and nature, and emerging issues such as responsible consumption and production (SDG 12). One of the conclusions of the authors of this survey is that

> tracking progress should not leave blank spots – indicators should cover not only what is easy and convenient for policy, but also those issues, such as sustainable consumption and production, that are conflicted, where entrenched interests may work against transparency or change but that may lie at the heart of the SDG challenge.

In situations where there is a lack of trust about data, block chain technology might become important in the future, for example in order to achieve database consistency and integrity in a context of a distributed decentralized database. A recent pilot study on using block chain technology to increase transparency and trust in agrifood chains has demonstrated that it is feasible to put basic information concerning certificates on a block chain, to ensure that different parties share the same layer of information on the validity and provenance of food integrity certificates (Lan et al., 2017).

10.5 Conclusions

- Public sector reform should be 'tested' before it starts, on (a) the sensitivity for governance style interactions and metagovernance, (b) on the appropriateness of the normative assumptions – which model or mixture?, on (c) having a sense of direction which should at least be not detrimental to the implementation of Agenda 2030, but preferably promoting this.
- Promoting policy and institutional coherence for the SDGs should be inclusive, well-coordinated in e.g. national programmes, based on a range of available approaches, supported by dedicated reforms, accompanied by peer learning programmes, training and networks of practitioners.
- Developing metagovernance capacity requires focusing on skills and capabilities linked to multi-perspective thinking, and on developing willingness, discretionary power and capability to apply metagovernance.

- Indicators for measuring progress on the implementation of the SDGs are always a trade-off between technical and normative aspects; this implies that indicators sets are partially coloured by governance style preferences and a metagovernance view would mean trying to be sensitive to this: progress reports are the result of a whole chain of selection and making trade-offs, and their merits can only be assessed when this chain is transparent.

Notes

1. www.oecd.org/gov/innovative-government/a-framework-for-public-sector-innovation.htm (retrieved on 03.02.2018)
2. http://ec.europa.eu/research/participants/docs/h2020-funding-guide/cross-cutting-issues/climate-sustainable-development_en.htm retrieved on 3 February 2018
3. The paper on which Sections 10.2–10.6 are based was prepared in consultation with five other members of the United Nations Committee of Experts on Public Administration, whom I am very grateful for their suggestions: Cristina Duarte, Geraldine Fraser-Moleketi, Bridget Katsriku, Ma Hezu and Joan Mendez.
4. https://unstats.un.org/sdgs/indicators/indicators-list/

References

Aagaard, P. 2016. How to make the mix matter: A case study of post-transformational leadership in hybrid public management. *International Journal of Public Administration*, 39, 1171–1179.

Allen, C., Nejdawi, R., El-Baba, J., et al. 2017. Indicator-based assessments of progress towards the sustainable development goals (SDGs): A case study from the Arab region. *Sustainability Science*, 12, 975–989.

Bizikova, L. & Pinter, L. 2017. Indicator preferences in national reporting of progress toward the sustainable development goals. *IISD Briefing Note*. IISD. Available: https://www.iisd.org/sites/default/files/publications/indicator-preferences-national-reporting-progress-toward-sdgs.pdf.

European Commission. 2017a. *The EU Environmental Implementation Review: Common challenges and how to combine efforts to deliver better results*. COM(2017) 063 final. Brussels: European Commission.

European Commission. 2017b. *TAIEX-EIR Peer 2 Peer. A new tool bringing together environmental implementation policy experts*. Available: http://ec.europa.eu/environment/eir/p2p/index_en.htm. Brussels: European Commission.

European Commission. 2017c. Sustainable development in the European Union. Monitoring report on progress towards the SDGs in an EU context. Brussels: Eurostat.

De Ruijter, A. 1995. Cultural pluralism and citizenship. *Cult Dynam*, 7(2), 215–231.

Derkx, B. & Glasbergen, P. 2014. Elaborating global private meta-governance: An inventory in the realm of voluntary sustainability standards. *Global Environmental Change*, 27, 41–50.

Friedland, R., & Alford, R. R. 1987. Bringing back. In: *Symbols, Structure and Institutional Contradiction*. Paper presented at the Conference on Institutional Change, Center for Advanced Study in Behavioral Sciences, Stanford, CA, May 15–16.

Hartley, J. 2005. Innovation in governance and public services: Past and present. *Public Money & Management*, 25, 27–34.

Hersey, P. & Blanchard, K. H. 1988. *Management of Organizational Behavior: Utilizing Human Resources*. New York: Prentice Hall.
Hill, C. J. & Lynn, L. E. 2005. Is hierarchical governance in decline? Evidence from empirical research. *Journal of Public Administration Research and Theory*, 15(2), 173–195.
Howard, C. & Bakvis, H. 2015. Conceptualizing interagency coordination as metagovernance: Complexity, dynamism, and learning in Australian and British statistical administration. *International Journal of Public Administration*, 1–12.
Jessop, B. 2003. Governance and metagovernance: On reflexivity, requisite variety, and requisite irony. In: Bang, H. (ed.). *Governance as Social and Political Communication*. Manchester: Manchester University Press.
Kooiman, J. 2008. Exploring the concept of governability. *Journal of Comparative Policy Analysis: Research and Practice*, 10, 171–190.
Kroemer, J. 2010. *Meta-Governors and Their Influence on Network Functioning. A Study of Meta-Governance in the Case of the European City Network Eurocities*. Master International Public Management and Public Policy, Erasmus University Rotterdam.
Lan, G., Brewster, C., Spek, J., et al. 2017. *Blockchain for Agriculture and Food; Findings from the Pilot Study*. Wageningen: Wageningen Economic Research.
Laske, O. 2006. *Measuring Hidden Dimensions. The Art and Science of Fully Engaging Adults*. Medford, MA: Interdevelopment Institute Press.
Lipsky, M. 1980. *Street-Level Bureaucracy: Dilemmas of the Individual in Public Services*. New York: Russell Sage Foundation.
Meuleman, L. 2008. *Public Management and the Metagoverance of Hierarchies, Networks, and Markets*. Heidelberg: Springer.
Meuleman, L. 2011. Metagoverning governance styles – broadening the public manager's action perspective. In: Torfing, J. & Triantafillou, P. (eds.). *Interactive Policy Making, Metagovernance and Democracy*. Colchester; ECPR Press.
Meuleman, L. 2013. Cultural diversity and sustainability metagovernance. In: Meuleman, L. (ed.). *Transgovernance – Advancing Sustainability Governance*. Berlin/Heidelberg: Springer Verlag.
Meuleman, L. 2018. *Promoting Policy and Institutional Coherence for the Sustainable Development Goals*. New York: United Nations, Committee of Experts on Public Administration.
Meuleman, L. & Niestroy, I. 2015. Common but differentiated governance: A metagovernance approach to make the SDGs work. *Sustainability*, 12295–12321.
Neshkova, M. I. & Kostadinova, T. 2012. The effectiveness of administrative reform in new democracies. *Public Administration Review*, 72, 324–333.
Niestroy, I. & Meuleman, L. 2016. Teaching silos to dance: A condition to implement the SDGs. *IISD SD Policy & Practice*. Guest Article. posted on 21 July 2016 [Online]. Available: http://sd.iisd.org/guest-articles/teaching-silos-to-dance-a-condition-to-implement-the-sdgs/.
O'Connor, D., Mackie, J., Van Esveld, D.,.Hoseok, K., Scholz, I. & Weitz, N. 2016. *Universality, Integration, and Policy Coherence for Sustainable Development: Early SDG Implementation in Selected OECD Countries*. Washington DC: World Resources Institute.
OECD. 2006. *Whole of government approaches to fragile states*. DAC Guidelines and Reference Series. Paris: OECD.
OECD. 2015. *Lead-Engage-Perform. Public Sector Leadership for Improved Employee Engagement and Organisational Success*. Discussion Paper. Paris: OECD.
Pintér, L., Kok, M. & Almassy, D. 2017. Measuring progress in achieving the Sustainable Development Goals. In: Kanie, N. & Biermann, F. (eds.) *Governing Through Goals. Sustainable Development Goals as Governance Innovation*. Cambridge, MA: MIT Press.
Pirsig, R. M. 1991. *Zen and the Art of Motorcycle Maintenance: An Inquiry Into Values*. London: Vintage Classics.

Pollitt, C. & Bouckaert, G. 2011. *Public Management Reform: A Comparative Analysis: New Public Management, Governance, and the Neo-Weberian State*. Third Edition. Oxford: Oxford University Press.

Porter, M. E. 2000. Location, competition, and economic development: Local clusters in a global economy. *Economic development quarterly*, 14, 15–34.

Rayner, J. 2015. The past and future of governance studies: From governance to metagovernance? In: Capano, G., Howlett, M. & Ramesh, M. (eds.). *Varieties of Governance: Studies in the Political Economy of Public Policy*. Basingstoke: Palgrave Macmillan.

Ricard, L. M., Klijn, E. H., Lewis, J. M., et al. 2017. Assessing public leadership styles for innovation: A comparison of Copenhagen, Rotterdam and Barcelona. *Public Management Review*, 19, 134–156.

Sachs, J., Schmidt-Traub, G., Kroll, C., et al. 2016. SDG Index & Dashboards, A global report. *Sustainable Development Solutions Network*, 58–58.

Scott, W. R. 1987. The adolescence of institutional theory. *Administrative Science Quarterly*, 32, 493–511.

Susskind, L. & Field, P. 1996. *Dealing with an Angry Public: The Mutual Gains Approach to Resolving Disputes*. New York: Free Press.

Taylor, I. & Kelly, J. 2006. Professionals, discretion and public sector reform in the UK: re-visiting Lipsky. *International Journal of Public Sector Management*, 19, 629–642.

Ugyel, L. 2016. Convergences and divergences of public sector reform in Bhutan: dynamics of incremental and transformational policies. *Asia Pacific Journal of Public Administration*, 38, 118–130.

United Nations. 2015. Committee of experts on public administration. Report on the fourteenth session (20–24 April 2015). New York: United Nations.

United Nations. 2016. *2016 Synthesis of Voluntary National Reviews Synthesis of Voluntary National Reviews*. New York: United Nations

United Nations. 2017. *2017 Synthesis of Voluntary National Reviews Synthesis of Voluntary National Reviews*. New York: United Nations.

United Nations. 2018. Supporting civil service restoration and reform in fragile and conflict-affected settings. New York: United Nations.

Wackernagel, M., Hanscom, L. & Lin, D. 2017. Making the sustainable development goals consistent with sustainability. *Frontiers in Energy Research*, 5, 18–18.

Ziekow, J. & Bethel, R. 2017. *Institutional Arrangements for the Sustainable Development Goals*. New York: United Nations, Committee of Experts on Public Administration (CEPA).

11 Conclusions
Metagovernance as framework for SDG implementation

> Sustainable development is the pathway to the future we want for all. It offers a framework to generate economic growth, achieve social justice, exercise environmental stewardship and strengthen governance.
> – Ban Ki-moon, former UN Secretary-General (2013)

11.1 How the story has developed

This book has told a story about two very abstract topics. Sustainable development and metagovernance are both holistic concepts that offer a framework for policies ('what to do?') to achieve the over-arching set of goals (a meta-policy) of sustainability, and for governance of governance (i.e. metagovernance) to implement these policies ('how to do?'), respectively.

Part 1 introduced sustainability governance as a complex challenge (Chapter 1), covering problems emerging from tensions between governance styles (Chapter 2) or governance failures (Chapter 3), and how metagovernance as governance of governance styles has been developed to prevent or tackle such challenges (Chapter 4). Part 2 put several features of sustainability (meta)governance in the spotlight, starting with a catalogue of 50 'shades of governance' with each three distinct operational forms (Chapter 5). Chapter 6 discussed how to deal with the cultural dimension of governance, and was followed by a chapter on the alluring 'best practice' recipes of New Public Management (Chapter 7). Part 3 focused on the applicability of metagovernance on complex challenges such as sustainable development. Chapter 8 hinted at application of metagovernance with regard to each of the SDGs. In Chapter 9, a sketch of a practical method for the use of metagovernance was presented, followed by introductions on public sector reform, promotion of coherence and capacity building (Chapter 10).

This last chapter brings the storylines together that I consider most relevant for SDG implementation. Section 11.2 reiterates the principle of 'common but differentiated governance' as an imperative for sustainability. Then several preconditions for the use of metagovernance as framework concept for analysis, design and management of the implementation of the SDGs will be suggested (11.3). Two important cross-cutting themes, namely transitions (11.4) and partnerships

(11.5) are described from a metagovernance perspective. Finally, we return to the central hypothesis and three questions raised in Chapter 1 (11.6).

11.2 Metagovernance as 'common but differentiated governance' for the SDGs

The Agenda 2030 clearly excludes blueprint-thinking as a solution for the implementation of the SDGs: "We recognize that there are different approaches, visions, models and tools available to each country, in accordance with its national circumstances and priorities, to achieve sustainable development" (United Nations, 2015, p. 13). The more concrete sustainability governance becomes, the more the challenge arises between, on the one hand, universally applicable goals and (quantified) targets, and, on the other hand, the need for diversification at the levels of implementation. Already Principle 7 of the Rio Declaration (United Nations, 1992), 'Common but Differentiated Responsibility' (CBDR) implied that different levels of administration should develop their own, specific approach to translate globally agreed policies and targets. This principle has become focused on financing and in general on a North-South dichotomy. However, since the adoption of Agenda 2030, governments of developed countries have a responsibility beyond providing financial means. Like all other countries, they need to reduce their ecological footprint, achieve a higher Human Development Index (HDI), and make their economies sustainable. Therefore, the principle of CBDR remained only in the introduction of Agenda 2030 but was not integrated into the Goals: "Not placing CBDR in the goals or in their targets maintained the universality of the goals instead of differentiating them by level of development" (Kamau et al., 2018, p. 143).

As a consequence of the adoption of the set of SDGs and targets at the UN Summit on the Agenda 2030 in New York, countries will not only need to define specified targets and timelines reflecting their situation, but also to design corresponding processes for implementation. Implementation of the SDGs needs 'mindfulness' (in the sense of Weick and Sutcliffe, 2009, see also Chapter 7). It requires systemic thinking, which implies using comprehensive approaches (taking into account all relevant aspects). SDG implementation therefore requires having a holistic view, which means trying to keep in mind the importance of the whole and the interdependence of its parts, both horizontally and vertically. Horizontal coordination should promote spill-over effects from progress in one goal area in others, and vertical coordination should ensure coherence between the priorities and actions at different levels of government (Pintér et al., 2014). Maintaining the holistic view is challenging. From the very beginning of the implementation of the Agenda 2030, countries have argued that they need to be selective and set priorities among the Goals, because of their limited resources. This has already resulted in 'cherry-picking': giving priority to the Goals and targets which are within reach without too much effort.

Mindfully and holistically implementing the SDGs in differentiated ways in different situations is not enough. Sustainability governance at the same time needs to ensure the universal applicability of the Goals and their indivisibility, while not compromising universal principles such as rule of law. In order to give this insight a prominent place, 'Common But Differentiated Governance' (CBDG) has been proposed as guiding principle for the implementation of the SDGs (Meuleman and Niestroy, 2015). Applying this principle might help overcoming the dichotomy between developed and developing countries that has evolved as connotation for the Rio principle of 'Common but Differentiated Responsibilities'. CBDG-based governance cannot be the exclusive domain of one of the three basic governance styles. It requires metagovernance (of hierarchies, networks and markets) because only such an approach make it possible to 'orchestrate' SDG implementation frameworks in ways that take into account the full context, including cultures, history, geography, existing skills, capacity and resources of public authorities, in relation to the type of problems and the feasibility of using certain instruments.

11.3 Feasibility of using metagovernance

Sustainability metagovernance is not under all circumstances feasible. Ten years ago, I formulated five framework conditions for metagoverance, based on five case studies with successful metagovernance on environmental policy in three Western European Countries and at the European Commission (Meuleman, 2008). These framework conditions co-determine whether and to which extent governance style combinations can be designed and managed in situational ways. In addition, I mentioned three other conditions to make metagovernance feasible. These eight conditions – and it is certainly not a complete list – are below formulated as questions to be dealt with to enable 'common but differentiated governance' for the SDGs.

Are public sector organizations ready to think outside of their favourite governance style?

An organizational culture may be 'open' or 'closed', 'professional' or 'task-oriented', and the dominant style of leadership may be 'command and control', 'coaching' or 'enabling'. In order to be able to apply metagovernance, public sector organizations should have some degree of open-mindedness, awareness of what is going on in the governance environment, reflexive thinking, flexibility and openness to innovation. Thinking 'outside of the box' is a necessity when a holistic concept should be applied. However, public sector organizations almost by definition have an underlying hierarchical governance style, which has many strengths but among them are not the ones I listed above. A classical, Weberian bureaucracy has an internally oriented, closed culture which is typically not innovative or reflexive. For example, at the European Commission, it has taken a dedicated political intervention around 2000, culminating in a 'White Paper on Governance' (Grote and Gbikpi, 2002) to add two 'new modes of governance' (Héritier, 2002) to the repertoire.

In many cases, the situation is not so straightforward. In every organization there are more and less open-minded people. Policy, project or programme managers who are convinced that a metagovernance approach would result in having governance frameworks which are less deemed to fail, need to find the right 'sponsors' or may need to wait until a 'window of opportunity' opens with a change of senior management or political leadership.

Is cross-sectoral collaboration enabled?

Each and every policy sector has its own tradition and preferred governance approach. A ministry of economic affairs may have a market governance tradition, whereas environmental ministries, in most countries, may have a preference for hierarchy, as this implies enforcement, with the help of legislation, norms and standards. Stakeholders may have the same diversity concerning their preference for a governance style: ministries are often mirroring the societal sectors they represent. This hinders cross-sectoral collaboration. Such collaboration is also hampered by the silo-thinking of hierarchical public sector organizations. Overcoming these obstacles is a precondition for implementation of the SDGs and can be achieved with a combination of central leadership and decentral informal measures such as suggested in Chapter 7 under the metaphor of 'teaching silos to dance' (Niestroy and Meuleman, 2016).

Are line, project and process management combined?

Project management is usually the first organization principle beyond the hierarchical line organization, with which public sector organizations become familiar. The ideas behind project management are a combination of hierarchical (control) and market governance (efficiency of resource uses) concept. Process management is a very different approach, where the main objective is not control of resources, but dealing with uncertainties, unpredictability, the views and interests of all stakeholders, the history of an issue, coalitions between actors, etcetera. This requires network governance thinking. As mentioned in Chapter 5.4 (feature 33), metagovernance requires a dynamic balance of these two approaches.

Is cultural compatibility an option?

The culture of people and organizations influences the feasibility of metagovernance at all levels of administration. Understanding that there is often one predominant governance style (or hybrid) – of which people may not be aware because it is part of what they conceive as 'normal' – creates opportunities but also limitations. A hierarchical style is difficult to implement in a consensus society like Denmark, Sweden, Finland or the Netherlands, even when the problem at hand is of the urgent type. However, a consensus culture is a good fundament for network governance, a governance style which is especially useful for dealing with complex and unstructured problems. In Chapter 6, I have suggested that making a governance framework culturally compatible is likely to be more effective than, for example, forcing a governance style upon people that does not relate to their values and traditions.

Is individual capability in line with the requirements?

In Chapter 10, metagovernance capacity and capability were touched upon. Capability is about individual qualities. The individual who is to apply metagovernance would have to have a notion of efficacy (a rationale); a reasonable level of willingness; discretion and capability; should be experienced in applying strategies such as switching (styles), linking, and conflict resolution; and first and foremost, should understand the logics of all governance styles. Jessop (2009, p. 96) adds what he calls the hardest condition for metagovernance, namely "ironic, experimental approach that relies on collective intelligence to overcome tendencies towards scepticism, cynicism, opportunism and spin".

Does the personal conviction of politicians or public managers promote or prevent metagovernance?

Each governance framework is normative and therefore different political parties tend to design different governance approaches. A metagovernance perspective does not obstruct the normativity but points at strengths and weaknesses. One might say that metagovernance of hierarchies, networks and markets 'softens' the normative reflex of politicians to go for one type of solutions, or at least shows the negative externalities of such a preference.

Personal convictions, beliefs, experiences influence the extent to which a political leader or a public manager is willing to apply metagovernance strategies. For a politician (or a politically appointed public servant) it may be very difficult (and risky) to switch from one to another governance style, as this can be seen as betrayal of the programme of the political party. I once worked for a minister who believed in hierarchical governance as the way to achieve his goal, while I believed that we needed to cooperate with stakeholders in order to join forces to achieve the goal. He won the dispute, of course – but I also won something: the insight that the government could not prevent that the other (public and non-public) stakeholders engaged in an interactive network process from which we, the ministry, benefited strongly (Meuleman, 2003).

Do stakeholder expectations regarding the role of public authorities support or hinder metagovernance?

As a rule of thumb, I would say that when civil society and/or private sector actors demand influence and offer co-responsibility with regard to public policies, this increases the feasibility of metagovernance, as this approach is able to deal better with different perspectives than any of the governance styles as such. Stakeholder interests may align with a stronger or a weaker role of the state, with less or more legal standards, and with different roles of public authorities, such as ensuring oversight and rule of law, being a partner in joint activities, or being a service provider and market regulator. The other way around, where (organized) stakeholder organizations encounter difficulties to establish or maintain a serious/ influential relation with government authorities, these authorities may not be triggered enough to design situationally adapted governance.

Is the problem type solvable with the preferred governance approach?

The problem definition co-determines which governance style will serve best as the dominant style. If the policy problem is defined as an urgent matter (a crisis or emergency), the rationale would be to choose a hierarchical approach; if it is a routine issue that should be dealt with as efficiently as possible, market governance seems to work best, and for wicked problems, using network governance mostly is the best start. The match between a governance approach and the type of problem it should deal with is an important factor determining success or failure. However, I have put this condition as the last in the list, because a mismatch of problem and governance can only be solved by reframing the problem, changing the governance framework or both – and in all cases the fulfilment of the above-mentioned conditions helps a lot, to use an understatement.

11.4 Metagovernance and systemic sustainability challenges

System thinking originated in natural sciences but has been also become important in social sciences. This started with the discovery – or construction? – of analogies between natural and social systems (in terms of complexity and unpredictability) and the hunch (or hope) that causal relations observed in natural science might also work in social systems. As social systems are without exception very complex, causality turned out to be difficult to prove and dealing with complexity became a key challenge in many social disciplines (except in economics, where there is still a strong school of thinking around causal effects between economic principles and (rational) human behaviour).

The Sustainable Development Goals are not only designed for their achievement, but also to account for the systemic changes that the Goals and ways of achieving them imply (Ziekow and Bethel, 2017, p. 19). It is, however, to be seen whether the SDGs will trigger deep systemic changes in all areas where this is necessary to achieve sustainability, such as how economic development is measured (qualitative or quantitative growth?), and the transitions from fossil to renewable energy, from resource-intensive agriculture to sustainable food production, and from throw-away to circular economy.

Context matters: understanding sustainability in a changing world

When Klaus Töpfer came back to Germany after having served as executive director of UNEP in Nairobi, he founded the Institute for Advanced Sustainability

Studies (IASS) in Potsdam, which would be a transdisciplinary institute, combining academic and practical knowledge. The very first project of the new institute (2010–2012) was an international project about sustainability governance of which I had the honour to be the project leader. The Transgov project was a search for insights from various social sciences including not-so-common ones in governance studies such as cultural anthropology and history, as well as practical knowledge. It analysed what implications recent and ongoing changes in the relations between politics, science and media – together characterized as the emergence of a knowledge democracy – may have for governance for sustainable development, on global and other levels of societal decision making, and vice versa. How can concepts such as second modernity, reflexivity, configuration theory, (meta)governance theory and cultural theory contribute to a 'transgovernance' approach which goes beyond mainstream sustainability governance?

The project resulted in a monograph (In 't Veld et al., 2011) and an edited book (Meuleman, 2013) with explorations of topics such as international relations, governance and metagovernance theory, (environmental) economics and innovation science, institutions and transformation processes, and the paradigms behind contemporary sustainability governance. After a series of international, transdisciplinary workshops, a number of building blocks for this approach were identified (Figure 11.1).

The six concepts are:

Knowledge democracy (In 't Veld, 2010). This concept reflects the observation that pattern of interactions (including conflicts) between politics, science and media have emerged in which more interactive forms co-exist with the old forms.

Second modernity (Beck, 1992). Our societies are so complex and unpredictable that it is wise not to choose for single solutions, but for plurality: 'and' instead of 'or'. It is generally better not to throw away existing institutions

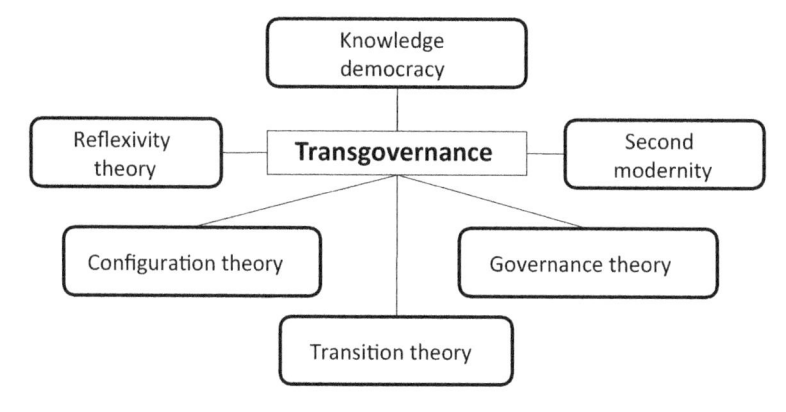

Figure 11.1 Transgovernance: the quest for sustainable development (Meuleman, 2012)

too quickly, as they may be needed again. Redundancy has a negative connotation, but should be revaluated: it has clear advantages.
- **Reflexivity** (Beck et al., 1994). Social systems are reflexive and therefore change all the time and adapt to new circumstances. This implies that solutions to societal problems need to be resilient.
- **Configuration theory** (Van Twist and Termeer, 1991). Innovation does not begin at the centre of groups and organizations, but at the borders, where people are working who are not fixed in one configuration of people with the same mind, but are 'multiple included' in various configurations.
- **Transition theory** (Grin et al., 2010). Sustainability transitions follow patterns. They emerge on the project level in 'niches'. When these innovations are successful, associated rules and other supporting mechanisms may constitute 'regimes'. Gradually, a transition 'landscape' will emerge in which niches and regimes are loosely linked.
- **Governance and metagovernance theory** (Jessop, 2002, Kooiman, 2003, Meuleman, 2008). It is important to accept that governance in practice is always normative; this knowledge should be used for designing and managing situational governance approaches/mixtures (metagovernance).

Using this 'basket' of concepts which, including their interlinkages, constitute the transgovernance framework, could help overcome some of the typical misconceptions and often-repeated dead end solutions, like the idea that only legally binding agreements can be successful, or that cultures and traditions are a hindrance rather than part of the solution. In the discussion on whether the often-observed 'fragmentation' of global environmental governance is a sign of duplication or complementarity (Ivanova and Roy, 2007), a transgovernance perspective would probably favour the latter.

Metagovernance and transition management

There is a similarity in thinking between metagovernance and transition management, as the latter also "distinguishes between different types of governance that relate directly to different patterns of change in societal systems" (Loorbach, 2010). Transition management is "a metagovernance approach made operational by a series of process and content combining steps" (Frantzeskaki and Loorbach, 2014). There are, however, also differences. For instance, I agree with the critical comments Meadowcroft (2011, p. 547) made, namely that promoters of transition management often suggest that experts should be the central players, while the political system is as seen as exogenous (see also Section 2.3). Moreover, transition management is sometimes – for me counter-intuitively – presented as the only road leading to sustainable development: "Policy Makers are interested in transitions because incremental, technical changes based on end-of-pipe solutions, cleaner products or eco-efficiency, are not believed to lead to sustainability" (Wieczorek, 2017). A similar conviction that only radical change will do the trick is expressed by Loorbach, stating that "Sustainable Development itself has become part of the problem", because it has become part of the established regime and has primarily served to make it a little less unsustainable (Loorbach, 2014, p. 32). The transition management approach seems difficult to replicate in other countries than in which it emerged (i.e. Western Europe). Transition theory cannot be applied without adaptation in e.g. developing countries, where it is important "to avoid reproducing ill-functioning institutions that continue benefitting the privileges of a few, while undermining the well-being of many" (Ramos-Mejía et al., 2017).

Personally, I am not convinced that an expert-based approach replacing representative democracy will work more effectively and will ensure the principle of inclusiveness agreed in the Agenda 2030. In addition, not all sustainability challenges are about systemic transitions. Part of the SDGs and their targets will be achievable with incremental change within existing systemic conditions and paradigms. In general, the "easier" roads towards sustainability are about optimizing and modernizing existing pathways, where they are not unsustainable.

In metagovernance terms, it is relevant that the governance design for sustainability transitions follows the same logic we have seen throughout this book: there are different approaches which can be characterized along governance styles: top-down (hierarchical), bottom-up/deliberative (network) and self-regulatory (market governance); these approaches all have their benefits and weaknesses. The overall characteristics of national energy transition approaches in the Germany, the Netherlands and the UK seem to reflect the national preferences for, respectively, hierarchical, network and market governance (Figure 11.2, after Meuleman, 2015). However, in all three examples, the other governance styles were used as auxiliary tools. The German energy transition ("Energiewende") initiated in 2011 is a good example. It started with a central government initiative establishing a dedicated Commission, which organized stakeholder and citizens' consultations. The Commission's conclusions were adopted by the German government with legal provisions and an action programme, followed by financial interventions (investments).

11.5 Metagovernance and partnerships

Participation and partnerships are key terms in contemporary governance, and play an important role in SDG 17. How actors – governmental or non-governmental – are involved in (sustainability) governance may vary: from not even being informed, via being kept up-to-date, involved as participant, co-creator up to co-decision-making. This 'stairway' of participation is not a hierarchy in itself, comparable with the 'waste hierarchy' in environmental policy. I would not ask the general and uninformed public to decide through a referendum on a decision of war and peace.

Partnerships to support public policy design and implementation can be designed as both a structure and a process, and as a means to an end and an end in itself; they can be meant as instruments and have an intrinsic value in cementing social capital (Stott, 2017). Such partnership types are often characterized

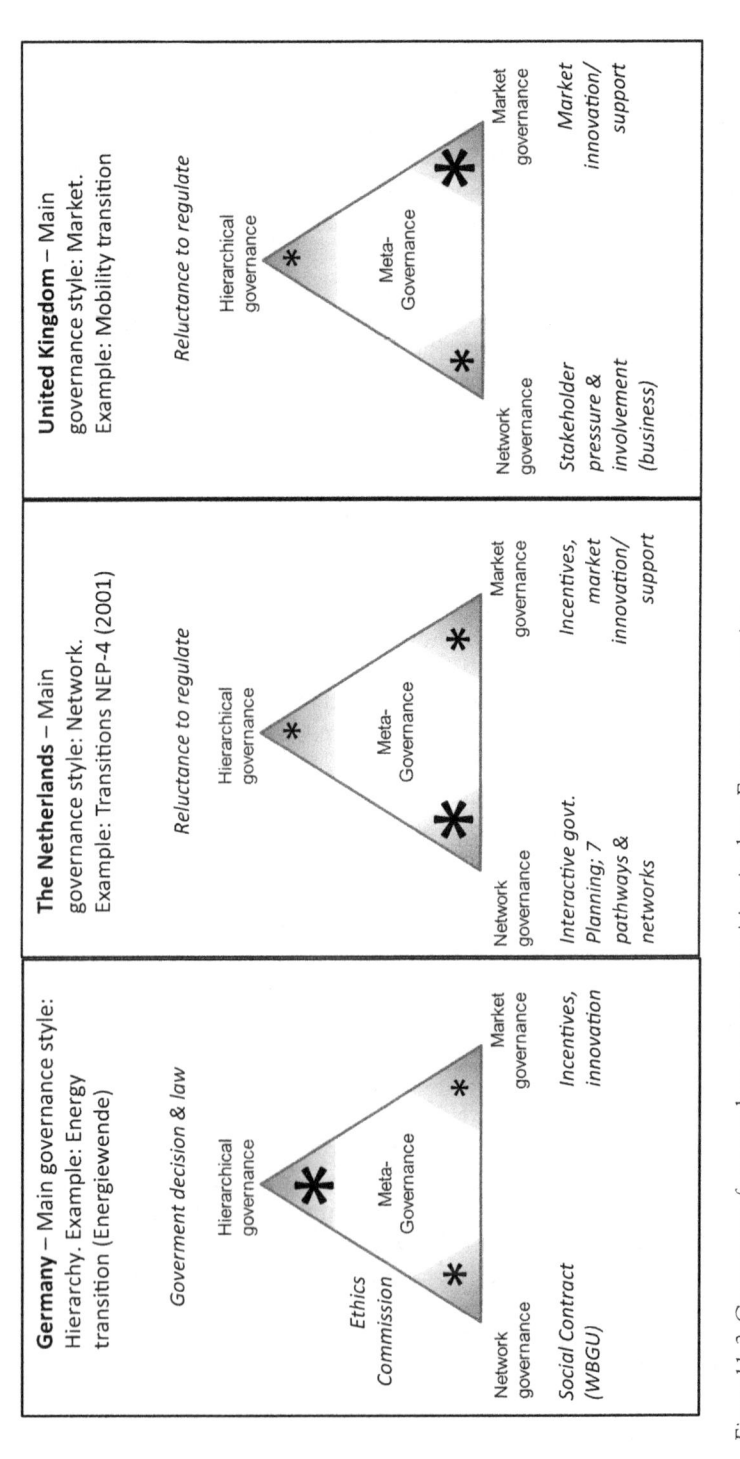

Figure 11.2 Governance frameworks on energy transition in three European countries
(Meuleman 2015)

according to who are the partners: public-public partnerships, public-private partnerships or broad alliances of administrations, business and civil society.

Stakeholders of a policy-making or implementation process are defined by having something a stake. Their interests are supported by assumptions, values, traditions and sometimes by alliances with other stakeholders. Stakeholders have interests and positions or statements, which are two very different things. A key element in the Mutual Gains Approach to negotiation (Susskind and Field, 1996) is to find out about the deeper interests of stakeholders, which are often hidden by statements. Working effectively with stakeholders in a metagovernance approach requires in the first place an analysis of the stakeholder environment, because metagovernance combines elements of different but always value-based governance styles, and stakeholders may have an implicit or explicit preference, underpinned by their values and situation, for a specific governance style. Hierarchy requires rulers and subjects, networks build on partners and market governance frames actors as producers and consumers. The appropriate use of such terms is helps the management of expectations. A 'partner' expects a rather equal position, whereas a 'customer' expects a product or service with a good price. In Section 9.2, a series of methods was presented to analyse the stakeholder environment.

Strengthening partnerships is central to the entire implementation approach of the SDGs. Goal 17 includes targets to enhance a 'Global Partnership for Sustainable Development' complemented by multi-stakeholder partnerships to encourage and promote effective public, public-private and civil society partnerships, building on the experience and resourcing strategies of partnerships. The chapter on 'Means of Implementation' asserts that the entire 2030 Agenda will be judged on the success of partnership constructs and their implementation of every Goal.

In a critical discussion on public-private partnerships (Section 7.7) I raised the question whether the envisioned revitalized global partnerships for the SDGs should be fundamentally different from the most popular existing arrangement, namely public-private partnerships (PPP). The term public-private partnership defines who owns the partnership – public actors and private actors – and not what its purpose is. Partnerships for the SDGs should instead be described in a way that reflects the actors as well as their purposes. To reflect this, they could be named 'Administration – Business – Civil society (ABC) Partnerships' (Meuleman et al., 2016).

ABC partnerships will require a goal-reorientation of all three parties. For administrative partners, the goal could be achieving concrete targets in alliance with societal partners while achieving mutual gains, instead of cost-saving or downsizing government. For business partners, corporate social responsibility could become an integrated objective, in addition to creating added value. For civil society organizations, the goal could be taking co-responsibility for solving societal challenges. Usually, however, they interact with governments and businesses as advocates for the common good. Advocacy includes lobbying, convincing, fund-raising, campaigning, protesting, as well as being a "watchdog".

Engaging in partnerships on equal footing with their classical "opponents" will be a new challenge. Civil society organizations who could see a role for themselves in ABC partnerships should be encouraged to learn from existing good practices in development programmes, where more experience has been gathered than for example in environmental programmes.

PPPs and multi-stakeholder partnerships (MSPs) for sustainability have traditionally been implemented on a "North-South axis", within the aid-development paradigm. Dedicated CSOs have been involved in carrying out such "North-South" development projects. However, as the SDGs call for universal application, there is a need to develop a new implementation basis, with a new partnership philosophy and narrative, which will be different than in the past as regards purpose (better, not just cheaper results), vision (keep implementation holistic, inclusive and long-term oriented), scope (not only North-South, but also North-North and South-South), and roles (each of partners could take up a leadership role).

ABC partnerships could be used to address a wide variety of issues, such as administrative decentralization, small-scale and direct democracy, access and participation in a transparent manner, green-growth, steady-state economy and the challenge of the growth paradigm, the implications of the precautionary principle, and the polluter pays principle.

An important question is whether CSOs are interested, willing and capable of taking up new responsibility as actors in ABC partnerships. The challenge of developing CSOs' capacity to engage in ABC partnerships should not be underestimated. On the other hand, the UN member states have agreed to SDG 17 and other references to partnerships in the 2030 Agenda, including a specific target (17.9) on support for capacity building in developing countries. It should be expected that means to build capacity for new partnerships will be made available.

To conclude, partnerships for the 2030 Agenda should be inclusive, with in principle an equal weight of CSOs, business and government organizations. PPPs should be transformed, or a separate type of partnership must be developed. Adding civil society to PPP as an after-thought is not enough: putting wings on a car doesn't ensure that it will fly; it is still a car.

Metagovernance can be used to manage and balance the different interests of actors in partnerships. Experiences from the European Union may be useful to learn lessons from (Petersen, 2010). Beisheim and Simon (2015) see a limited role for metagovernance to support partnerships for sustainable development, but we have to take into account that these authors use a narrow definition of the term, namely metagovernance of network governance. In a case study on Kenya, they conclude that existing attempts to use metagovernance are rather weak and fragmented, but also that respondents believe that sustainability partnerships could benefit from such a framework, which could promote local ownership of the partnerships and increase the potential to scale up successful cases (Beisheim et al., 2017). An earlier case study in the UK (Wales) shows that 10 partnerships at the local level suffered from "the dysfunctional effects of hierarchical and market coordination" (Entwistle et al., 2007) . Here the authors suggest that the more a partnership depends on one governance style as basis for coordination, the more likely is it that dysfunctions of that style will influence the success. In my view, a broad metagovernance approach would exactly focus on remediating such problems.

11.6 Wrapping up, conclusions and outlook

In Chapter 1, I formulated three guiding questions which should now be answered.

1   How have metagovernance theory and practice developed during the last decade (2008–2018), based on academic and other publications?

The use of metagovernance as analytical concept has increased sharply over the last decade, across the world, on very different policy and implementation challenges. There is, however, not yet much research on cases where metagovernance was used as design and management of policies. At the same time, the fact that metagovernance is a heuristic concept – the practice was there before the theory – should lead us to assume that there is ample empirical material for such research.

In addition, theory development has been limited. Like 10 years ago, there are two main schools of thinking on metagovernance: the broad version dealing with hierarchical, network and market governance, and the narrower version focusing on network management. Unfortunately, there is little exchange between the two schools.

2   Is it plausible that unsuccessful public managers are, generally speaking, not using the metagovernance lens to prevent and mitigate governance failures?

This question remains difficult to answer. In addition to the examples on governance failures I gave in Chapter 3, and especially those resulting from inherent weaknesses of each of the three governance styles, I have presented some examples earlier (Meuleman, 2008), while also Christopoulos et al. (2012) have formulated what can happen in the absence of metagovernance, as this "enables customized approaches based on endogenous knowledge, equally important for all three pillars of sustainability. Approaches not sensitive to what is previously mentioned are bound to fail".

3   How can the holistic concept of metagovernance be applied to the most holistic of contemporary policies: striving for sustainable development, and in particular for implementation of the 2015 UN Sustainable Development Goals (SDGs)?

I have argued that much of the existing studies and research papers published about the governance needed for the implementation of the SDGs is of a stocktaking or prescriptive kind (see Section 1.2). In this book we have touched upon some of the main 'root' causes of obstacles to implementation of the SDGs, using the analytical lens of metagovernance and three basic, ideal-typical governance styles. Metagovernance has the potential to guide the design of governance for a transformative agenda, while being grounded in existing cultures and traditions. It offers contextualized solutions and suggests that learning from peer nations (considering "good practices" or "inspiring examples") should be a priority.

In Chapter 1, I combined the triangle of metagovernance and the triangle of the sustainable development dimensions (Figure 1.1), which brought about the question what the affinity is of each of the three sustainability dimensions with each of the three governance styles. Which values do they share? The easiest link is the one between market governance and the economic dimension of sustainability, although one could imagine a different type of economic development than the classical, GDP- and growth-driven version that we know has dominated, at least at the global level, for decades. Hierarchical governance seems to link best to the environmental dimension, because of the need to have at least minimum standards for protecting the quality of the living environment. However, the same can be said about social conditions: without legal standards for equal treatment and decent jobs, sustainability remains a faint light on the horizon. Network governance reflects the wide range of stakeholders involved in environmental and social policies. To conclude, inking the three dimensions of sustainability and the three governance styles is not so straightforward, and probably national cultures and history may make that the situation will be different on different countries.

I do not claim that my views on governance and metagovernance are universally the 'best'. They are, as all views, normative. For instance, I have the normative standpoint that there are universal norms such as rule of law, which should inform all governance practice. At the same time, the different dimensions of metagovernance discussed in this book have resulted in the conclusion that (meta)governance for sustainability and the SDGs should be 'common but differentiated'. I suppose it is also a normative position (namely about science) that I believe that academic work is only credible when it is transparent about its assumptions and the values behind them. I have criticized the blind trust in facts and numbers as in 'evidence-based policy making' but I do think that scientific rigour is essential.

This brings me to two conclusions on how governance is treated in academic publications. Having seen the hundreds of publications referred to in this book, I cannot escape the idea that network (meta)governance is the new 'New Public Management', which I have critically assessed as cause of many governance failures in Chapter 7. The promotion of network governance has some of the same flaws as NPM (market governance):

- Claims not founded by empirical research (or only by research on specific topics where network governance is expected to be dominant; therefore e.g. claiming that hierarchical governance is disappearing).

- Lack of academic rigour (such as frequent lack of transparency about the definition of (meta)governance and how this relates to other definitions). Part of network governance literature is self-referential – only authors are cited which think along the same lines – which does not add to the credibility of social sciences across society.

Moreover, a recurrent but often between-the-lines theme of network governance literature is that democracies are not able to tackle the great governance challenges of our time and that network arrangements organized and managed by experts would be better placed to take the lead.

Outlook: maintaining a holistic view while differentiating on the ground

The Agenda 2030 is the most over-arching but not the only important global agreement of the last years – the Paris Climate Agreement is another crucial step forward. Of all these agreements, the Agenda 2030 with its Goals and targets is the one with the strongest focus on governance: the SDGs not only give both a direction but also the tools to get there. For the implementation of the Goals, many lessons can be learned from the negotiation processes preceding the 2015 adoption of the SDGs, as analysed in a recommendable book written by three closely involved key experts (Kamau et al., 2018).

The analysis in this book has highlighted the need for 'common but differentiated governance' for the SDGs, as suggested in an earlier publication (Meuleman and Niestroy, 2015). I have tried to show that in order to succeed on this pathway, it is necessary to combine stimulating diverse approaches that take into account their contexts, with keeping the broad view and being able to link Goals and targets, to revise concrete governance frameworks when existing ones are not effective, by using various metagovernance strategies.

I have emphasized that metagovernance or governance of governance exists in practice: it is not only a theoretical concept. Those who use the approach – the 'metagovernors' – are not necessarily the ones in charge in a hierarchical sense. Metagovernance as analytical approach and practical method can be used by anyone who has some kind of coordinating responsibility for public policy making, implementation, and/or management. Also coordinators of non-public organizations can take up such an attitude.

Part of the challenge of maintaining the holistic view can be addressed by improving multi-level governance. A metagovernance view shows that there are several ways to organize effective multi-level governance for sustainability – top-down, bottom-up, with broad involvement or focused on public actors. In any case, sub-national authorities will be key actors for implementing the SDGs and it needs to be recognized that they are, often more than national governments, the ones that can connected the Goals with the needs of citizens. Implementing the SDGs is very much about creating synergies between the various levels, sectors and actors, while recognizing the distribution of power across these levels, sectors and actors and making this more transparent.

Last but not least, the nexus approach of linking different Goals will remain on the political and research agendas. While many of the SDGs themselves are largely sectoral in their approach (Stevens and Kanie, 2016), addressing their integration also in nexus programmes will at least show areas of easy synergy and topics where severe dispute can be expected. It may turn out that the nexus approach will be more valuable as strategic agenda-setter than as concrete solution-provider. In any case, nexus themes such as the linkages between food, energy and water, allow us to see them as complex, 'wicked' problems for which there are no simple solutions, and which require "careful analysis to define situationally optimal, yet implementable and acceptable governance style mixtures" (Venghaus and Hake, 2018) – in other words, a nexus approach would benefit from applying metagovernance.

This last chapter discussed the central hypothesis and three questions raised in this book, and concluded that governance for sustainability and in particular for implementing the Sustainable Development Goals, will benefit from the holistic and situational approach supported by using metagovernance of hierarchical, network and market governance styles. At the same time, there is a need to further analyse the experiences with the practical application of metagovernance: some questions are still open, such as which guidance could be developed for practitioners of (sustainability) metagovernance, to deal with the tensions between universality and differentiation.

The chapter introduced the principle of 'common but differentiated governance' as an imperative for sustainability because of the balance needed between common goals and differentiated implementation. It addressed several preconditions for the use of metagovernance as framework concept for analysis, design and management of the implementation of the Sustainable Development Goals. An important cross-cutting theme, namely systemic transitions towards sustainability, was described from a metagovernance perspective.

I need to end with a big caveat. By far not all relevant and interesting topics could be discussed in the context of this book, and I will have missed or underrepresented important work done by scholars and practitioners. Still, I hope that what could be presented has been worthwhile enough to raise new research questions and help practitioners of sustainability (meta)governance to at least ask the relevant questions – even if some answers may still be out of reach. One of my personal mottos is that asking the right questions is more important than giving the right answers.

References

Ban, K.-M. 2013. *New Agenda for Development Must Be 'Tuned' to Leading Challenges of Decent Jobs, Inclusive Growth, Governance, Peace, Climate Change, Says Secretary-General*. New York: United Nations.

Beck, U. 1992. *Risk Society: Towards a New Modernity*. London: Sage.

Beck, U., Giddens, A. & Lash, S. 1994. *Reflexive Modernization: Politics, Tradition and Aesthetics in the Modern Social Order*. Cambridge: Polity Press.
Beisheim, M., Ellersiek, A., Goltermann, L., et al. 2017. Meta-governance of partnerships for sustainable development: Actors' perspectives from Kenya: Kenyan metagovernance for partnerships. *Public Administration and Development*. https://doi.org/10.1002/pad.1810.
Beisheim, M. & Simon, N. 2015. Meta-governance of partnerships for sustainable development. Actors' perspectives on how the UN could improve partnerships' governance services in areas of limited statehood. SFB-Governance Working Papers Series. Berlin: Collaborative Research Center (SFB).
Christopoulos, S., Horvath, B. & Kull, M. 2012. Advancing the governance of cross-sectoral policies for sustainable development: A metagovernance perspective. *Public Administration and Development*, 32, 305–323.
Entwistle, T., Bristow, G., Hines, F., et al. 2007. The dysfunctions of markets, hierarchies and networks in the meta-governance of partnership. *Urban Studies*, 44, 63–79.
Frantzeskaki, N. & Loorbach, D. 2014. Transition management as a meta-governance experiment to build energy resilience in European cities. Paper presented at the Resilience Conference 2014, Montpellier, France, May 4–8, 2014,
Grin, J., Rotmans, J., Schot, J., et al. 2010. *Transitions to Sustainable Development: New Directions in the Study of Long Term Transformative Change*. New York: Routledge.
Grote, J. R. & Gbikpi, B. 2002. Participation and metagovernance: The white paper of the EU commission. In: Grote, J. R. & Gbikpi, B. (eds.) *Participatory Governance: Political and Societal Implications*. Wiesbaden: VS Verlag für Sozialwissenschaften.
Héritier, A. 2002. New modes of governance in Europe: Policy making without legislating? *IHS Political Science Series – No. 81*. [Working Paper]. Vienna: Institute for Advanced Studies.
In 't Veld, R. J. 2010 (ed.). *Knowledge Democracy: Consequences for Science, Politics, and Media*. Heidelberg: Springer.
In 't Veld, R. J., Töpfer, K., Meuleman, L., et al. 2011. *Transgovernance: The Quest for Governance of Sustainable Development*. Potsdam: Institute for Advanced Sustainability Studies (IASS).
Ivanova, M., & Roy, J. (2007). The architecture of global environmental governance: pros and cons of multiplicity. In: Speth, J. G. and Haas (eds.). *Global Environmental Governance*. Delhi: Pearson Education India.
Jessop, B. 2002. Governance and meta-governance in the face of complexity: On the roles of requisite variety, reflexive observation, and romantic irony in participatory governance. In: Heinelt, H., Getimis, P., kafkalas, G., Smith, R. & Swyngedouw, E. (eds.) *Participatory Governance in Multi-Level Context*. Wiesbaden: VS Verlag für Sozialwissenschaften.
Jessop, B. 2009. From governance to governance failure and from multi-level governance to multi-scalar meta-governance. In: Arts, B. & Al, E. (eds.) *The Disoriented State: Shifts in Governmentality, Territoriality and Governance*. Heidelberg: Springer.
Kamau, M., Chasek, P. & O'connor, D. 2018. *Transforming Multilateral Diplomacy: The Inside Story of the Sustainable Development Goals*. London: Routledge.
Kooiman, J. 2003. *Governing as Governance*. London: Sage.
Loorbach, D. 2010. Transition management for sustainable development: A prescriptive, complexity-based governance framework. *Governance*, 23, 161–183.
Loorbach, D. 2014. *To Transition! Governance Panarchy in the New Transformation*. Rotterdam: DRIFT.

Meadowcroft, J. 2011. Sustainable development. *In:* Bevir, M. (ed.). *The Sage Handbook of Governance.* London: Sage.

Meuleman, L. 2003. *The Pegasus Principle – Reinventing a Credible Public Sector.* Utrecht: Lemma.

Meuleman, L. 2008. *Public Management and the Metagoverance of Hierarchies, Networks, and Markets.* Heidelberg: Springer.

Meuleman, L. 2012. Transgovernance: Sustainability governance in knowledge democracies. *Guest lecture, University of Lüneburg, 1 March 2012* Lüneburg.

Meuleman, L. (ed.) 2013. *Transgovernance: Advancing Sustainability Governance.* Berlin, Heidelberg: Springer.

Meuleman, L. 2015. Transition governance: It takes two to tango. *6th International Sustainability Transitions (IST-6) Conference.* Brighton, UK.

Meuleman, L. & Niestroy, I. 2015. Common but differentiated governance: A metagovernance approach to make the SDGs work. *Sustainability,* 12295–12321.

Meuleman, L., Strandenaes, J.-G. & Niestroy, I. 2016. From PPP to ABC: A New Partnership Approach for the SDGs. *IISD SD Policy & Practice. Guest Article, posted on 11 October 2016* [Online]. Available: http://sdg.iisd.org/commentary/guest-articles/from-ppp-to-abc-a-new-partnership-approach-for-the-sdgs/

Niestroy, I. & Meuleman, L. 2016. Teaching silos to dance: A condition to implement the SDGs. *IISD SD Policy & Practice. Guest Article, posted on 21 July 2016* [Online]. Available: http://sd.iisd.org/guest-articles/teaching-silos-to-dance-a-condition-to-implement-the-sdgs/.

Petersen, O. H. 2010. Emerging meta-governance as a regulation framework for public-private partnerships: An examination of the European Union's approach. *International Public Management Review,* 11, 1–21.

Pintér, L., Almássy, D., Antonio, E., et al. 2014. *Sustainable Development Goals and Indicators for a Small Planet – Part I: Methodology and Goal Framework.* Singapore: Asia-Europe Foundation (ASEF).

Ramos-Mejía, M., Franco-Garcia, M.-L. & Jauregui-Becker, J. M. 2017. Sustainability transitions in the developing world: Challenges of socio-technical transformations unfolding in contexts of poverty. *Environmental Science & Policy,* 84, 217–223.

Stevens, C. & Kanie, N. 2016. The transformative potential of the Sustainable Development Goals (SDGs). *International Environmental Agreements: Politics, Law and Economics,* 16, 393–396.

Stott, L. 2017. *Partnership and Social Progress: Multi-Stakeholder Collaboration in Context.* PhD, The University of Edinburgh.

Susskind, L. & Field, P. 1996. *Dealing with an Angry Public: The Mutual Gains Approach to Resolving Disputes.* New York: Free Press.

United Nations. 1992. *Rio Declaration on Environment and Development.* New York: United Nations General Assembly.

United Nations. 2015. Transforming our world: The 2030 agenda for sustainable development. New York: United Nations.

Van Twist, M. & Termeer, C. 1991. Introduction to configuration approach: A process theory for societal steering. *In:* In 't Veld, R. J., Schaap, L., Termeer, C. J. A. M. & Van Twist, M. J. W. (eds.). *Autopoiesis and Configuration Theory: New Approaches to Societal Steering.* Dordrecht: Springer.

Venghaus, S. & Hake, J. F. 2018. Nexus thinking in current EU policies – The interdependencies among food, energy and water resources. *Environmental Science & Policy*. https://doi.org/10.1016/j.envsci.2017.12.014.

Weick, K. E. & Sutcliffe, K. M. 2009. *Managing the Unexpected: Resilient Performance in an Age of Uncertainty*. 2nd Edition. Hoboken, NJ: John Wiley.

Wieczorek, A. J. 2017. Sustainability transitions in developing countries: Major insights and their implications for research and policy. *Environmental Science & Policy*. 84, 204–216.

Ziekow, J. & Bethel, R. 2017. *Institutional Arrangements for the Sustainable Development Goals*. New York: UN Committee of Experts on Public Administration (CEPA).

Index

Aagaard, P. 269
ABC partnerships 202, 287
Abdallah-Pretceille, M. 91
access to information 137
access to justice 197
accountability 137–138, 206
Ackerman, F. 177
actor analysis 228
adhocracy 140
Adler, P. S. 131
Agenda 2030 4, 8, 49, 91, 169, 241, 278, 291
Agenda 21 49
Agu, S. U. 84
Albareda, L. 216
Alford, R. R. 254
Ali, S. H. 215
Allen, C. 272
Allmendinger, P. 96
Amore, A. 75, 84, 211
Andersen, M. S. 46
Arentsen, M. J. 126
argumentation analysis 229
Arnouts, R. 54
Arts, B. 54
Assens, C. 121, 137
Atapattu, S. 8
Atkinson, R. 109
Australia 48, 76, 84, 209, 272

Bakker, K. 61, 65, 209
Bakvis, H. 272
Ban Ki-moon 277
Barabási, A.-L.
Barcelona 270
Barnard, C. 177
Baroncelli, A. 121, 137
Bartels, G. 228
Beck, U. 24, 183, 231, 284

Beisheim, M. 15, 78, 84, 180, 198, 202, 288
Belgium 78, 84
Belize 205
Bell, S. 22, 76, 78, 84, 209
Benington, J. 114, 119, 140
Berger, P. 177
Bernstein, S. 35
best practices 174–175
Bethel, R. 38, 252, 258, 282
better regulation 63, 179–180
Bevir, M. 4, 22, 113, 224
Bhutan 35, 93, 248, 256
Biermann, F. 35
big data 264
Bizikova, L. 203, 273
Blanchard, K. H. 3, 144, 250
block chain technology 273
blueprint 11, 49
Boas, I. 207
Boin, A. 99
Bolivia 180
Bonivento, J. 84
Bouckaert, G. 25, 27, 141, 183, 248
Bovens, M. xii, 47, 50
Bowen, K. J. 199
Boxall, A. M. 29, 48
Brammer, S. 153
Brandtner, C. 79, 119
Branwell, B. 84
Brazil 84, 95
Brundtland, G. xiii, 7
Bulgaria 235
Burau, V. 33, 84

Cabo Verde 260, 262
Canada 84, 138, 203
Candel, J. 57, 88
Capano, G. 79

Cape Town 58
Caviedes, A. 86
Centre for Public Impact 53
CEPA 11, 35, 83, 91, 201, 247, 251
Challe, S. 204
China 83, 162, 205, 216, 217
Christopoulos, S. 37, 82, 97, 210, 289
Cini, M. 86
circular economy 215
cities 212–213
climate action 217
Clubb, D. O. 46
cockpit-ism 35
Cohen, M. D. 150
Collier, P. M. 175
Colombia 81, 84
common but differentiated governance 75, 92, 184, 250, 278–279
common but differentiated responsibility 278
common pool resource 34
communities 212–213
competences 145
compliance assurance: EU measures 59; lack of 59
conceptual crowd 25
configuration theory 284
conflict resolution 147–149
Considine, M. 27, 115
Copenhagen 270
corporate social responsibility 81, 215
corruption 53, 197
Costa Rica 201
cost-benefit analysis 177
Council of Europe 58
Covenant of Mayors 213
Croatia 83, 210
cultural: assimilation 167; compatibility 168; diversity 167–168; relativism 167; theory 111
culture 111; definition 161

Damian, H. 82
Daugbjerg, C. 73, 79
Davies, J. S. 196, 202, 216
Davis, G. 127
De Aguiar, T. R. S. 84, 95
De Bruijn, H. 35, 90, 140
decent work 211
De Grauwe, A. 138, 206
Demil, B. 31
Denmark 24, 36, 206, 269
Derix, G. 25

Derkx, B. 15, 78, 83, 268
De Ruijter, A. 167, 260
devolved agencies 55
De Wit, B. 119
De Zeeuw, A. 135, 177
Dimitrakopoulos, D. 129, 233
Dixon, J. 52, 54, 113, 119, 178
Doberstein, C. 84, 138, 213
Dodds, F. 180
Dogan, R. 52, 54, 113, 119, 178
Dommett, K. 54, 73, 84
Donnelly, J. 167
dossier teams 64, 183
Driessen, P. 22, 31
Dunleavy, P. 114
Dunn-Cavelty, M. 181
Dunsire, A. 15, 68

Eckert, S. 109
Ecological Footprint 9, 272
economy of scale 58
education 205–206
Egeberg, M. 66
Ehren, M. 206
Elkins, D. J. 168
empowerment 144
endogenous development 212
energy 210–211
Entwisle, T. 52, 55, 288
Environmental Implementation Review 50, 180, 226–227, 231, 235, 238, 240, 242
Eshuis, J. 32
Estonia 8, 98
Eurocities 96, 213
European Commission 6, 39, 48, 51, 57, 59, 66, 133, 142, 179, 200, 219; as centre of metagovernance 86–89
European Environment Agency 46
European Union: education governance 206; fisheries governance 209
Eurostat 114
evidence-based policy making 176–179

factor analysis 228
Fawcett, P. 73, 79
Ferretti, J. 136
Festinger, L. 178
Field, P. 128, 230, 287
Finland 35, 84
Fioretti, G. 150
Fisher, R. 130
Fleming, J. 72
Flinders, M. 54, 73, 84

flip-flop governance 168
focus or unfocus 185
food production 208
fragile states 196, 264
framing 52, 62, 92–94, 152, 184
France 126
Frances, J. 27
Fransen, L. 82
Frantzeskaki, N. 8, 213, 284
Freire, F. D. S. 84, 95
Friedland, R. 254
Fund for Peace 197, 217

Gabizon, S. 207
Gaebler, T. 12
Gailing, L. 54
garbage can theory 150
Gbikpi, B. 86, 279
Gee, D. 177
gender equality 207
geography and governance 169
Georgia 53, 197
Germany 35, 73, 84, 131
Geyskens, I. xiii, 28
Gini coefficient 165
Glasbergen, P. 15, 78, 82, 216, 268
glocalization 24
Good Governance 9, 23
governability, definition 55
governance: definition 10, 25; Goals 10; good enough 23; savviness 52; sustainability 6
governance capacity, definition 55
governance environment: context 79; definition 11; mapping 227
governance failure 45; capacity failure 51; definition 47; design failure 51; and government failure 47; of hierarchical governance 52–53; management failure 51; of market governance 54–55; meta-causes of 68; of network governance 53–54; root causes of 50
governance framework: definition 10; mismatch 57; principles 242
governance styles 21; adaptive management 32; bazaar governance 31; community governance 33; control and coordination mechanisms 127; convergence or co-habitation 30; dealing with power 147; default style 73; definition 27; differences between 109–159; distinct logics 110; fourth style 31; hierarchical governance 28; hybrids 31; flexibility 128; incompatibility of 59, 61; inherent weaknesses 52; institutional logic 124; institutions 122; interactive governance 31; leadership styles 144; knowledge governance 32; market governance 29; metaphors 117; modes of calculation 114; motives 115; network governance 28; organizational concertation 32; private interest government 31; professional self-regulation 33; provider-based governance 33; public-private governance 31; response to resistance 120; roles of government 116; self-governance 29; solidarity governance 33; styles or modes 27; theoretical background 113; typical problems 38; virtues 115
Granovetter, M. S. 146
Graves, C. W. 145
green deals 126
Grin, J. 36, 284
Grindle, M. S. 23
Grint, K. 52
Groetelaers, D. A. 137
Grote, J. R. 86, 279
Guatamala 205
Gulbrandsen, L. H. 37
Gupta, J. 90

Hague, The 67
Hajer, M. 22, 24, 35
Hake, J. E. 88, 219, 292
Hall, C. M. 75, 84, 211
Hall, E. T. 161
Hartley, J. 31, 114, 119, 126, 140, 142, 248
Haughton, G. 96
Haveri, A. 84
He, Z. 88
Head, B. W. 46
health and well-being 204–205
Heinzerling, L. 177
Hemmati, M. 180
Hendriks, F. 169
Héritier, A. 30, 39, 86, 109, 279
Hersey, P. 3, 144, 250
Hertin, J. 177
heuristic 73
hierarchical lock-in 149
Higdem, U. 84
High-Level Political Forum 13, 35, 82, 251
Hill, C. J. 30, 77, 79, 263

Hindmoor, A. 78
Hisschemöller, M. 237
Hofstede, G. 86, 164–166, 224–225, 233–234
Hofstede, G. J. 233
holistic 5
Hölscher, K. 213
Holzscheiter, A. 205
Höpner, M. 109
Hoppe, R. 134, 151, 161, 229, 237
Howard, C. 272
Howlett, M. xii, 31, 46, 52, 59
Hull, R. 124
Hulst, R. 57
Human Development Index 9
Huntington, S. P. 122
Hutchcroft, P. D. 127

ICLEI 213
Ide, T. 219
impact assessment 134–136
INCSR 216
India 99, 205
indicators 115, 203
Indonesia 65
inequality 208
infrastructure 211
Ingraham, P. W. 129
innovation 142
institutional resilience 183
institutional void 22
Inter Municipal Cooperation (IMC) 58
In 't Veld, R. J. 9, 23, 37, 97, 111, 132, 134, 162, 228, 283
Israel 219
Ivanova, M. 284

Jann, W. 67–68
Japan 84, 209, 218
Jayasuriya, K. 197, 201
Jentoft, S. 89, 98
Jessop, B. 4, 15, 27, 33, 45, 68, 74–75, 78, 80, 86, 96, 114, 117, 152, 168, 224, 269, 281, 284
Jordan, A. 5, 27, 238
Jungcurt, S. 179

Kahneman, D. 186
Kamau, M. 5, 13, 193, 195, 278, 291
Kanashiro, P. 185
Kang, V. 137
Kanie, N. 292
Kao, G. Y. 162, 167
Kaplan, R. 169

Kaufman, F.-X. 27, 127
Kay, A. 29, 48
Keech, W. 47
Kelly, J. 58, 206, 269
Kemp, R. 32, 36
Kenya 84, 219
Kickert, W. J. M. xiii, 25, 27, 29, 59, 143, 162–163, 169
Kim, S. E. 31, 84
Kirschbaum, C. 84
Klausen, J. E. 109
Klijn E.-H. 24, 78
Knill, C. 88
knowledge: democracy 132; roles of 131–132
Kolar-Planinsic, V. 148
Kooiman, J. 10, 15, 27, 38, 55, 68, 76, 89, 98, 143, 268, 284
Koppenjan, J. F. M. 78
Kostadinova, T. 248
Kroemer, J. 96, 270
Kroll, C. 38, 114
Kuenkel, P. 37
Kull, M. 84, 96

Lampropoulou, M. 84
Lan, G. 273
Lange, P. 95
Larsson, O. 4, 15, 24, 54, 74, 78, 84, 197, 213
Laske, O. 145, 174, 268, 269
Latvia 98
Lecocq, X. 3
Lees-Marshment, J. 144
Le Galès, P. 47
Lehmkuhl, D. 39
Lenschow, A. 88
less is more 175
Levin, M. 230
Lewis, J. M. 27, 115
Li, J. 84
Lieder, M. 214
life on land 218
Lindblom, C. E. 29, 177
Lipsky, M. 206, 269
Lodge, M. 127
logic: of hierarchies 124; of markets 125; of networks 125
Lomi, A. 150
Loorbach, D. 32, 213, 284
Lorvi, K. 98
Lowndes, V. 27, 94
Lucas, P. 194
Luckmann, T. 177

Lujie, W. 82
Lynn, L. E. 30, 77, 79, 263

Maas, W. 86
Magelhaes, A. 84, 206
majoritarian system 97, 168
Malinoski, M. 166
Mamadouh, V. 168
management development 146
Marshall, T. 169
Mason, M. 137
Mayntz, R. 129
McConnell, A. 47
McCullough, A. 136
McHugh, A. 140
Meadowcroft, J. 5, 6, 8, 12, 32, 36, 119, 225, 284
meta-exchange 80
metagovernance: coordination potential of 82; definition 15, 74; dissemination of 83; first order and second order 77; as governance of governance 38, 72–74; of hierarchies 77; internal and external 76, 95; long-term 38, 97; of markets 80; as method 225; multi-perspective approach 37; of networks 78; poly-centred 81; power dimension of 98–99; principles 89–92; private 15, 81–82, 216; public 15
metagovernance strategies 91–96; combining 94; framing 92; maintenance 95; switching 73, 94, 213
metagovernor 73
meta-hierarchy 78
metamarkets 80
meta-organization 78
meta-solidarity 80
Metzger, J. 94
Mexico 84, 205, 209
Meyer, E. 166, 233
Meyer, R. 119
mindfulness 174
mindsets 174
Mininberg, L. 181
Mintzberg, H. 56, 118, 140
Mongolia 83, 210
Monteiro, M. B. 135
Moomaw, M. 93, 218
Moore, G. 114
Morgan, G. 117
Mori Junior, R. 215
Mostert, E. 147
Moynihan, D. 198
muddling through 29

multi-level governance 63, 66, 86, 260
multi-stakeholder partnerships 180, 202, 288
Mundle, L. 82
Munger, M. 47
Murakami, I. 83
mutual gains approach 58, 147, 186, 230, 287

negotiation methods 230
Nepal 83
Neshkova, M. I. 248
Netherlands, the 24, 25, 36, 53, 58, 62, 65, 72, 93, 95, 120, 126, 133, 135, 163, 198, 233, 255
network analysis 229
Newman, J. 46
New Public Governance 12
New Public Management 11, 30, 143, 173
New Zealand 84
nexus: air-mobility 219; approach 66, 88, 185, 219; climate-conflict 219; energy-climate 88, 255; energy-poverty 207; energy-transport 258; environment-poverty 204; food-energy-water 88, 219, 292; gender-energy-poverty 207; inequality-gender-work-justice 88; nature-rural land use-urbanization 219; politics-institutions 196; water-agriculture 258; water-nature-food 219
Niestroy, I. 11, 37, 66, 75, 86, 90, 111, 119, 125, 162, 174, 177, 181, 194, 226, 241, 263, 279, 291
Nigeria 83, 201
Niksic Radic, M. 8
Noordegraaf, M. 237
Nooteboom, S. 78
Norway 84, 206, 209, 218
Nurse, K. 8

O'Connor, D. 253
OECD 153, 249, 264
Oikonomou, G. 84
Olsen, J. P. 37
Olsen, S. H. 37
Open Method of Coordination (OMC) 78, 86, 96
Open Working Group 4
organizing meetings 230
Osborne, D. 12
Osborne, S. 12, 23
Ostrom, E. 34, 216

Page, E. 129, 233
Pahl-Wostl, C. 162

302 Index

Painter, M. 55, 162
Palestine 219
Papa, M. 93, 218
Park, A. 22, 76, 84, 209
parliaments 97, 265
Partidário, M. 134, 135
partnerships: multi-stakeholder 121; in Wales 52
peer-to-peer learning 266, 271
Perryman, J. 206
Peter, L. J. 124
Peters, B. G. 27, 46, 48, 127, 145, 176, 206
Peterse, A. 229
Petersen, O. H. 288
Pierre, J. 29
Pintér, L. 271, 273, 278
Pirsig, R. M. 269
Pisano, U. 8
planetary boundaries 177
Plastics Strategy (EU) 200, 233–234, 236
policy: cycle 67; failure 46; implementation 63; principles 61; theories 255
policy and institutional capacity building 261; alignment 259; coherence 251–267; compatibility 260; coordination 256; definitions 253–254; empowerment 262; integration 258; intervention types 257; multi-level governance 260; principles 262–266; public sector reform 261; reconciliation 261
political cultures 168–169
political skill 56
political will 55–56
polity/politics/policy 48
Pollit, C. 25, 27, 141, 183, 248
Pomerantz, P. R. 53
Porras-Gómez, A.-M. 86, 98
Porter, M. E. 259
Portugal 84, 135
poverty 203–204
Poverty and Environment Initiative 204
Powell, W. 27, 34, 128, 130
precautionary principle 178
process and project management 140–141
public: choice 114; goods 114; value 114
public managers, roles of 143
public-private partnerships 180, 202, 287
public procurement 152–153
public sector reform 141, 248–251

Quality of Public Administration (Toolbox) 78, 174, 175

Rachel, P. 82
Radaelli, C. 86
Ramesh, M. xii, 31, 46, 52, 59, 205
Ramos-Mejía, M. 285
Rashid, A. 214
Rayner, J. 4, 22, 79, 124, 179, 263
redundancy 186
reframing *see* framing
relation: management 96; types 14
relational values 25, 111, 162–163
Renaud, E. N. C. 84, 209
Rhodes, R. A. W. 28, 36, 113, 127, 225
Ricard, L. M. 270
Rijnja, G. 130
rings of influence 230
risk analysis 229
risk-based inspections 58
Rittel, H. W. J. 53, 176, 236
RMNO 133
Röhring, A. 54
Roth, M. 197
Rotmans, J. 32, 36
Rotterdam 270
Roy, J. 284
Roy, M. N. 99
rule of law 195

Sachs, J. 272
Salingaros, N. A. 178
Scharpf, F. W. 113
Schedler, K. 113
Schiffler, M. 180
Schmidt, F. 177
Schmitt, P. 94
Schmitter, P. C. 27, 31, 116, 120
Scholz, I. 38
Schout, A. 27
Schuppert, G. F. 79, 83
science-policy relations 132–133
Scott, W. R. 254
shades of governance 109–159
SIDS 169
silos 38, 66, 86, 125; political, mental and institutional 181–184
Simon, H. A. 146, 177
Simon, N. 15, 198, 202, 288
situational leadership 144, 201, 250
Skelcher, C. 27, 94
Smolović Jones, O. 144
Sørensen, E. 13, 24, 54, 78, 80, 81, 88, 91, 128, 174, 202, 216
South Africa 84, 196, 212
South Korea 31, 84

Sowman, M. 84
Spangenberg, J. H. 8
Stacey, R. 118
Stakeholder Forum 49
stakeholders, mapping 122, 228–233
Starik, M. 185
Stark, A. 83, 86
statehood 81
Steenkamp, J.-B. E. M. xiii, 28
Steurer, R. 8, 15, 73, 81, 109
Stevens, V. 84, 141, 292
Stewart, F. 185
Stott, L. 285
strategy styles 118
Streeck, W. 27, 31, 116, 120
street-level bureaucrats 115, 206, 269
strength analysis 229
Struyven, L. 78
Susskind, L. 128, 230, 287
sustainability 12; schools of governance 35
sustainable consumption and production 213–217
sustainable development: definition 7; strategies 119
sustainable development councils 64
Sutcliffe, K. M. 174, 178, 278
Suter, M. 181
Sweden 24, 36, 53, 84, 197, 206, 235

Tacoli, C. 47
Tang, Y. 216
Taylor, I. 206, 269
Teisman, G. 67, 78
Tenbensel, T. 33
Termeer, C. 116, 284
Thailand 205
Thang, H. V. 84, 209, 218
't Hart, P. 50
Thijs, N. 48, 141, 165, 234
Thøgersen, M. 206
Thompson, G. 127
Thompson, M. 24, 32, 111, 127, 161
Thorelli, H. B. 27
Thrift, C. 203
Throsby, D. 161
Tidwell, C. H., Jr. 165
Timmermans, F. x, 3, 180
Tömmel, I. 86
Töpfer, K. 282
Torfing, J. 24, 78, 174, 216
transaction types 128
transdisciplinary 16, 283
transgovernance 283

transition management 32, 36
transition theory 284
Treadway, D. C. 55
Triantafillou, P. 24, 78
Trinidad and Tobago 165, 265
Tromp, H. 131
Trondal, J. 66
Tsheola, J. P. 84

Ugyel, L. 248
Uitto, J. I. 204
Ulbert, C. 197
UNCSD 119
UNDESA 11
UNDP 58
UNECE 61, 137
United Kingdom 24, 73, 84, 99, 175, 196, 206, 209, 213, 272, 288
UNOSD 83
UNPAN 83
USA 84, 209

values, of civil servants 145
Vanclay, F. 136
Van der Meer, F. M. 57
Van Kersbergen, K. 27
Van Montfort, A. 57
Van Parys, L. 78
Van Twist, M. 116, 284
Van Waarden, F. 27
Venghaus, S. 88, 219, 292
Verbong, G. 129
Verhoest, K. 84, 141
Verweel, P. 167
Vietnam 84, 162, 209, 218
Voets, J. 84, 92
Voluntary National Reviews 13

Waage, J. 194
Wackernagel, M. 170, 272
Waddock, S. 216
Wagenaar, H. 24
Walker, G. 98
Walker, H. 153, 209
Walter, T. 34
Wan, Y. K. P. 84
Warner, J. 92–93
water 209
Webber, M. M. 53, 176, 236
Wegrich, K. 67–68, 127
Weick, K. E. 174, 179, 278
Wesselink, A. 92–93
Whitehead, M. 24, 77, 80, 213

wicked problems 53, 61, 93, 151, 176, 179
Wieczorek, A. J. 284
Wollmann, H. 30
World Bank 12, 23, 98, 147, 162, 180, 196
Wu, X. 46

Yoo, J. W. 31, 84
Young, O. R. 6
Yu, J. 88

Ziekow, J. 38, 252, 258, 282

Made in the USA
Las Vegas, NV
21 June 2021